Instability and Transition

Volume I

ICASE/NASA LaRC Series

Stability of Time Dependent and Spatially Varying Flows
D.L. Dwoyer and M.Y. Hussaini (eds.)

Studies of Vortex Dominated Flows
M.Y. Hussaini and M.D. Salas (eds.)

Finite Elements: Theory and Application
D.L. Dwoyer, M.Y. Hussaini and R.G. Voigt (eds.)

Instability and Transition, Volumes I and II
M.Y. Hussaini and R.G. Voigt (eds.)

M.Y. Hussaini R.G. Voigt
Editors

Instability and Transition

Materials of the workshop held
May 15–June 9, 1989 in Hampton, Virginia

Volume I

With 119 illustrations, 2 in color

Springer-Verlag
New York Berlin Heidelberg
London Paris Tokyo Hong Kong

M.Y. Hussaini
R.G. Voigt
Institute for Computer Applications
in Science and Engineering (ICASE)
ICASE
NASA Langley Research Center
Hampton, VA 23665
U.S.A.

Library of Congress Cataloging-in-Publication Data
Instability and transition / [edited by] M.Y. Hussaini, Robert Voigt.
p. cm. – (ICASE/NASA LaRC series)
Proceedings of a workshop on Instability and Transition, held May 15–June 9, 1989, sponsored by the Institute for Computer Applications in Science and Engineering (ICASE) and the Langley Research Center (LaRC).
1. Fluid dynamics – Congresses. 2. Transition flow – Congresses. 3. Stability – Congresses. I. Hussaini, M. Yousuff. II. Voigt, Robert G. III. Institute for Computer Applications in Science and Engineering. IV. Langley Research Center. V. Series.
TA357.I55 1990
532′.0526 – dc20 90-9806
ISBN-13: 978-1-4612-8008-8

Printed on acid-free paper.

Camera-ready copy provided by the editors using TeX.

9 8 7 6 5 4 3 2 1

ISBN-13: 978-1-4612-8008-8 e-ISBN: 978-1-4612-3430-2
DOI: 10.1007/978-1-4612-3430-2

PREFACE

These two volumes contain the proceedings of the workshop on Instability and Transition, sponsored by the Institute for Computer Applications in Science and Engineering (ICASE) and the Langley Research Center (LaRC), during May 15 to June 9, 1989. The workshop coincided with the initiation of a new, focused research program on instability and transition at LaRC. The objectives of the workshop were to (i) expose the academic community to current technologically important issues of instability and transition in shear flows over the entire speed range, (ii) acquaint the academic community with the unique combination of theoretical, computational and experimental capabilities at LaRC and foster interaction with these facilities, (iii) review current state-of-the-art and propose future directions for instability and transition research, (iv) accelerate progress in elucidating basic understanding of transition phenomena and in transferring this knowledge into improved design methodologies through improved transition modeling, and (v) establish mechanisms for continued interaction. The objectives (i) to (iii) were of course immediately met. It is still premature to assess whether objectives (iv) and (v) are achieved.

The workshop program consisted of tutorials, research presentations, panel discussions, experimental and computational demonstrations, and collaborative projects. There were tutorials on Applications of Transition Research (Bushnell), Challenging Research Topics in Stability and Transition (Morkovin), Initiation and Evolution of Disturbances in Shear Flows (Morkovin), Flight Transition Research (Holmes), Theoretical Tools (Hall), Measurement Techniques (Gaster), Modeling of Laminar-Turbulent Transition (Narasimha), Laminar Flow Control in Flight Research (Wagner), 3 Ft. Transonic Pressure Tunnel Laminar Flow Control Research (Bobbitt and Harvey), Computational Tools (Zang), and Effects of Flow Quality on Transition (Bobbitt). There were also numerous demonstrations on experimental techniques. Special sessions were devoted to research presentations on Gortler vortices, crossflow vortices, compressible boundary layer stability.

These proceedings contain only panel discussions and the research papers resulting from the collaborative projects started during the workshop and completed shortly thereafter. There were six panels with three to six members in each – the Theory Panel was chaired by Sir James Lighthill, the Low-Speed Experiment Panel by Percy Bobbitt, the High-Speed Experiment Panel by Dennis Bushnell, the Computation Panel by Steven Orszag, the Receptivity Panel by Eli Reshotko, and the Roughness Panel by Mark Morkovin. The participants of the workshop were divided into several focused groups, each hosted by either a Langley or an ICASE scientist: Experiments (Wilkinson, Dagenhart), Receptivity (Gatski), Advanced Asymptotics (Hall), Boundary Layer Stability (Erlebacher), Free Shear Flows (Macaraeg), Compressible Stability and Transition (Malik), Transition Simulation (Zang, Streett), and Transition Modeling (Speziale).

The first volume contains the panel summaries and the position papers by the panelists as well as the research papers of the Experiment Group and the Receptivity Group. The second volume contains the research papers from the remaining groups. The success of the workshop was in no small part due to the contributions of the individuals noted above. We thank all the participants of the workshop for making it such an exciting event.

We are deeply grateful to Lee Beach and Roy Harris whose support and encouragement was essential from the conception through the execution of the workshop; we are thankful to Richard Barnwell and Edwin Prior for their continuous assistance. Then there are many other Langley scientists and logistical staff whose assistance and cooperation contributed to the smooth operation of the workshop. Our thanks are due to them all, in particular Percy Bobbitt, Dennis Bushnell, Douglas Dwoyer, and Manuel Salas.

It is a pleasure to express our sincere appreciation to Linda Johnson and Emily Todd who handled all the preliminary arrangements for the workshop, to Holly Joplin, who served as the workshop secretary; and to Etta Blair, who handled reimbursement for the participants. We are also appreciative of Emily Todd who coordinated the collection of manuscripts and facilitated their editing, and to Barbara Stewart who typed the majority of manuscripts. Thanks are also due to Zvi Ruder and other staff of Springer-Verlag for their cooperation in bringing out these volumes.

M. Y. Hussaini
R. G. Voigt

WELCOME ADDRESS

We at Langley consider the Instability and Transition Workshop extremely important in helping us to assess the state of knowledge in the area of transition from laminar to turbulent flow, to help us identify promising future areas of research, and to help us build future interactions with the world research community as our panels discuss the status of the major aspects of transition research.

As I look to the future, I see our ability to predict and control viscous flow phenomena becoming increasingly important to the successful design of aircraft. Of course, a major feature of these viscous flows is the transition process. Our major aeronautical technology initiatives all require advanced knowledge of transition for their success. Future transport aircraft, both subsonic and supersonic, will require laminar flow control to increase aerodynamic efficiency. The National Aero-Space Plane's design requires an accurate prediction of the hypersonic transition process in order to properly estimate drag, installed engine performance, and aerodynamic heating. Advanced military aircraft with high agility require a knowledge of flow separation and control of separation location in order to execute the design.

In my view, now is the right time to give increased emphasis to transition research. From my previous discussion, there is clearly a need to move ahead in this arena. At the same time, significant new research tools give us the opportunity to make unprecedented progress over the next several years. These new tools are analytical, computational, and experimental. In the analytical arena, new nonlinear theoretical techniques give us the ability to understand the fundamental nature of nonlinear interactions in the transitioning flow. Modern computational fluid dynamics techniques give us the ability to directly simulate the transition process, yielding information on details of the flow well into the nonlinear stage previously unattainable. In the experimental area, new facilities and instrumentation allow us to probe transitioning flows under conditions more closely matching the flight environment than previously possible. These facilities include not only quiet low-speed facilities, but now a family of quiet supersonic and hypersonic wind tunnels. Additionally, many of the advanced instrumentation techniques developed for wind tunnel use in the recent past are now being successfully applied in flight testing.

Langley has played a significant role in the development of all of these techniques and facilities. It is our intention to continue to play a leading role in furthering the state of knowledge of transition in the future. This workshop represents a new beginning for us in helping to shape our future research program.

Roy V. Harris, Jr.
Director for Aeronautics
NASA Langley Research Center

Contents

Contents
Volume II

The research of the following authors who contributed to this volume was supported by the National Aeronautics and Space Administration under NASA Contract No. NAS1-18605 while they were in residence at the Instability and Transition Workshop.

Henrik Alfredsson	Sophie Jallade
Daniel Arnal	J. M. Kendall, Jr.
R. Balasubramanian	Edward Kerschen
Fabio Bertolotti	Leonhard Kleiser
Sedat Biringen	Eckart Laurien
Ndaona Chokani	Patrick Leehey
Thomas Corke	Sir James Lighthill
Jeffrey Crouch	Thomas S. Lund
A. Demetriades	S. M. Mangalam
Peter Duck	Mark Morkovin
Stuart Edwards	Marc C. Mousseux
H. Fasel	B. Mueller
Jerzy M. Floryan	Xavier Normand
Mohamed Gad-el-Hak	Demetrius Papageorgiou
J. S. B. Gajjar	Ugo Piomelli
Michael Gaster	Eli Reshotko
Philip Hall	Daniel Riahi
Roland A. E. Heinrich	William Saric
Dan Henningson	Frank T. Smith
Thorwald Herbert	Terrance Tritz

CONTRIBUTORS

Ridha Abid
Vigyan Research Association
30 Research Drive
Hampton, VA 23666, USA

Henrik Alfredsson
Department of Mechanics
Royal Institute of Technology
S-10044
Stockholm
SWEDEN

Daniel Arnal
ONERA/CERT - DERAT
2 Avenue Edouard Belin
31055 Toulouse, Cedex
FRANCE

David Ashpis
Mail Stop 5-11
NASA Lewis Research Center
Cleveland, OH 44135, USA

S. Balachandar
High-Technology Corporation
28 Research Drive
Hampton, VA 23666, USA

R. Balasubramanian
Spectrex Inc.
P. O. Box 707
Gloucester, VA 23061, USA

Promode R. Bandyopadhyay
Mail Stop 163
NASA Langley Research Center
Hampton, VA 23665, USA

Ivan E. Beckwith
Mail Stop 163
NASA Langley Research Center
Hampton, VA 23665, USA

Fabio Bertolotti
Department of Mechanical Engineering
The Ohio State University
Columbus, OH 43201, USA

H. Bippes
Deutsche Forschungsanstalt für Luft-und Raumfarht 3400
Göttingen, WEST GERMANY

Sedat Biringen
Department of Aerospace Engineering Sciences
Engineering Center
University of Colorado
Campus Box 429
Boulder, CO 80309-0429, USA

A. E. Blanchard
Department of Mechanical Engineering & Mechanics
Old Dominion University
Norfolk, VA 23529, USA

Percy J. Bobbitt
Mail Stop 285
NASA Langley Research Center
Hampton, VA 23665, USA

Dennis M. Bushnell
Mail Stop 197
NASA Langley Research Center
Hampton, VA 23665, USA

Fang-Jenq Chen
High Technology Corporation
28 Research Drive
Hampton, VA 23666, USA

Ndaona Chokani
Department of Mechanical and Aerospace Engineering
North Carolina State University
Box 7910
Raleigh, NC 27695-7910, USA

Meelan Choudhari
Aerospace and Mechanical Engineering
University of Arizona
Tucson, AZ 85721, USA

F. Collier
Mail Stop 261
NASA Langley Research Center
Hampton, VA 23665, USA

Thomas Corke
Fluid Dynamics Research Center
Mechanical & Aerospace Engineering Department
Illinois Institute of Technology
Chicago, IL 60616, USA

Jeffrey Crouch
Code 4420
Laboratory for Computational Physics and Fluid Dynamics
Naval Research Laboratory
Washington, DC 20375-5000, USA

G. Dannabasoglu
Department of Aerospace Engineering Sciences
Engineering Center
University of Colorado
Campus Box 429
Boulder, CO 80309-0429, USA

A. Demetriades
Department of Mechanical Engineering
Montana State University
Bozeman, MT 59717, USA

Surya Dinavahi
Analytical Services & Materials, Inc.
107 Research Drive
Hampton, VA 23666, USA

Peter Duck
Department of Mathematics
University of Manchester
Manchester M139PL
ENGLAND

Stuart Edwards
Center for Nonlinear Dynamics and Department of Physics
University of Texas
Austin, TX 78712, USA

Nabil M. El-Hady
Analytical Services &
Materials, Inc.
107 Research Drive
Hampton, VA 23666, USA

Gordon Erlebacher
Mail Stop 156
NASA Langley Research Center
Hampton, VA 23665, USA

H. Fasel
Institute of Mechanics
University of Stuttgart
Prassenwaldring #9
700 Stuttgart 80
GERMANY

Jerzy M. Floryan
Department of Mechanical
Engineering
The University of Western
Ontario
London N6A 5B9
CANADA

Mohamed Gad-el-Hak
Aerospace and Mechanical
Engineering Department
University of Notre Dame
Notre Dame, IN 46556, USA

J. S. B. Gajjar
Department of Mathematics
Iowa State University
400 Carver Hall
Ames, IA 50011, USA

Michael Gaster
Department of Engineering
Cambridge University
Cambridge 6B2 1PZ
ENGLAND

Thomas B. Gatski
Mail Stop 163
NASA Langley Research Center
Hampton, VA 23665, USA

William P. Gilbert
Mail Stop 286
NASA Langley Research Center
Hampton, VA 23665, USA

A. A. Godil
High Technology Corporation
Hampton, VA 23666, USA

Marvin Goldstein
NASA Lewis Research Center
2100 Brookpark Road
Cleveland, OH 44135, USA

Chester E. Grosch
ICASE
Mail Stop 132C
NASA Langley Research Center
Hampton, VA 23665, USA

Philip Hall
Department of Mathematics
Exeter University
North Park Road
Exeter EX 4 4QE
ENGLAND

Roland A. E. Heinrich
University of Arizona
Aerospace and Mechanical
Engineering Department
Tucson, AZ 85721, USA

Dan Henningson
Department of Mathematics
Massachusetts Institute of
Technology
Room 2-339
Cambridge, MA 02139, USA

Thorwald Herbert
Department of Mechanical
Engineering
Ohio State University
206 W. 18th Street
Columbus, OH 43210, USA

Lennart Hultgren
NASA Lewis Research Center
2100 Brookpark Road
Cleveland, OH 44135, USA

Thomas L. Jackson
ICASE
Mail Stop 132C
NASA Langley Research Center
Hampton, VA 23665, USA

Sophie Jallade
ONERA/CERT
2 Avenue Edouard Belin
31055 Toulouse Cedex
FRANCE

J. M. Kendall, Jr.
Jet Propulsion Laboratory
California Institute of
Technology
4800 Oak Grove Drive
Pasadena, CA 91109, USA

Edward Kerschen
Department of Aerospace
and Mechanical Engineering
University of Arizona
Tucson, AZ 85721, USA

Leonhard Kleiser
DFVLR
Institute for Theoretical
Fluid Mechanics
Bunsenstr 10
D-3400 Goettingen
FEDERAL REPUBLIC OF
GERMANY

L. R. Kubendran
Analytical Services &
Materials, Inc.
107 Research Drive
Hampton, VA 23666, USA

Eckart Laurien
Technical University of
Braunschweig
Institut for Fluid Mechanics
Bienroder Weg 3
33 Braunschweig
FEDERAL REPUBLIC OF
GERMANY

Patrick Leehey
Department of Mechanical Engineering
Room 3-262
Massachusetts Institute of Technology
Cambridge, MA 02139, USA

Sir James Lighthill
Provost, University College
Gower Street
London WCIE 6BT
UNITED KINGDOM

Thomas S. Lund
Purdue University
School of Aeronautics and Astronautics
West Lafayette, IN 47907, USA

Michele G. Macaraeg
Mail Stop 156
NASA Langley Research Center
Hampton, VA 23665, USA

Lucio Maestrello
Mail Stop 359
NASA Langley Research Center
Hampton, VA 23665, USA

Mujeeb R. Malik
High Technology Corporation
28 Research Drive
Hampton, VA 23666, USA

S. M. Mangalam
Analytical Services and Materials, Inc.
107 Research Drive
Hampton, VA 23666, USA

Robert J. McGhee
Mail Stop 359
NASA Langley Research Center
Hampton, VA 23665, USA

F. Meyer
DFVLR
Institute for Theoretical Fluid Mechanics
Bunsenstr 10
D-3400 Goettingen
FEDERAL REPUBLIC OF GERMANY

Mark Morkovin
Professor Emeritus
Illinois Institute of Technology
Chicago, IL 60616, USA

Marc C. Mousseux
Department of Mechanical and Aerospace Engineering
Arizona State University
Tempe, AZ 85257-6106, USA

B. Müller
Deutsche Forschungsanstalt für Luft-und Raumfarht 3400
Göttingen, WEST GERMANY

Lian Ng
Analytical Services and Materials, Inc.
107 Research Drive
Hampton, VA 23666, USA

Xavier Normand
Institut de Mecanique de Grenoble
Equipe de Modelisation de la turbulence
B.P. 53 X, 38 041 Grenoble Cedex
FRANCE

Steven A. Orszag
Program in Applied Mathematics
218 Fine Hall
Princeton University
Princeton, NJ 08544, USA

Demetrius Papageorgiou
Department of Chemical Engineering
City University of New York
140th Street
New York, NY 10031, USA

Samuel Pate
Technical Support Division
Sversrup Technology Inc.
MS 500
Arnold Air Force Base Station
Manchester, TN 37389, USA

Ugo Piomelli
Department of Mechanical Engineering
University of Maryland
College Park, MD 20742, USA

Eli Reshotko
Department of Mechanical and Aerospace Engineering
Case Western Reserve University
Cleveland, OH 44106, USA

Daniel Riahi
Department of Theoretical and Applied Mechanics
215 Talbot Laboratory
University of Illinois at Urbana-Champaign
104 South Wright Street
Urbana, IL 61801, USA

William Saric
Department of Mechanical Engineering
Arizona State University
Tempe, AZ 85287-6106, USA

Gregory V. Selby
Department of Mechanical Engineering and Mechanics
Old Dominion University
Norfolk, VA 23529-0247, USA

Bart A. Singer
High Technology Corporation
28 Research Drive
Hampton, VA 23666, USA

Frank T. Smith
Department of Mathematics
University College London
Gower Street
London WC1E 6BT
ENLAND

Philippe Spalart
Mail Stop 202A-1
NASA Ames Research Center
Moffett Field, CA 94035, USA

Charles G. Speziale
ICASE
Mail Stop 132C
NASA Langley Research Center
Hampton, VA 23665, USA

John P. Stack
Mail Stop 359
NASA Langley Research Center
Hampton, VA 23665, USA

K. F. Stetson
AFWAL/FIMG
Wright-Patterson Air Force
Base
Dayton, OH 45433, USA

Craig L. Streett
Mail Stop 159
NASA Langley Research Center
Hampton, VA 23665, USA

Terrance Tritz
Mechanical Engineering
Department
Montana State University
Bozeman, MT 59717, USA

Andrew Walton
Department of Mathematics
University College London
Gower Street
London WC1E 6BT
ENGLAND

Stephen P. Wilkinson
Mail Stop 163
NASA Langley Research Center
Hampton, VA 23665, USA

Thomas A. Zang
Mail Stop 159
NASA Langley Research Center
Hampton, VA 23665, USA

1. Percy Bobbit
2. Dan Henningson
3. Bob Voigt
4. M. Y. Hussaini
5. Holly Joplin
6. Bernhard Mueller
7. Michael Gaster
8. Ian Poll
9. Ndaona Chokani
10. Sharon Seddougui
11. Fabio Bertolotti
12. Peter Duck
13. Nicola Horseman
14. Andrew Walton
15. Jeffrey Crouch
16. Xavier Normand
17. Jon Lee
18. Dennis Bushnell
19. Eckart Laurien
20. Ged Gasperas
21. David Ashpis
22. S. Balachandar
23. Meelan Choudhari
24. Sophie Jallade
25. Ugo Piomelli
26. Gordon Erelbacher
27. J. S. B. Gajjar
28. Charles Speziale
29. Jerzy Floryan
30. Tom Gatski
31. G. Dannabasoglu
32. Chet Grosch
33. Mujeeb Malik
34. Roland Heinrich
35. Alric Rothmayer
36. Robert Bowles
37. David Samuels
38. Stephen Cowley
39. Stuart Edwards
40. Scott Eberhardt
41. D. Papageorgiou
42. P. Blennerhassett
43. Terrance Tritz
44. Marc Mousseux
45. Nicholas Blackaby
46. Phil Hall
47. M. Gad-el-Hak
48. Stephen Wilkinson
49. Stan Berger
50. Henrik Alfredsson
51. Mehdi Khorami

1. Lucio Maestrello
2. Edward Kerschen
3. Albert Braslow
4. Phillipe Spalart
5. Bud Bobbit
6. Frank Smith
7. H. Fasel
8. Phil Hall
9. Thorwald Herbert
10. Sir James Lighthill
11. J. M. Kendall
12. Mark Morkovin
13. K. F. Stetson
14. Lennart Hultgren
15. Ivan Beckwith
16. Dennis Bushnell
17. Eli Reshotko
18. Daniel Joesph
19. Daniel Arnel
20. Steven Orszag
21. Doug Dowyer
22. Anthony Patera
23. Tom Zang
24. William Saric

PROLOGUE

M. Y. Hussaini

ICASE, NASA Langley Research Center
Hampton, VA 23665

and

T. A. Zang

NASA Langley Research Center
Hampton, VA 23665

The charter for the various panels was to review the state-of-the-art, emphasizing the progress made during the eighties, and to propose future directions for instability and transition research. For both the novices and the experts this was the highlight of the workshop.

The discussion began in grand style with the Theory Panel, chaired by Sir James Lighthill. Among the singular accomplishments have been secondary instability theory and strongly nonlinear theories of transition. The compelling challenge for the latter is to establish its validity via experiment or computation just as Schubauer and Skramstad confirmed the controversial theory of Tollmien and Schlichting. There exists a marvelous opportunity to apply these new tools to the terrain of transition in compressible flows. Indeed the lack of uniquely compressible stability theories was conspicuous by its absence from the panel discussions.

Two panels surveyed the state of experimental research, with the low-speed regime chaired by Percy Bobbitt and the high-speed regime by Dennis Bushnell. New developments in experimental research include supersonic quiet tunnel facilities, three-wire hot wire and thin film array instrumentation, and the bispectral data analysis. Challenges for the experimentalists include extracting transition physics data and not just transition Reynolds number from quiet tunnels, providing meaningful clues to nonlinear phenomena in low-speed flows, and providing at least enough data from flight experiments to permit mean flow computations.

The Computation Panel, graced by the presence of its chairman and coach Steven Orszag, focussed on the role that computation has played in stability (rather than transition) research: it has provided the first clues of phenomena such as secondary instability which has later proven amenable to much simpler analyses and it has established its credibility by reproducing many observed features of the early nonlinear stages of transition. The challenges are to relax the existing geometric constraints, to include important physical effects such as shock waves, and to provide data and insight for modeling the transition region.

The Receptivity Panel, chaired by Eli Reshotko, indicated that the foundation of a theory has been laid. To date predominantly a single tool – linear asymptotic methods with a heavy emphasis on triple deck-theory – has been applied. The principle challenges are to verify the predictions by experiment and computation, to furnish detailed experimental data on the disturbance background,and to include compressibility effects particularly shock waves.

The Roughness Panel, in which Mark Morkovin served as prosecution, defense, judge and jury, emphasized the primitive stage of this crucial subject. Although the experimental and computational difficulties are formidable, there is a major opportunity for substantial progress.

The Panels contributions to this workshop were immensely appreciated. The true test of their success will be the state of transition research in five years from now.

THEORY

PANEL SUMMARY: THEORY

Sir James Lighthill

University College London
London, England

The chairman began the proceedings by welcoming members of the panel as follows:

Dr. Marvin Goldstein, NASA Lewis Research Center
Professor Philip Hall, Exeter University
Professor Thorwald Herbert, Ohio State University
Professor Daniel Joseph, University of Minnesota
Professor Frank Smith, University College London.

He emphasized that the work of the panel would need to be linked closely with the work of other panels. It would be essential to stress connections between theoretical analyses on the one hand and experimental and/or CFD studies on the other.

The workshop as a whole was giving opportunities for researchers in every area of instability and transition studies to become aware of the impact of many new research tools developed during the 1980's. Just as other panels would make known the results derived from newly introduced experimental techniques or computational schemes, so the panel on theory had the responsibility of indicating the conclusions – and, especially, the new physical insights – that are emerging from a selection of the most modern and promising <u>theoretical</u> methods now available for studying instability and transition.

A common feature of most of these modern theoretical approaches was "strong" nonlinearity; which represented a major step forward from early methods that merely went to a second or third approximation beyond a simple linear stability analysis. At the same time, the discussion would concern itself with a wide range of different types of undisturbed flow: two-dimensional or three-dimensional, incompressible or compressible, parallel or nonparallel.

Like the workshop as a whole, the panel on theory had partly "tutorial" objectives, and would wish to make the richness of these new strongly nonlinear methods widely known to the workshop's large audience. Beyond all that, the panel would be seeking to define new directions and goals. The objective, in short, was to map out the future of transition theory.

The proceedings of the panel lasted for three hours, out of which the first two took the form of a discussion amongst panel members based on short presentations by individual members. The last hour was devoted to a general discussion in which the audience as a whole took part.

Panel members, beginning with Dr. Goldstein, recalled that much of the theoretical work was directed towards analyzing the results of transition experiments in which a shear layer was excited at a particular frequency by a vibrating-ribbon technique or by acoustic means, in such a way that the resulting disturbance was initially two-dimensional and periodic but progressively developed a chaotic character. Theoretical methods along the lines of matched asymptotic expansions had to "marry" the initial linear evolution of the disturbance to a later nonlinear phase of evolution.

Despite the crucial importance of this nonlinear phase, it was necessary to remember that the character of the theory as a whole could be strongly influenced by particular characteristics of the linear stability analysis. Several examples of this were cited, including characteristics of the stability analysis for weakly nonparallel flows; the importance of Görtler vortices for curved flows; and at higher supersonic Mach numbers, the special features of the second-mode instability discovered by Mack.

The earlier theories of nonlinear evolution of disturbances, associated with work by Benney & Lin; Craik, Stuart, & Watson; Stewartson & Stuart, Davey, Di Prima, & Stuart and many others, had emphasized evolutionary equations for the development of disturbance amplitudes (e.g., in a resonant triad, or in other groupings of disturbances). The work of Stuart had been particularly influential in the fruitful development of these nonlinear wave evolution theories.

However, the main panel discussion centered upon the newer, more strongly nonlinear, theoretical developments of the past decade. Some of these allowed the disturbances to bring about a complete restructuring of the mean flow regime itself. Among those discussed were:

(a) triple-deck theories (Smith);

(b) vortex-wave interaction theories (Hall and Smith);

(c) nonlinear vortex-flow theories, including restructuring of the basic flow by Görtler-type vortices (Hall);

(d) a new nonlinear treatment of weakly nonparallel flows (Herbert);

(e) a Floquet theory of secondary or parametric instability of the nearly periodic flow created by finite-amplitude primary disturbances (Herbert);

(f) Euler-stage theories, requiring the inviscid analysis of the developing turbulent flow in the main part of the boundary layer, with sporadic eruptions of vorticity into it from an inner viscous sublayer (Smith).

The interdependence of different methods among these half-dozen was strongly emphasized.

Essentially, the theories were concerned with the spatial evolution of disturbances. Thus they were not easily comparable with CFD analyses of the temporal evolution of streamwise periodic flows. Again, they did not lend themselves yet to the application of techniques from dynamical-systems theory and "chaos" theory.

The most exciting feature of the new theories was an increasingly apparent capability to span the entire transition process, bridging the gap between the initial instability and the final well-developed turbulent motion. En route, they allowed for bursting and intermittency. Whatever the formal nature of some of the theories as "asymptotic" analyses for "infinite" Reynolds number, they could in practice be applied successfully for just "moderately" large Reynolds numbers.

Amongst panel members and among those contributing to the discussion from the general audience, there were many diverse views of the relative merits of approaches (a) to (f); different speakers found different approaches relatively more valuable and more productive of good physical insights than others. Some constructive discussion of all of them ensued.

Finally, the chairman summed up on a note of optimism. After a period of highly ingenious innovation in the 1980's, the theoreticians "had their tails up"; they possessed a half-dozen exciting new theoretical approaches which, between them, seemed likely in the 1990's to lead to a greatly deepened understanding of the phenomena of instability and transition.

POSITION PAPER FOR THE PANEL ON THEORY

M. E. Goldstein

NASA Lewis Research Center
Cleveland, OH 44135

1. Subsonic Flows

(a) Experimental observations. Since the early days of the Schubauer and Skramstad experiment, most experimenters have approached the boundary layer transition problem by artificially exciting their flows with relatively two-dimensional, small amplitude, single frequency excitation devices, such as vibrating ribbons or acoustic speakers. They often go to great lengths to reduce the background disturbance level to an absolute minimum in order to make the external forcing in their experiments as small as possible. The initial disturbances should then exhibit harmonic time dependence, be well described by linear stability theory, and be nearly two-dimensional for the low Mach numbers at which most of the experiments have been carried out. This two-dimensional linear behavior can persist over long streamwise disturbances when the excitation levels are sufficiently small, but eventually becomes three-dimensional, as evidenced by the appearance of Λ-shaped structures in experiments where smoke-flow visualization is used. These structures, which are arranged in rows, can either be aligned or staggered in alternating rows. The unstaggered arrangement, which was originally observed in 1962 by Klebanoff, Tidstrom, and Sargent, is now commonly referred to as "peak valley" splitting.

(b) Theoretical description. The staggered arrangement, which usually appears at low excitation levels, is now believed to be the result of a resonant triad interaction between a pair of oblique subharmonic modes (which originate from the background disturbance environment) with the basic fundamental two-dimensional mode. This type of interaction was originally analyzed for the case of viscous-dominated Tollmien-Schlichting type instabilities by Raetz, and later on by Craik, who proposed that the Klebanoff peak-valley splitting could also result from a resonant triad interaction, which would then involve a pair of oblique fundamental harmonic modes interacting with the small two-dimensional instability mode that is invariably generated at the harmonic of the excitation frequency. All of the relevant modes can then be generated by the excitation device and do not have to originate from the background disturbance environment. However, the observed gradual transition from a two- to three-dimensional flow structure requires that the (common) amplitude of the oblique modes greatly exceed that of the (small) two-dimensional first harmonic that produces their enhanced growth in the Craik resonant triad model. Such behavior would obviously be favored if the oblique modes were unable to react back to suppress the initial linear growth of the first harmonic until they became very large.

(c) Adverse pressure gradient boundary layer. Many authors have analyzed this resonant-triad interaction for viscous-dominated Tollmien-Schlichting type instabilities on flat-plate-type boundary layers. But transition usually occurs in regions of adverse pressure gradient in most technological devices, and three-dimensionality is usually confined to about five wavelengths of the resulting transition point. I will therefore concentrate on flows where the adverse pressure gradients are strong enough to make the linear instability inviscid.

It is still appropriate to require that the adverse pressure gradient be small, because even relatively weak gradients can separate the boundary layer. The linear growth rates will then be small (O (pressure gradient squared)), and the instability waves will have a well-defined critical layer, but it will be of the nonequilibrium (or growth dominated) type, rather than of the equilibrium (or viscous dominated) type associated with Tollmien-Schlichting waves. This brings in a new nonequilibrium effect that does not occur in the viscous-type analyses and leads to a different type of amplitude equation that involves upstream history effects and is therefore an integro-differential equation, rather than an ordinary differential equation (or more accurately, a set of ordinary differential equations) as in the viscous-type analyses.

It is appropriate to suppose that the nonlinear interactions arise from the continued downstream growth of a resonant triad of initially linear instability waves (a single two-dimensional mode and two oblique modes with half the frequency and streamwise wavenumber of the two-dimensional mode and appropriate equal and opposite spanwise wavenumbers). The three modes can then interact nonlinearly with the nonlinearity first coming into play in the common critical layer of these three modes. The nonlinearity has a first-order effect on the (common) oblique mode growth rate when the order of magnitude of the transverse velocity fluctuation of the two-dimensional mode becomes equal to the (small) growth rates to the seven-halves power. But this initial nonlinearity will be weak in the sense that it will enter through an inhomogeneous term in a higher-order problem rather than as a coefficient in the lowest-order equation, with the result that the instability wave amplitude can then be explicitly determined from a set of amplitude equations.

These remarks apply to both the subharmonic and peak-valley transition processes. First, suppose that the initial amplitude of the oblique modes do not exceed that of the two-dimensional first harmonic. Then the subsequent nonlinear interaction produces no critical layer velocity jump at the frequency of the two-dimensional mode, which means that this mode can continue to grow at its initial linear growth rate even when the oblique modes become very large. This, in turn, allows these latter modes to exhibit faster than exponential growth, even when they are much larger than the more slowly growing two-dimensional mode. In this regard, it is worth noting that most adverse pressure gradient experiments seem to exhibit (Klebanoff) peak-valley splitting-type transitions – in which case, the oblique mode amplitudes must exceed the (frequently very small) two-dimensional first harmonic mode in order to be observable above the (usually much larger) two-dimensional fundamental mode at the frequency of the excitation device. This type of behavior is obviously promoted by the suppressed back reaction alluded to above.

The oblique modes eventually react back on the two-dimensional mode, but *only* when their (common) amplitude is of the order of the two-dimensional modal amplitude divided by the cube of the small streamwise wavenumber. The corresponding back reaction term is then quartic in the oblique mode amplitudes rather than quadratic (as in the flat-plate-type analyses), and – perhaps even more importantly – the mutual interaction between the two oblique modes then has a first-order effect on their mutual growth rates. This latter type of interaction produces a singularity at a finite downstream position – indicating an explosive growth of the instability waves there. This then leads to a new stage of evolution, which is described by the full three-dimensional nonlinear Euler's equations. The motion will therefore be exceedingly complex and may even be indistinguishable from turbulence.

2. Supersonic Flows

(a) Experimental considerations. The nonlinear dynamics can be quite different at supersonic speeds. The most rapidly growing linear mode can then be an oblique wave in the appropriate Mach-number range, and it might then be very difficult to generate the two-dimensional modes experimentally (Maslov and Levchenko). There are then two possibilities: One can generate a single oblique mode by placing the excitation device at the appropriate oblique angle to the flow (corresponding to the direction of maximum instability wave growth), or one can produce a pair of oblique waves with the same streamwise wavenumber and frequency, but equal and opposite spanwise wavenumbers (corresponding to a standing wave in the spanwise direction), by placing the excitation device perpendicular to the flow.

(b) Theoretical description. In the latter case, the two (initially linear) oblique modes can interact nonlinearly in their mutual critical layer, causing the nonlinearity to come into play at very small instability wave amplitudes (i.e., when the amplitude is of the order of the asymptotically small growth rate cubed). The critical layer nonlinearity is again weak, with the result that the instability wave amplitude can again be determined from a single integro-differential equation.

The inviscid solution to this equation always ends in a singularity at a finite downstream position – indicating an explosive growth of the instability wave there.

The so-called second mode has the largest growth rate when the mean flow Mach number is sufficiently large. The plane wave will then exhibit the fastest growth. Here the temperature fluctuations, which (on linear theory) develop an algebraic singularity in the critical layer, cause the critical layer nonlinearity to again occur at very small instability wave amplitudes. Nonlinearity now occurs when the instability wave amplitude is of the order small growth rate to the five halves power. But the nonlinearity is again weak, so the amplitude can again be determined from a single amplitude equation, which is similar to the equation for the single oblique mode case discussed above. We again worked out the solution to this equation and found that it ends at a singularity at a finite downstream position when the appropriate scaled viscous parameter is sufficiently small. But there is now a certain range of parameters in which the singularity is bypassed when this viscous parameter exceeds a certain finite value – in which case the solution just goes to a finite amplitude equilibrium state further downstream.

SOME THOERETICAL ASPECTS OF BOUNDARY LAYER STABILITY THEORY

Philip Hall

Exeter University
Exeter, England

1. Introduction

Our increased understanding in recent years of boundary layer transition has been made possible by the development of strongly nonlinear stability theories. Thus, after some twenty or so years when nonlinear stability theory was restricted to the application of the Stuart-Watson method or less formal amplitude expansion procedures, there now exist strongly nonlinear theories which can describe processes which have an $0(1)$ effect on the basic state. Most of the discussion in this paper will concern these strongly nonlinear theories and their possible role in pushing our theoretical understanding of transition ever further into the nonlinear regime. However, we will first review the nonparallel theories of boundary layer stability theory which enabled the strongly nonlinear theories to be formulated.

2. Linear Stability Theory

In recent years the nonparallel stability problems for incompressible boundary layers have yielded to different asymptotic methods. Thus, following the work of Bouthier (1973) and Gaster (1974) a reliable and efficient successive approximation procedure has been available to describe the nonparallel linear growth of Tollmien-Schlichting waves in developing boundary layers. At zeroth order the Orr-Sommerfeld equation is obtained and must be solved numerically. Nonparallel effects then introduce forcing terms on the right hand sides of the Orr-Sommerfeld equations obtained by this procedure. Gaster showed that the approach was capable of significantly improving the agreement between stability theory and the experimental results of for example Schubauer and Skramsted (1947) and Ross et al (1970).

An alternative approach based on more formal asymptotic methods was developed by Smith (1979a) and Smith and Bodonyi (1981). Here the asymptotic expansion is developed in an appropriate manner to capture either the lower or upper branch of the neutral stability curve. Typically it is found that the lower branch structure, based on triple-deck theory, is the most successful in reproducing the available experimental results. Since the successive approximation procedure is capable of producing both branches of the neutral curve, it has been argued elsewhere this is the most efficient method of determining growth rates in developing boundary layers. This argument has been reinforced by the development of high speed computers which make Orr-Sommerfield calculations a routine task. However, since the most important problems in boundary layer stability theory concern nonlinear effects, and the multi-deck theories provide the only rational framework for nonlinear studies, it is clear that, if anything other than linear growth rates are required, the successive approximation procedure is of limited use. Furthermore the multi-deck theories, Smith (1989), Cowley and Hall (1988) have recently been used to elucidate many of the viscous stability properties of compressible boundary layers. It appears that the Cowley-Hall calculation for the stability of wedges flows in the presence of shocks cannot yet be tackled using the successive approximation procedure.

In compressible boundary layers inviscid modes are possible causes of instability. The wavelength of these modes typically scales on the boundary layer thickness so that a multiple scale approach leads to a Rayleigh equation at zeroth order in an asymptotic description of these modes. Recent asymptotic investigations of the high Mach number structure of the neutral curves for inviscid compressible modes, see Brown and Smith (1989), Cowley and Hall (1988), have revealed much of the complex structure found by Mack (1984) in his extensive study of compressible stability theory. The high Mach number structure found by these investigators points to the possibility of complex multi-mode interactions in the nonlinear regime and provides a rational framework for its study.

The other instability mechanism which can occur in boundary layers is the Görtler vortex mechanism which is a centrifugal instability associated with wall curvature. Here the status of a quasi-parallel approach (which is the zeroth order problem for either of the two approaches to Tollmien-Schlichting instabilities) is now clear. Since the vortices grow on a lengthscale comparable to the boundary layer streamwise variable it is not a self-consistent approximation to replace streamwise derivatives by constants in parallel flow stability calculations of the type performed by Görtler (1940), Hammerlin (1956) and many subsequent authors. At large wavenumbers Hall (1982a) showed that a quasi-parallel approach can be formally justified and that the vortices become trapped in internal viscous shear layers. At order one vortex wavenumbers the vortex structure must be found by the numerical solution of the parabolic partial differential equations obtained by linearizing the Navier-Stokes equations. The parabolic nature of the equations means that the downstream position of neutral stability must depend on the nature of the upstream initial conditions. However, Hall (1983) showed that the different neutral curves appropriate to different initial conditions are in much closer agreement to experimental observations than those given by the parallel flow theories. Much more recently different authors have tried to restore some credibility to the parallel flow theories by arguing that they should be used to provide the initial conditions for the full partial differential equations. The justification for this procedure is somewhat unclear and merely confuses the accepted view that there is no such thing as a unique growth rate or neutral curve in the Görtler problem. If an engineer must use some measure of vortex growth in order to predict transition the most accurate method would be to use the small wavelength theory developed by Hall (1982a). The justification for this statement is that in a growing boundary layer almost all of the growth of a vortex occurs in this regime. Recent work on the hypersonic Görtler vortex problem, Hall and Fu (1989) , has generalized the incompressible work of Hall (1982a, 1983). Surprisingly it is found that that the parallel/non-parallel nature of the stability problem depends on the viscosity-temperature dependence of the fluid. Further work being carried out at Exeter concerns the role of real gas effects in the Görtler mechanism in both the linear and nonlinear regimes.

Other asymptotic investigations of the linear stability properties of three-dimensional boundary layers concern the structure of inviscid modes; see for example Bassom and Gajjar (1988). These problems are particularly suitable for asymptotic investigations because of the large variety of modes available. These calculations are of course relevant to swept-wing flows and have recently been generalized into the compressible situation by Gajjar(1989).

3. Nonlinear Theories

For many years nonlinear stability theory in hydrodynamics was almost exclusively based on the ideas and expansion procedure of Stuart and Watson. This procedure describes how the energy of a wave cascades into the harmonics and the mean flow and it enables us to calculate the amplitude of the wave at Reynolds numbers close to the neutral one. The application of the method to boundary layers is complicated by boundary layer growth and it was not until Smith (1979b) applied triple-deck theory to the problem that a self-consistent derivation of the amplitude equation for Tollmien-Schlichting waves was available. This formulation is valid near to the lower branch of the neutral curve and later Hall and Smith (1984) showed that the approach could be used to explain the three-dimensional secondary instability of a two-dimensional Tollmien-Schlichting wave.

Less formal expansion procedures have been used to investigate the secondary instability mechanism, see for example Herbert (1988) for a review of these theories. Here the shape assumption is made in order to specify arbitrarily the two-dimensional Tollmien-Schlichting amplitude before investigating its instability to three-dimensional waves. It appears that this type of approach can successfully explain some aspects of transition. Indeed, in view of the agreement between the results of this approach and parallel flow numerical simulations, it appears that it captures some of the physics of transition even though a justifiable derivation of the theory from the Navier-Stokes equations has not been given.

Even at this stage when we are discussing small but finite amplitude Tollmien-Schlichting disturbances there is already a significant difference from the corresponding state for Görtler vortices in growing boundary layers. Here the full numerical simulations of Hall (1988) and the weakly nonlinear theory of Hall (1982b) show nonlinear effects set in as an interaction between the mean state and the fundamental component of the vortex. A strongly nonlinear description of Görtler vortices was subsequently given by Hall and Lakin (1988) who were able to describe vortices so large that they completely restructure the basic state over most of the flow field. Hall and Lakin showed that a strongly nonlinear vortex state induces a new mean flow which makes all possible vortices neutrally stable , this idea is of course closely related to the marginal theory of turbulence proposed by Malkus. The subsequent three-dimensional unsteady breakdown of Görtler vortices was described by Hall and Mackerrell (1987) who showed that this secondary instability occurs in a localized region at the top or bottom of the vortex. This latter result had previously been found experimentally by Peehossaini and Wesfried (1987) so it appears that asymptotic terms can describe the linear, weakly nonlinear, strongly nonlinear, and the breakdown states of

Görtler vortices.

The first strongly nonlinear description of Tollmien-Schlichting waves was given by Smith and Burggraf (1985). This approach is valid as the frequency of the Tollmien-Schlichting wave approaches that of the upper branch of the neutral curve and Smith and Stewart (1987) later showed that the approach was remarkably successful in predicting the resonant triad states found by Kachanov and Levechenko (1982). The approach also points to the existence of a hierarchy of nonlinear states possibly relevant to the later stages of transition. A more recent development is the theory of vortex-wave interactions in boundary layers and indeed many other flows. Here the basic idea is that the significantly different velocity scales for the streamwise and the other two velocity components in a longitudinal vortex allows a large amplitude vortex to be driven by a remarkably small amplitudes wave system. (See for example Hall and Smith (1988, 1989)). Again there is a hierarchy of possible nonlinear states with the most nonlinear one being capable of restructuring the incoming boundary layer. The interaction equations are capable of describing a secondary instability of a two-dimensional Tollmien-Schlichting wave and the growth of the induced vortex state. In fact the wave system can also describe Rayleigh waves so future work will concern the simultaneous interaction between Tollmien-Schlichting waves, Rayleigh waves and longitudinal vortices. The approach will hopefully be capable of describing the "spike" stages of transition. The interaction theory is also relevant in the compressible case and in particular the dominant interactions in the high Mach number regime concern inviscid waves and vortices trapped in the temperature adjustment layer at the edge of the boundary layer. The vortex-wave interaction theories also provide an alternative framework to describe the secondary instability of Görtler vortices. Other applications of the theory concern such diverse flows as plane Couette flow, free surface flows, and free surface flows. It can be argued that the vortex-wave interaction is probably the most dangerous one possible in many flows because of the remarkably small wave system amplitudes required to sustain it.

4. Future Research

The 'frontier' problems of transition exclusively concern the strong nonlinear development of different instability mechanisms in the presence of non-parallel effects. These must either be studied using numerical simulation techniques or the recent fully nonlinear formulations described above. Since it is not yet possible to accurately simulate fully 3-D unsteady disturbances in growing boundary layers the strongly nonlinear theories offer perhaps the best chance for us to understand the later stages of transition. The computational requirements of some of the strongly nonlinear theories are significant, however it is clear that they are typically an order of magnitude smaller than those appropriate to unsteady 3-D simulations of the Navier Stokes equations; this is because the Reynolds number is scaled out of the problem and the time dependence simplified. Of course the approaches should compliment each other and identify the scales and structures in the flow field of transitional boundary layers. Some obvious areas where the strongly nonlinear theories should prove invaluable are:

1. The 'spike' stage of transition where Tollmien-Schlichting and Rayleigh instabilities co-exist with longitudinal vortex structures.

2. All of the nonlinear stages of inviscid and viscous instabilities of compressible boundary layers, particularly the hypersonic regime where localized vortex and wave structures interact in a manner governed by the vortex-wave interaction theory.

3. Wave interaction problems in three-dimensional boundary layers; the strongly nonlinear theories provide a rational basis for Tollmien-Schlichting crossflow interactions.

4. Tollmien-Schlichting-Görtler crossflow interactions in curved geometries.

5. Görtler vortex structures in compressible boundary layers; as yet there are no nonlinear solutions of the Görtler equations for compressible boundary layers.

References

[1] Bassom, A. and Gajjar, J. 1988, *Non-stationary crossflow vortices in three-dimensional boundary layer flows*, PRS(A),417,179.

[2] Bennett, J., Hall, P. and Smith, F. T. 1988, *The strong nonlinear interaction of Tollmein-Schlichting waves and Taylor-Görtler vortices in curved channel flows*, ICASE Report No. 88-45, submitted to J.Fluid Mech.

[3] Bouthier, M. 1973, *Stabilite' line'aire des e'coulements presque paralle'les. Partie II. La Couche limite de Blasius*, J.Mecanique, 12, 75-95.

[4] Brown, S. N. and Smith, F. T. 1989, ICASE Report in preparation.

[5] Cowley, S. and Hall, P. 1988, *On the instability of hypersonic flow past o ledge*, ICASE Report No. 88-72, submitted to J.Fluid Mech.

[6] Gajjar, J. 1989, ICASE Report in preparation.

[7] Gaster, M. 1974, *On the effects of boundary layer a growth on flow stability*, J. Fluid Mech. 66, 465.

[8] Görtler, H. 1940, *Über eine driedimensionale Instabilität laminarer Grenzschichten an konkaven Wänden*, Tech. Mem. Nat. Adv. Comm. Aero Rep 1375.

[9] Hall, P. 1982a, *Taylor-Görtler vortices in growing boundary layers*, J. Fluid Mech. 124, 475, 494.

[10] Hall, P. 1982b, *On the nonlinear evolution of Görtler vortices in growing boundary layers*, IMA J. App. Math. 29,173-192.

[11] Hall. P. 1983, *The linear development of Görtler vortices in growing boundary layers*, J. Fluid Mech. 130, 41-58.

[12] Hall, P. 1988, *The nonlinear development of Görtler vortice in growing boundary layers*, J. Fluid Mech. 193, 243-266.

[13] Hall, P. and Mackerrell, S. 1987 , *On the onset of three-dimensinality and time dependence in the Görtler vortices problem*, ICASE Report No. 87-71, J.Fluid Mech. 1989, 204, 405-620.

[14] Hall, P. and Lakin, W. 1988, *The fully nonlinear development of Görtler vortices in growing boundary layers*, PRS(A), 415, 421-444.

[15] Hall, P. and Smith, F. T. 1984, *On the effects of non-parallelism, three-dimensionality and mode interaction in boundary layer stability*, Studies in Applied Math, 70, 91-120.

[16] Hall, P. and Smith, F. T. 1988, *The nonlinear interaction of Tollmien-Schlichting waves and Taylor-Görtler vortices in curved channel flows*, PRS(A), 417, 255-282.

[17] Hall, P. and Smith, F. T. 1989, *Interaction of longitudinal vortices and Tollmien-Schlichting waves in boundary layers*, to appear in Euro.J.Mech.

[18] Hall, P. and Fu, Y. B. 1989, *On the Görtler vortex instability mechanism at hypersonic speeds*, to appear in Theoretical and Computational Fluid Dynamics.

[19] Hämmerlin, G. 1955, *Über das Eigenwestproblem der dreidimensionalen Instabilität laminarer Grenzschichten an Konkaven Wand*, J. Rat. Mech. Anal. 4, 279-321.

[20] Herbert, T. 1988, *Secondary instability of boundary layers*, Ann. Rev. Fluid Mech, 20, 487-526.

[21] Kachanov, Yu. S. and Levechko, V. Ya. *The resonant interaction of disturbance at laminar-turbulent transition in a boundary layer*, IUTAM paper 10-82 ,USSR Acad.Sci.

[22] Mack, L. 1984, *Boundary layer stability theory*, AGARD 709.

[23] Peerhossaini, H. and Wesfried, J. 1987, *An experimental investigation of Görtler vortices in channel flows*, paper presented at Taylor Vortex Working Party Tempe Arizona.

[24] Ross, J. Barnes, F. H., Burns, J. G. and Ross, M. A. 1970, *The flat plate boundary layer. Part 3 comparison of theory with experiment*, J. Fluid Mech. 43, 8, 19-832.

[25] Schubauer, G. B. and Skramsted, H. K. 1948, *The flat plate boundary layer. Part 3 comparison of theory with experiment*, NACA 909.

[26] Smith, F. T. 1979a, *Laminar boundary layer oscillations and transition on a flat plate*, PRS(A) 366, 91-109.

[27] Smith, F. T. 1979b, *On the nonparallel flow stability of the Balsius boundary layer*, PRS(A) 368, 573-589.

[28] Smith, F. T. 1989, *Nonlinear stability of boundary layers for disturbances of various sizes*, 1987 UTRC Report, to appear in J. Fluid Mech.

[29] Smith, F. T. and Stewart, P. 1987, *On the first mode instability in subsonic, supersonic or hypersonic boundary layers*, J. Fluid Mech. 179, 227-252.

THEORY OF INSTABILITY AND TRANSITION

Thorwald Herbert

The Ohio State University
Columbus, OH

1. Introduction

The theory of instability and transition has seen remarkable progress over the past decade. For certain classes of flows such as the Bénard and Taylor problem, the progress is largely due to the application of new analytical techniques and the ability to describe the dynamics by relatively simple amplitude equations. This analytical work is supported by current experimental and computational studies.

The attempt to model, simplify, and analyze the dynamics of through-flow systems, especially boundary and mixing layers, with similar techniques has not yet succeeded. Some of the reasons are the lack of a true Reynolds number (as in boundary and mixing layers), the unavailability of nonlinear equilibrium motions (in boundary and mixing layers, pipe flow, plane Couette flow), and the nonlinear subcritical instability and consequent "snap-through" transition in the few remaining prototype flows (e.g., plane Poiseuille flow). Once the flow becomes unstable, it develops through some stages of distinguished character into a turbulent flow without settling into any regular motion or stable equilibrium. Nevertheless, equilibrium states, bifurcation points, and symmetry breaking have been revealed for plane Poiseuille flow and these results have strongly influenced the theoretical developments for other flows, in particular for boundary layers. Owing to the complexity of the dynamical equations, the theoretical work on through-flow systems has strong computational aspects and is in many cases closely related to spectral simulations of nonlinear instability and transition.

New results have been obtained and methods of analysis have been developed for various classes of flows that permit mathematical rigor:

- Parallel flows with velocity vector $\mathbf{V} = (U(y), 0, W(y))$ or $\mathbf{V} = (U(y), 0, 0)$, such as plane Poiseuille flow. Here, U is in the streamwise x direction.

- 3D flows with $\mathbf{V} = (U, V, W)$ and $\mathbf{V} = \mathbf{V}(y, z)$, as in rectangular ducts, or $\mathbf{V} = \mathbf{V}(r, \phi)$, as in curved or elliptic pipes.
- Periodic flows with periodicity in different variables.

The class of periodic flows has found broad interest owing to their frequent occurrence in practical applications and as a source of parametric secondary instabilities. The main groups studied are:

- Time periodic with $\mathbf{V} = (U(y,t), 0, 0)$ and $U(y,t) = U(y, t+\lambda_t)$ (Davis 1976),
- Periodic in the cross-stream direction with $\mathbf{V} = (U(y), 0, 0)$ and $U(y) = U(y + \lambda_y)$ (Gotoh & Yamada 1986),
- Periodic in the streamwise direction with $\mathbf{V} = (U(x,y), V(x,y), 0)$ and $\mathbf{V}(x,y) = \mathbf{V}(x + \lambda_x, y)$.

The last group incorporates 2D equilibrium states in parallel flows as they develop in the nonlinear stage of primary instability (Orszag & Patera 1980, Herbert 1981, 1983, Nagata & Busse 1983, Pierrehumbert & Widnall 1982). Even the states associated with traveling waves are steady in a Galilean frame that moves with the phase velocity of the waves.

With appropriate approximations, new analytical and numerical tools have also been developed for the technologically important class of weakly nonparallel flows where the basic flow involves the boundary layer approximation (Hall 1983, Herbert & Bertolotti 1987). The local approach (leading to ordinary differential equations) is complemented or replaced by solving parabolic partial differential equations for given initial conditions. Besides nonparallelism, the nonlinear evolution of the disturbance field can be studied. The need for initial conditions, however, poses a new challenge to clarify the receptivity issue.

Many of the tools developed and established for incompressible flows are currently ported to tackle the stability of compressible flows. In this area, stability theory has barely progressed beyond the classical linear analysis, and the physical mechanisms of instability still await clarification and deeper understanding.

2. Weakly and Strongly Nonlinear Theories

The charge to this panel is to place emphasis on the strongly nonlinear area of theory. Given that the phenomena in this area are governed by the Navier-Stokes equations in more or less complex form, one has to compromise. Direct numerical simulation is feasible but is more a replacement for laboratory experiments, not theory. Asymptotic theories for large Reynolds (or other) numbers lead to simplified equations that allow closed-form solutions (in exceptional cases) or ease numerical solutions to nonlinear problems (see panel discussion, F. T. Smith). These theories based on early asymptotic studies (Lin 1955) enlist impressive results for linear problems and local (e.g., receptivity) phenomena that can be captured within a single asymptotic structure. However, the structures are different for solutions near branch I (triple deck) and branch II (five decks) of the neutral curve. Therefore, studies of physically relevant nonlinear mode interactions (as those causing secondary instability) or the streamwise evolution of the disturbance field seem outside any single structure of the asymptotic theory. Moreover, various phenomena in the transition process critically depend on the combination of physical parameters and may not be accessible to asymptotic methods for high Reynolds numbers.

Linear and "weakly" nonlinear (i.e., perturbation) theories are not as ineffectual as the notation suggests. "It is not the process of linearization that limits insight. It is the nature of the state we choose to linearize about." (E. T. Eady). Moreover, perturbation methods can solve strongly nonlinear problems provided they are rational in the sense of Van Dyke (1975), consider higher-order terms, and are combined with the arsenal of tools for analyzing, improving, and extending the convergence of perturbation series (Van Dyke 1984). The weaknesses of the weakly nonlinear theory are essentially the inappropriate formulation in earlier work, (Herbert 1983), and the lack of guidance for the choice of the lowest-order basis. When properly applied, there should be in fact no difference between a weakly nonlinear or perturbation theory, a nonlinear theory, and a truly nonlinear theory. Perturbation methods can be very relevant to obtain insight into the nonlinear transition process.

In the following, we discuss three areas of theoretical/numerical development that contribute to understanding the transition mechanisms, and moreover, provide new means for quantitatively analyzing and predicting transition. The incompressible flow over a flat plate with zero-pressure gradient is chosen as an example while applications range to other shear flows, including three-dimensional and compressible boundary layers.

3. Nonlinear Stability of Nonparallel Flows

Boundary layers, mixing layers, jets, and wakes are interesting subjects to stability studies, but are solutions to the Navier-Stokes equations only within the boundary-layer approximation. Usually, the stability analysis at a fixed streamwise position (and Reynolds number) neglects the streamwise variation and small transverse velocity, and assumes a locally parallel flow. While this assumption can be tolerated for disturbance waves with sufficiently large streamwise wavenumbers α (Herbert & Morkovin 1980), it is incorrect for small wavenumbers, especially for $\alpha = 0$. The most affected disturbances with $\alpha = 0$ are longitudinal vortices, including Görtler vortices, and nonlinear distortions of the mean flow.

For Görtler vortices, the effect of the transverse velocity is of first order, not of order $O(Re^{-1})$ (Herbert 1976). It is necessary, therefore, to pose the stability problem as an initial-boundary-value problem for parabolic partial differential equations (Hall 1983), in essence the 3D boundary-layer equations for a spanwise periodic flow. For initial conditions in the form of traditional normal modes, some features of earlier parallel-flow and weakly-nonparallel-flow results are qualitatively reproduced (Day, Herbert, & Saric 1988), however, the local growth rates and integrated N factors are quite different and depend on the initial conditions. The concept of a branch I of the neutral curve, the traditional go-sign for instability and the integration of N factors, does not carry over to the new approach. Considered that N factors are a cornerstone of transition prediction in engineering practice, theory should not only push forward into virgin territory but also scrutinize all those convenient assumptions that were made along the way.

TS waves with $\alpha > 0$ cannot be described by the boundary-layer equations, but the effect of non-parallelism on the linear stability can be captured by perturbation methods. Earlier work suffers criticism since the perturbation approach is not rational in Van Dyke's sense. Moreover, the approach cannot incorporate nonlinear effects of finite TS amplitude. To overcome these deficiencies, Herbert & Bertolotti (1987) introduce a multiplicative decomposition of the waves in a nonparallel flow into a strictly periodic wave function and an amplitude function. Within the boundary-layer approximation, the amplitude function is governed by a parabolic differential equation that can be solved by a marching scheme for given initial wavenumber and initial amplitude distribution. The streamwise change of the complex wavenumber can be extracted from the amplitude function. A method to obtain the local solution to the stability problem similar to earlier work has been derived without further approximation. This method requires simultaneously solving for the local wavenumber, amplitude function, and streamwise amplitude variation, thus solving for all terms of order $O(Re^{-1})$ at once.

The parabolized stability equations (PSE) permit accurate and very efficient calculation of spatial growth rates and N factors, but moreover, they offer various capabilities not previously available. The equations contain both the unsteady boundary-layer equations and the Orr-Sommerfeld equation as special cases, and therefore govern the link between their solutions (Goldstein 1983) without using matched asymptotic expansions. This link is key to the receptivity for sound near the leading edge. The PSE can maintain nonlinear terms and describe the evolution of harmonics and mean flow distortion. The PSE can also be utilized to analyze mode interactions and the linear and nonlinear stages of secondary instability without the need for a downstream boundary condition.

The results of both local and marching solution of the PSE for linear TS waves agree and largely confirm Gaster's (1974) results. In contrast to the findings of Saric & Nayfeh (1977) and Smith (1979), the incorporation of nonparallel effects does not improve the agreement of the neutral curve with the experimental data. The neutral curve shown by Saric & Nayfeh is based on a different measure for growth than that used in the experiments. We conclude that the disagreement between theoretical and experimental results is caused by experimental inaccuracies combined with the sensitivity of the problem (see panel discussion, W. S. Saric). The results of the marching scheme for the nonlinear development of TS waves are consistent with the weakly nonlinear parallel-flow results of Herbert (1974) and Itoh (1974), and in perfect agreement with numerical solutions to the Navier-Stokes equations (Bertolotti, Herbert, & Spalart 1989).

4. Linear Secondary Instability

Significant progress in the analysis of the transition process has been achieved by the theory of linear secondary instability of shear flows (Bayly, Orszag, & Herbert 1988, Herbert 1988). The concept of a three-dimensional secondary instability parametrically excited by the primary TS waves in a boundary layer was first used in a largely unnoticed paper by Maseev (1968) and revived by Herbert & Morkovin (1980).

In a coordinate system moving with the wave speed, a parallel flow with a superposed wave of fixed amplitude is strictly periodic and its linear stability is governed by a Floquet system of differential equations with periodic coefficients. General form, properties, and classification of disturbances can be derived with mathematical rigor. For cases like the 2D periodic equilibrium motions in a plane channel that are associated with nonlinear TS instability, the "primary instability," parametric resonance can lead to "secondary instability" with respect to a variety of 2D or 3D, subharmonic, fundamental (peak-valley splitting), or combination modes. The physical mechanism of this instability rests on tilting and stretching of vortices periodically arranged by the primary wave and is essentially of inviscid nature (Bayly, Orszag, & Herbert 1988). Accordingly, the growth of secondary modes on a convective scale can be much stronger than the slow, viscous growth of TS amplitudes.

Physical mechanism and classes of secondary modes are common to a wide variety of unstable flows. Unstable boundary layers exhibit the same types of secondary instabilities (Herbert 1988). For boundary layers, applicability of the Floquet analysis requires some approximations: the parallel-flow assumption and the neglect of the TS-amplitude growth. Both these approximations are well justified in regions of strong secondary disturbance growth. (These approximations need scrutiny, however, for other primary disturbances such as Görtler vortices or cross-flow vortices.) With the new concepts for incorporating streamwise changes discussed above, the assumptions of earlier work can be relaxed.

The Floquet theory of secondary instability provides convincing explanations for numerous puzzling observations and establishes a framework for quantitatively studying the later stages of transition. Formal and numerical results are in encouraging agreement with detailed experimental data for subharmonic modes (Kachanov & Levchenko 1984) and fundamental modes (Klebanoff, Tidstrom, & Sargent 1962, Cornelius 1985) in the Blasius flow. The predicted characteristics of combination modes are consistent with the scarce set of observations.

Combined with numerical results, the theory also permits asymptotic studies that provide guidance for properly modeling the primary-wave interaction in weakly nonlinear theories. Craik's (1971) resonant triad appears as a special case of subharmonic resonance as the TS amplitude tends to zero. Other weakly nonlinear theories have not been found relevant to the explanation or quantitative analysis of the secondary instability. Smith & Stewart (1987) have developed an asymptotic theory of the subharmonic resonant-triad interaction based on a triple-deck structure. Their results for the experimental conditions of Kachanov & Levchenko disagree with those provided by the Floquet theory (and the experiment) in various aspects such as the wave angle, growth rate, and streamwise variation of the growth rate. The discrepancies cannot be removed by incorporating the nonlinear interaction between TS wave and 3D subharmonic mode. A possible explanation may be the difference in the physical mechanisms. In the experiments and in the Floquet theory, the subharmonic secondary instability originates from the interaction of an upper-branch TS wave with a lower-branch subharmonic wave, while the triple-deck structure is appropriate only for lower-branch waves.

Both temporal and spatial growth concepts have been pursued with the Floquet theory. Other than in Gaster's transformation for primary disturbances, the leading term of the temporal-to-spatial transformation of growth rates for secondary disturbances is the phase speed of the primary wave and higher terms are small. Hence, transformed temporal and directly calculated spatial growth rates are very similar. This result explains some of the striking similarities between temporal Navier-Stokes simulations and spatial experiments. The results of the Floquet theory have not only stimulated numerous transition simulations but agree with their results and permit their interpretation up to the stage where nonlinear coupling between TS wave and 3D secondary modes comes into play. The theory yet bears the advantage of not being bound to a pre-selected computational domain that restricts the wavenumbers of the 3D disturbance field. The work of Spalart & Yang (1987) with a large spanwise domain is a notable exception.

5. Nonlinear Wave Interaction

In the absence of a nonlinear interaction of the 3D secondary modes with the primary TS wave, the secondary modes decay as their parametric excitation fades away. Analysis of the energy balance between mean flow, 2D wave, and 3D waves shows, however, that the 3D wave may feed energy into the 2D field (Herbert 1988). This contribution to the 2D field increases quadratically with the amplitude of the 3D mode and may halt the decay of the primary wave, thus leading to a feedback loop for self-sustained growth of both 2D and 3D disturbance field. A more detailed analysis of the processes involved (Crouch & Herbert 1989) is based on a perturbation method expanding simultaneously in the amplitudes of primary and secondary modes to maintain the mechanism of parametric excitation. Subharmonic and peak-valley-splitting route to transition turn out to be quite distinct in their sequence of events. In both cases, however, the nonlinear interaction can lead to either an ultimate decay of the disturbance field or to a simultaneous growth of 2D and 3D components. The threshold conditions for such simultaneous growth can be exploited to predict breakdown and transition in a given disturbance environment (Herbert & Crouch 1989).

The results for fundamental modes are in close agreement with the experimental data of Cornelius (1985) up to the stage where an increasing number of spikes heralds the breakdown of the laminar flow. Peak-valley splitting can lead to breakdown only if the 3D disturbance reaches a threshold amplitude upstream of branch II. In contrast, subharmonic modes can cause self-sustained growth even downstream of branch II as observed by Kachanov & Levchenko (1984) and Corke & Mangano (1987). The strong growth of the 2D component is owing to a purely forced 3D-3D interaction while the TS amplitude reaches an almost constant level. The two main contributions to the 2D component can be clearly identified in the velocity profiles at different streamwise positions measured by Kachanov, Kozlov, & Levchenko (1977). The theoretical results are in good agreement with direct simulations of transition, although the number of Fourier modes in the streamwise and spanwise direction is relatively small. The ability to represent the essence of the transition process with crude Fourier approximations was also noted by Rozhdestvensky & Simakin (1984) and Kleiser & Gilbert (1989).

6. Future Directions

Building on the encouraging results of the theoretical concepts, present work aims at their integration into an efficient and reliable tool to analyze a broader variety of flows and wave interactions, especially those affected by nonparallelism. To further develop understanding and predictive capabilities for transition, emphasis will be on

- Receptivity, in particular leading-edge receptivity, to provide the relation between environment and the initial and boundary conditions for the transition analysis.
- Breakdown and evolution of small scales, to provide the link between transitional and turbulent flows.
- Compressible flows, to gain insight, predictive capabilities, and support for numerical simulations.

In view of the difficulties of gathering detailed experimental data at high Mach numbers under realistic conditions, the development of reliable theoretical and computational methods for studies on transition in compressible flows has high priority.

7. Acknowledgement

This work is supported by the Air Force Office of Scientific Research under Grant AFOSR-88-0186 and under Contract F49620-88-C-0082.

References

[1] Bayly, B. J., Orszag, S. A., and Herbert, Th. (1988) "Instability mechanisms in shear flow transition," *Ann. Rev. Fluid Mech.*, Vol. 20, pp. 359-391.

[2] Bertolotti, F. P., Herbert, Th., and Spalart, P. R. (1989) "On the stability of the Blasius boundary layer," *J. Fluid Mech.*, in preparation.

[3] Corke, T. C. and Mangano R. A. (1987) "Transition of a boundary layer: controlled fundamental-subharmonic interactions," in Proc. IUTAM Symposium on Turbulence Management and Relaminarization, Ed. H. W. Liepmann and R. Narasimha, Springer-Verlag.

[4] Cornelius, K. C. (1985) "Three dimensional wave development during boundary layer transition," Lockheed Georgia Research Report LG85RR0004.

[5] Craik, A. D. D. (1971) "Nonlinear resonant instability in boundary layers," *J. Fluid Mech.*, Vol. 50, pp. 393-413.

[6] Crouch, J. D. and Herbert, Th. (1989) "Nonlinear evolution of secondary instabilities in boundary layers," *J. Fluid Mech.*, in preparation.

[7] Davis, S. H. (1976) "The stability of time-periodic flows," *Ann. Rev. Fluid Mech.*, Vol. 8, pp. 57-74.

[8] Day, H. P., Herbert, Th., and Saric, W. S. (1988) "Comparing local and marching analysis of Görtler instability," *AIAA J.*, submitted for publication.

[9] Gaster, M. (1974) "On the effects of boundary-layer growth on flow stability," *J. Fluid Mech.*, Vol. 66, pp. 465-480.

[10] Goldstein, M. E. (1983) "The evolution of Tollmien-Schlichting waves near a leading edge," *J. Fluid Mech.*, Vol. 127, pp. 59-81.

[11] Gotoh, K. and Yamada, M. (1986) "Stability of spatially periodic flows," in Encyclopedia of Fluid Mechanics, pp. 589-610, Gulf Publ. Co., Houston.

[12] Hall, P. (1983) "The linear development of Görtler vortices in growing boundary layers," *J. Fluid Mech.*, Vol. 130, pp. 41-58.

[13] Herbert, Th. and Morkovin, M. V. (1980) "Dialogue on bridging some gaps in stability and transition research," in Laminar-Turbulent Transition, Ed. R. Eppler and H. Fasel, pp. 47-72, Springer-Verlag.

[14] Herbert, Th. (1974) "Über endliche Amplituden periodischer Störungen der Grenzschicht an der ebenen Platte," Deutsche Luft-und Raumfahrt, DLR FB 74-53. (Engl. Translat.: European Space Agency ESA TT-169, 1975)

[15] Herbert, Th. (1976) "On the stability of the boundary layer along a concave wall," *Arch. of Mech.*, Vol. 28, p. 1039.

[16] Herbert, Th. (1981) "A secondary instability mechanism in plane Poiseuille flow," *Bull. Am. Phys. Soc.*, Vol. 26, p. 1257.

[17] Herbert, Th. (1983) "Secondary instability of plane channel flow to subharmonic three-dimensional disturbances," *Phys. Fluids*, Vol. 26, pp. 871-874.

[18] Herbert, Th. (1988) "Secondary instability of boundary layers," *Ann. Rev. Fluid Mech.*, Vol. 20, pp. 487-526.

[19] Herbert, Th. and Bertolotti, F. P. (1987) "Stability analysis of nonparallel boundary layers," *Bull. Am. Phys. Soc.*, Vol. 32, p. 2079.

[20] Herbert, Th. and Crouch, J. D. (1989) "Threshold conditions for breakdown of laminar boundary layers," in Proc. IUTAM Symposium on Laminar-Turbulent Transition, Ed. R. Michel and D. Arnal, Toulouse, France.

[21] Itoh, N. (1974) "Spatial growth of finite wave disturbances in parallel and nearly parallel flows, Part 2.," *Trans. Japan Soc. Aeron. Space Sci.*, Vol. 17, pp. 175-186.

[22] Kachanov, Yu. S. and Levchenko, V. Ya. (1984) "The resonant interaction of disturbances at laminar-turbulent transition in a boundary layer," *J. Fluid Mech.*, Vol. 138, pp. 209-247.

[23] Kachanov, Yu. S., Kozlov, V. V. and Levchenko, V. Ya. (1977) "Nonlinear development of a wave in a boundary layer," *Izv. AN USSR, Mekh. Zhidk. i Gaza*, Vol. 3, pp. 49-53. (In Russian)

[24] Klebanoff, P. S., Tidstrom, K. D., and Sargent, L. M. (1962) "The three-dimensional nature of boundary-layer instability," *J. Fluid Mech.*, Vol. 12, pp. 1-34.

[25] Kleiser, L. and Gilbert, N. (1989) "Simulation of transition to turbulence," in Proc. 1989 Newport Conference on Turbulence, Newport, RI. To be published.

[26] Lin, C. C. (1955) *The Theory of Hydrodynamic Stability*, Cambridge University Press.

[27] Maseev, L. M. (1968) "Occurrence of three-dimensional perturbations in a boundary layer," *Fluid Dyn.*, Vol. 3, pp. 23-24.

[28] Nagata, M. and Busse, F. M. (1983) "Three-dimensional tertiary motions in a plane shear layer," *J. Fluid Mech.*, Vol. 135, pp. 1-28.

[29] Orszag, S. A. and Patera, A. T. (1980) "Subcritical transition to turbulence in plane channel flows," *Phys. Rev. Lett.*, Vol. 45, pp. 989-993.

[30] Pierrehumbert, R. T. and Widnall, S. E. (1982) "The two- and three-dimensional instabilities of a spatially periodic shear layer," *J. Fluid Mech.*, Vol. 114, pp. 59-82.

[31] Rozhdestvensky, B. L. and Simakin, I. N. (1984) "Secondary flows in a plane channel: their relationship and comparison with turbulent flows," *J. Fluid Mech.*, Vol. 147, pp. 261-289.

[32] Saric, W. S. and Nayfeh, A. H. (1977) "Nonparallel stability of boundary layers with pressure gradients and suction," in Laminar-Turbulent Transition, pp. 6/1-21, AGARD CP-224.

[33] Smith, F. T. (1979) "Nonlinear stability of boundary layers for disturbances of various sizes," *Proc. Roy. Soc. Lond. A*, Vol. 368, pp. 573-589. (And corrections 1980, A371, p. 439).

[34] Smith, F. T. and Stewart, P. A. (1987) "The resonant-triad nonlinear interaction in boundary-layer transition," *J. Fluid Mech.*, Vol. 179, pp. 227-252.

[35] Spalart, P. R. and Yang, K.-S. (1987) "Numerical study of ribbon-induced transition in Blasius flow," *J. Fluid Mech.*, Vol. 178, pp. 345-365.

[36] Van Dyke, M. (1975) *Perturbation Methods in Fluid Mechanics*, Parabolic Press, Stanford, CA.

[37] Van Dyke, M. (1984) "Computer-extended series," *Ann. Rev. Fluid Mech.*, Vol. 16, pp. 287-309.

THEORY ON INSTABILITY AND TRANSITION

Frank T. Smith

Department of Mathematics
University College London
Gower Street, London WC1E 6BT

1. Introduction

This research is concerned with the fundamental fluid dynamics governing instability and transition to turbulence in boundary layers. In particular, rational theory plays a key role in exploring the major aspects of *nonlinear* dynamics which are central to the transition process and to turbulent boundary-layer phenomena. These aspects include three-dimensionality, intermittency, bursting, gradual transition, by-pass transition, scale cascades, and interaction. The theory takes advantage of the fact that the Reynolds numbers of most practical interest in transition and turbulence are enormous, thus pointing immediately to the use and relevance of scaling analysis, the starting point of the viscous-inviscid interaction theory. The theory moreover appears at its most powerful, and useful, in the nonlinear regime which is the key area. In principle, all the major aspects above can then be captured and explained theoretically in terms of their scales and their interactions with different parts of the flowfield; the beginnings for such explanations are contained in some of the references cited below. The highlights of the research, mainly into the nonlinear three-dimensional (3D) unsteady fluid dynamics of transitional boundary layers, and the aims, are summarized below.

2. The Theoretical Approach

Our emphasis tends to be on truly nonlinear theories, in which the boundary layer's mean-flow profile is completely altered from its original form. This is as opposed to linear or weakly nonlinear analyses, of which there are many interesting examples (e.g., Orr-Sommerfeld based, Stuart-Watson-Landau, Benney-Lin, and the small-disturbance version of the present work), but in which the mean profile suffers little or no change, meaning that the relevance of such analyses to transition and turbulence is of course very limited. Truly nonlinear theories are most relevant, in contrast, and they appear to fall into the four main categories, namely *nonlinear Tollmien-Schlichting theory, nonlinear Görtler theory, nonlinear high-frequency/Euler theory*, and *vortex/wave nonlinear interaction theory*, aspects of which are addressed in the following subsections. There are few (if any) other truly nonlinear theories especially with a rational basis as here: the main alternative is large-scale DNS computation, apart perhaps from the interesting but *ad hoc* theories mentioned by Arnal [1], Morkovin [2], Sreenivasan [3], and Huerre [4] for example. Again, the present rational theory really excels in the nonlinear regime where it yields an alternative to DNS simulation at high Reynolds numbers Re and provides guidelines, theoretical and physical insight, parameter control, scales, and suggestions for improved numerical procedures (particularly desirable at high speeds). This type of approach, which involves computations hand-in-hand with analysis, has also been shown to be of practical value in comparisons [5-7, 26] with experiments and direct simulations.

2.1. Nonlinear TS transitions

These are governed typically by nonlinear triple-deck theory. In nonlinear unsteady triple-deck interactions, the response of the boundary layer to external or internal forcing, or to a sizeable initial input disturbance, is captured by the unsteady 3D interactive-boundary-layer equations (holding in the lower deck near the wall) with the unknown pressure and displacement coupled together [5-10]. The coupling is via the inviscid properties fed back from the rest of the boundary layer and the external motion.

The major features of this nonlinear system are as follows.

(i) The small-disturbance version [5-11] describes *linear stability* (TS waves) and *receptivity. Weakly nonlinear* analysis [5,6,12] accounts for certain mildly nonlinear phenomena observed such as initial bifurcation, resonant triads, secondary 3D instability and vortex induction. Both the linear and the weakly nonlinear work is in fair agreement with experiments.

(ii) In the *fully nonlinear* phase, the mean-flow profile in the lower deck can veer far away from its original form, producing an effect all across the boundary layer. Many types of 2D and 3D nonlinear response can then be triggered off [8,12], depending on the input amplitude, wavenumber and frequency; for example, the 3D-triple-deck contains, as a subset, one form of vortex/wave interaction [20,25]. The relations between these different forms of 3D nonlinear response are of much interest.

(iii) A significant feature as regards transition to turbulence is believed to be the recent prediction of finite-time *bursting* [7] in the full nonlinear system. This is felt to be associated ultimately with *intermittency* and a cascading in scales (much like that described in Section 2.2); the time scale involved is similar to empirical values [3] and the theory is in keeping with DNS computations [13]. Current work is extending the detailed analysis of the bursting process to encompass 3D motion, to discover how general the prediction is, and to find any other routes for the ultimate behavior of the nonlinear 3D flow, in contrast with the 2D one. Further, the bursting is predicted [7] as being possible for any physically sensible interaction law, including cases which correspond to by-pass transition in linearly stable flows. This also remains to be followed through in the 3D regime.

2.2. Euler-stage interactions

The Euler/high-frequency nonlinear theory stems from research [10,14] on the effects of enhanced disturbance size. It is found that relatively *large disturbances* to the boundary layer can evolve according to the unsteady 3D Euler equations but subject to sporadic eruptions of vorticity from a viscous sublayer at the wall. Weak versions include nonlinear critical-layer effects. The Euler stage results either from the downstream development of a nonlinear growing TS wave (gradual transition) or from increased input disturbance, e.g., incoming from the free stream (by-pass transition).

Here again the mean-flow profile is completely altered. The flow structure induced coincides to some extent with that underlying much of the on-going computational work on turbulence structures, e.g., by Hassan, Frisch, Lesieur, Herring, Orszag, and at Stanford, under various conditions [15-17], although here our goal is to incorporate viscous-inviscid influence in a rational way. The theory also appears to be promising so far as regards increasing understanding of the later developments in transitional/turbulent flow structure (e.g., see computations [23,24] by Zang, Erlebacher, Hussaini, Kleiser).

Thus, a recent extension [18] predicts rationally three principal length scales for a turbulent boundary layer, $\mathrm{Re}^{-\frac{1}{2}}\ell n\mathrm{Re}$, $\mathrm{Re}^{-\frac{3}{4}}$, $\mathrm{Re}^{-1}\ell n\mathrm{Re}$, times the airfoil chord. These are for the displacement thickness, the small-scale eddy interaction distance, and the turbulent wall-layer thickness, respectively. The last two scales agree exactly with, in turn, *the Kolmogorov interaction scale* and *the inner empirical thickness* of the Cebeci-Smith turbulence model for instance, which is very encouraging. The argument for the $\mathrm{Re}^{-1}\ell n\mathrm{Re}$ inner thickness pivots on a cascading of the length and time scales, locally, involving successive renormalization. Current research is extending the Euler-stage theory to 3D flow, first, and then considering the relevance of statistical methods in the scaling arguments to account for random influences present. The first step here requires analyses of 3D nonlinear critical layers erupting from the boundary layer, the 3D bursting of the viscous sublayer, and the possibility (suggested by computations but to date unconfirmed) of a finite-time singularity in the 3D Euler equations. The second step above would hope to provide an alternate rational account *inter alia* of the Kolmogorov $k^{-\frac{5}{3}}$ spectrum decay.

2.3. Vortex/wave interactions

The theory of vortex/wave nonlinear interaction, which is a form of short-scale-long-scale interplay, is a recent development [12]. It has become apparent that there are many such interactions, of different scales, and all of them are potentially powerful, altering the mean-flow profile even when the input wave is still of remarkably small amplitude. The nonlinear interactions can take place for instance for a *vortex* flow with nondimensional velocity vector $(u, v, w) \sim (1, \mathrm{Re}^{-\frac{1}{2}}, \mathrm{Re}^{-\frac{1}{2}})$, in cartesians (x, y, z) of order $(1, \mathrm{Re}^{-\frac{1}{2}}, \mathrm{Re}^{\frac{1}{2}})$, coupled with a TS-type *wave* in which (u, v, w) is small, within the $0(\mathrm{Re}^{-\frac{1}{2}})$ scaled boundary layer. The interaction may also arise, however, for Rayleigh-type nonlinear waves instead [25]. There are also connections with nonlinear Görtler transitions referred to earlier [12]. The typical controlling equations are parabolic in x and are given in some of the recent references [19,20], mainly for a single TS-type wave present, in which case the vortex flow is forced nonlinearly by amplitude-squared effects from the wave, while the wave is driven along nonlinearly by the skin friction of the strong induced vortex flow. As far as we know there is no alternate theory explaining the nonlinear interactions, although experimental observations of persistent vortex effects and waves have been made over many years.

Current research is proceeding as follows.

(i) Compute and analyze the behavior of the nonlinear interactions, including the suggested connection [20] with lambda-vortex eruption and collapse.

(ii) Multiple wave modes can arise, of TS or Rayleigh type, each controlled by a separate equation but acting in concert to force the vortex flow. It is possible that these modes can switch "on" and "off" as the interactive flow progresses downstream [25], thus giving a form of intermittency response.

(iii) The nonlinear interaction with Rayleigh waves excited may force the corresponding critical layer to erupt from the original boundary layer, linking with Subsection 2.2 to provide a new estimate for the increased boundary-layer thickness. The role of vortex/wave interactions throughout a turbulent boundary layer structure is unclear but again potentially powerful.

3. Compressible Boundary Layers

A large part of the nonlinear flow research described in Section 2 is focussed on incompressible boundary layers and channel flows, and it needs developing into the *compressible* range especially. The beginnings for this development are in recent theoretical work, or work in progress, on the flow structures concerning compressible/hypersonic mode-1 instability [22]; mode-2 instability; steady hypersonic flows and their instabilities; aspects of transonic instability and transition; forms of vortex/wave interaction; wall cooling; and compressible triple-deck interactions. These suggest many lines of further theoretical investigation, as scale analysis and parameterization are essential to shed light on compressible boundary-layer transition to turbulence. As before, and as in most new developments, linear and weakly nonlinear theory can provide some helpful groundwork, but the emphasis should move on eventually to the truly nonlinear features of transition. These include the progressions described above for gradual transition, e.g., from Section 2.1 to Section 2.2 or from Section 2.1 to Section 2.3 to Section 2.2, as transition intensifies.

New work is advancing the nonlinear theories into the compressible range, to analyze their major properties. The research noted in the previous paragraph, and other work, indicates that this should throw up many new features, rather than being merely a passive step. For example, the extension from References 12, 20 reveals the presence of significant nonlinear interaction between vortex flows and mode-2 (inviscid) waves [21]; while the full compressible/hypersonic version of the break-up singularity referred to in Section 2.1 suggests a by-pass transition route which remains to be studied.

Other flow regimes examined include the effects of *cross-flow* and *nonparallelism.*

4. Connections with Experiments and Computations

Various connections with experimental and computational findings exist [7,10,14] on the following essentially *nonlinear* phenomena: (i) sublayer bursting, (ii) intermittency, (iii) low-frequency responses for high amplitudes, (iv) turbulent Kolmogorov scale, (v) turbulent wall-layer scale, (vi) by-pass transition processes, produced well away from the linear neutral conditions, occurring due to increased input amplitudes, (vii) persistence of relatively long-scale vortices, and their subsequent break-up.

These encouraging links are described in the references and tend to confirm the physical insight provided by the theory.

5. Overall Trends

The emphasis here, attempting to define the general directions and certain short- and long-term goals, is as follows.

(i) Concentrate on extending *nonlinear* theory, especially in 3D, as this is the most relevant to transition.

(ii) Aim at understanding flow structures, to provide the necessary physical insight, e.g., for lambda vortices and their collapse, bursting, Klebanoff-type breakdown.

(iii) Continue the relevant computations alongside the theory.

(iv) Advance in the compressible and other flow regimes.

(v) Connect with the development of DNS and related numerical schemes (cf. Ref. 26) to capture flow structures, and variation with Reynolds number, accurately.

(vi) Continue comparisons with experiments and DNS computations, encouraged by the links so far, and aim in particular at more understanding of empirical transition criteria.

References

[1] Arnal, D., 1984, Stability and Transition of Laminar Flow, AGARD Report 709 (2-1 to 71).

[2] Morkovin, M. V., 1988, AIAA Paper No. 88-3675.

[3] Sreenivasan, K. R., in Turbulence Management and Relaminarization, Springer-Verlag, 1988.

[4] Huerre, P., in Instabilities and Non-Equilibrium Structures, D. Reidel Pub. Co., 1987.

[5] Smith, F. T., 1979, Proc. Roy. Soc., A*366*, 91, and A*368*, 573.

[6] Smith, F. T. and Stewart, P. A., 1987, J. Fluid Mech., *179*, 227 (and UTRC Report 86-26).

[7] Smith, F. T., 1988, Mathematika *35*, Part 2, 256-273. "Finite-time break-up can occur in any interacting unsteady boundary layer."

[8] Smith, F. T., 1985, UTRC Report 85-36 (and J. Fluid Mech., 1986, *169*, 353).

[9] Duck, P. W., 1985, J. Fluid Mech., *160*, 465.

[10] Smith, F. T. and Burggraf, O. R., 1985, Proc. Roy. Soc., A*399*, 25.

[11] Goldstein, M. E., 1984, J. Fluid Mech., *145*, 71.

[12] Hall, P. and Smith, F. T., 1988, Proc. Roy. Soc., A*417*, 255 and ICASE Report 88-46 (Euro. J. Mech., 1989, B*8*, 179-205). "Nonlinear Tollmien-Schlichting/vortex interaction in boundary layers."

[13] Fasel, H., 1984, in Turbulence and Chaotic Phenomena in Fluids, Elsevier Sci. Pub.

[14] Smith, F. T., 1988, "Interactions in boundary-layer transition." Invited Lecture, ICTAM Conference, Grenoble, August, published in "Theoretical and Applied Mechanics" (eds., P. Germain, M. Piau, and D. Caillerie), Elsevier Sci. Pub. B.V. (North-Holland), 1989.

[15] Frisch, U. She, Z.-S., and Sulem, P.-L., 1987, Physica *28D*, 382.

[16] Lesieur, M., Staquet, C., LeRoy, P. and Comte, P., 1988, J. Fluid Mech., *192*, 511.

[17] Moin, P., Leonard, A., and Kim, J., 1986, Phys. Fluids, *29*, 955.

[18] Smith, F. T., Doorly, D. J., and Rothmayer, A. P., 1989, Proc. Roy. Soc. A, submitted (based on Smith Report UTRC 87-43). "On displacement- thickness, wall-layer, and mid-flow scales in turbulent boundary layers, and slugs of vorticity in channel and pipe flows."

[19] Bennett, J., Hall, P., and Smith, F. T., 1988, ICASE Report 88-45 (submitted to J. Fluid Mech., 1989). "The strong nonlinear interactions of Tollmien-Schlichting waves and Taylor-Görtler vortices in curved channel flow."

[20] Smith, F. T., 1988, ICASE Report No. 88-66 (and Mathematika 1989, *36*, Part 2, with A. G. Walton). "Nonlinear interaction of near-planar TS waves and longitudinal vortices in boundary-layer transition."

[21] Mack, L. M., 1984, AGARD Report 709 (3-1 to -81).

[22] Smith, F. T., 1987, UTRC Report 87-52 (and J. Fluid Mech., 1989). "On the first-mode instability in subsonic, supersonic, or hypersonic boundary layers."

[23] Kleiser, L. and Laurien, E., 1985, "Three-dimensional numerical simulation of laminar-turbulent transition and its control by periodic disturbances," in Laminar-Turbulent Transition, ed. V. V. Kozlov, pp. 29-37, Springer-Verlag.

[24] Zang, T. A. and Hussaini, M. Y., 1985, "Numerical experiments on the stability of controlled shear flows," AIAA Paper No. 85-1698.

[25] Hall, P. and Smith, F. T., 1989, "Vortex/wave nonlinear interactions in boundary layers," ICASE Report, in preparation.

[26] Smith, F. T., Papageorgiou, D., and Elliott, J. W., 1984, "An alternative approach to linear and nonlinear stability calculations at finite Reynolds numbers," J. Fluid Mech., *146*, pp. 313-330.

HIGH-SPEED TRANSITION EXPERIMENTS

PANEL SUMMARY: HIGH-SPEED TRANSITION EXPERIMENTS

Dennis Bushnell

NASA Langley Research Center
Hampton, VA 23665-5225

The need for transition information at high speeds is both urgent and vital to the success of several national programs, most notably NASP. Several classes of experimental facilities can be used to provide such information, but each has problems vis-à-vis transition research. The usual litany of high-speed facility options includes: (a) conventional (blowdown, impulse, continuous) "cold" tunnels, (b) hot/high-energy tunnels (e.g., propulsion facilities), (c) ballistic ranges and, (d) atmospheric flight. Research in the late 1960's and early 1970's strongly indicated that sound radiation from turbulent nozzle wall boundary layers tended to dominate the transition physics in conventional wind tunnels. Especially in the high-energy facility case, stagnation enthalpy fluctuations can also create stream velocity fluctuations. The problems in the free-flight case, both range and atmospheric, tend to involve questions of data coverage and measurement capability, repeatability, and *expense.*

The experimental information of interest for high-speed transition research should obviously include details of 3-D wave dynamics, for which the current instrumentation capabilities are only partially suitable. At high speeds, normal hot wires measure total temperature and mass flow fluctuations and their correlation while slanted ones measure $(\rho u)'v'$ not $(\rho v)'u'$. However, sufficient frequency response is available so that, especially when employed using multiple wire arrays, significant dynamic information can be obtained. However, improved high-speed dynamic instrumentation is obviously required, particularly for receptivity studies where the initial disturbances can be of low amplitude. Also, the numerical simulations produce extraordinarily detailed information, which should be subjected to experimental scrutiny.

Reasonable experimental progress in the high-speed transition area requires development and utilization of low-disturbance quiet tunnels; a pilot version of such a facility, at $M_\infty = 3.5$, was developed over a ten-year period at NASA Langley Research Center by a group led by Ivan Beckwith. In this tunnel, extensive upstream treatments, suction slots upstream of the throat and ultra-smooth walls are used to delay transition on the nozzle walls up to high Reynolds numbers. Through tripping of the wall boundary layer, direct comparison can be made between quiet and noisy conditions for various transitioning flows. Results of such studies indicate that transitional region extent and the effects of model geometry, wall temperature, angle of attack, and bluntness are all influenced to first order by the noise fields in conventional facilities. The facility noise is far less important for case where the local flow itself produces significant or overwhelming disturbance fields such as large roughness effects or free-shear layers with adjacent surfaces and subsonic reverse flow. Currently, Langley has high-Reynolds-number quiet tunnels under development for Mach numbers of 3.5, 6, 8, and 20 with the Mach 20 version employing Helium as a test medium. NASA Lewis and NASA Ames plan low-Reynolds-number quiet tunnels primarily and specifically for receptivity studies where the actual transition location is not particularly required.

The key research issues for experimental high-speed transition studies include: (a) documentation of initial disturbance fields of all types including spectra, wave orientation, etc., (b) receptivity measurements and physics, especially the influences of subsonic nose regions and their attendant shock waves, (c) *detailed* transition studies of not only boundary layers but also separated and vortical flows including effects of adverse pressure gradients, shock interaction, flow curvature of all types, roughness, etc., (d) documentation of the transitional region (between transition and "fully turbulent" flow), again for multitudinous boundary conditions, (e) determination of flow chemistry effects and whether chemistry- specific instabilities are present, especially for exothermic reactions, (f) systems/practical aspects of supersonic laminar flow control and, (g) dynamics of the nonlinear end game. What is currently known is miniscule compared to the needs. The lack of quiet facilities has hampered the high-speed transition field for far too long. Sufficient facilities should be available within one to two years to allow significant advancements in the field.

SUGGESTED FUTURE DIRECTIONS IN HIGH-SPEED TRANSITION EXPERIMENTAL RESEARCH

Dennis Bushnell

NASA Langley Research Center
Hampton, VA 23665-5225

The substantiation in the late 1960's of the general dominance of radiated facility wall boundary-layer noise upon high-speed transition measurements in wind tunnels, along with the essential collapse of interest in high-speed aerodynamic research in the early 70's, had an understandably severe adverse effect upon the pace of experimental high-speed transition research. The community understood that further research progress required the invention and development of high-speed quiet (and otherwise low-disturbance) facilities. Due to funding actualities, the only sustained effort aimed at the development of such a facility was conducted (from 1972 - present) at the NASA Langley Research Center by a group headed by I. E. Beckwith. The quiet tunnel development problem is an extremely difficult one, being essentially a laminar flow control problem with the crippling condition that local suction cannot, in general, be used to accomplish the laminarization due to consequent production of finite strength waves at the suction sites. This wave system usually causes stream disturbances of a higher level than the turbulent wall radiated noise which one is trying to mitigate. The Langley effort indicated that mere reduction of the noise level (initially on the order of 1% in terms of $\left(\frac{P'_\infty}{P_\infty}\right)$) to the order of 0.3% (but still with turbulent wall flow) was just not sufficient, the high-frequency content of the disturbance field still affected transition to first order. The stream $P\prime$ level had to be reduced to 0(0.05%), which is the order of the flight level, to agree with flight (in terms of transition behavior). The only method of accomplishing this was to laminarize the facility wall boundary-layer flow. By 1983, after several false starts and blind alleys, Beckwith and his group were successful, by using a combination of techniques, in producing a viable pilot $M = 3.5$ quiet tunnel. This facility, still unique in the world, and several similar (but at other Mach numbers) follow-on facilities currently under design and construction at Langley allowed a revitalization of high-speed experimental transition research. This revitalization is particularly timely, as it is both aided and necessitated by present-day applications such as NASP and supersonic laminar flow control and spectacular developments in related/complementary research areas such as stability theory and the numerical simulation of transitioning flows. The status and comments on historical developments in the various subject areas of high-speed experimental transition research are contained in the position statements by the various panel members in the present volume. The following discussion concerns the suggested future directions in high-speed experimental transition research as determined by the panel membership. This discussion and listing of suggested research ares will undoubtedly and understandably be overtaken by future research developments. In all of the areas mentioned, there should be strong coupling with theory and/or numerical calculations.

Suggested Key Research Issues

1. Measurement and modeling of initial disturbance fields, both in ground facilities and flight, for all modes (acoustic, vorticity, entropy, particles, electrostatics, etc.) with the ultimate goal of true space-time specifications (including intermittency information). What is required is spectra (as a minimum) along with wave length and orientation.

2. Development of advanced high-speed instrumentation for disturbance field measurements with an emphasis upon nonintrusive techniques. Particle response delays in the presence of shock waves limits the applications of conventional LDV at high speeds and new techniques (LIF, electron beams, perhaps even ion beams) are needed to provide the detailed high-frequency experimental information on disturbance presence/instability required for both physical insight and detailed comparison with computation/theory.

3. Measurements of the details of receptivity in multitudinous flows (e.g., various types of 2-D/3-D leading edge and body geometries with various initial disturbances). The low level of initial disturbance fields typical of most flight transition situations and required in quality ground facilities makes such measurements extremely difficult.

4. Further development and use of high-speed quiet tunnels, including the development of hypersonic facilities ($M = 6,8$, and 0(15) tunnels currently under development at Langley), and, in particular, detailed measurements of the disturbance growth/physics over various bodies in both quiet conditions and in the presence of various imposed disturbance fields.

5. Stability and transition (physics, location) studies for multitudinous flows, including boundary layers, free flows, vortices, and separated and corner flows for various body geometries and including parametrics such as curvature, shock interaction, roughness, wall temperature, etc., i.e., flows beyond (and including) simple plates and cones should be explored in detail.

6. Detailed studies of the transitional region (downstream of the transition point) for boundary layers, free flows, vortices separated flows, corner flows, etc., including region extent, intermittency, turbulence amplification/overshoot, and resolution of several previous observations concerning the absence of conventional turbulence "spots" in high Mach number boundary-layer transitional regions. This information is again required for 2-D and 3-D body geometries and over parameter ranges typical of various flight applications.

7. Studies of flow chemistry affects upon the various transition phenomena. Such chemical effects are endemic in high hypersonic flight, but there is little realistic prospect, at this time, of the development of a high-speed quiet tunnel at the enthalpy levels required to simulate chemistry effects. Therefore, these experimental studies must, of necessity, be carried out either in ballistic ranges or free (atmospheric) flight.

8. Various applied research is required including the physics of perforated surface suction stabilization in the presence of swept (crossflow) disturbance growth and shock and simulated fuselage radiated noise interaction.

9. Various anomalies in the existing high-speed data base should be resolved, including: (a) Morkovin's "blunt-body paradox", (b) the apparent unit Reynolds number effect for slightly blunted cone ballistic range data, and (c) the source of the high-frequency

growth present in the Demetriatis, Stetsen, and Kendall hypersonic stability experiments.

HIGH-SPEED QUIET TUNNELS

Ivan E. Beckwith

NASA Langley Research Center
Hampton, VA 23665-5225

1. Introduction

The powerful adverse effect of wind tunnel noise on boundary layer transition was conclusively demonstrated about two decades ago by the remarkable correlations of Pate and Schueler at AEDC. They showed that transition Reynolds numbers were accurately correlated by noise dependent parameters in several different tunnels over the Mach number range from 3 to 20. Their correlations established that the high levels of wind tunnel noise were caused by acoustic noise radiation from the tunnel walls. Later at the Langley Research Center, Stainback showed that a quantitative relationship exists between wind tunnel rms pressure levels and transition.

Following Morkovin's comprehensive review in 1969 of the high-speed transition problem and his recommendation for the development of a low-noise hypersonic tunnel (based in part on the revelations of Pate and Schueler), the Boundary Layer Transition Study Group, chaired by Reshotko, formulated a program for transition research that included the development of quiet wind tunnels. Progress reports on this part of the BLTSG Program by a group of researchers at Langley Research Center are available.

By 1983, this groups had developed and tested Pilot Quiet Tunnels at Mach numbers from 3 to 5 that exhibited very low noise levels in upstream regions of the test rhombus to high values of the stream unit Reynolds number. Laminar wall boundary layers were observed on upstream regions of the nozzle walls only when the turbulent boundary layer in the subsonic approach to the nozzles was removed by suction slots upstream of the throat, although contributing factors were the very low acoustic and vorticity disturbances in the settling chamber and the highly polished walls of the nozzles. Detailed studies have indicated that accurate flight transition simulation requires the maintenance of laminar nozzle wall flows, and that simply altering the wall turbulence to a lower noise radiation state is not sufficient. The acoustic field produced by nozzle wall turbulence not only reduces the *levels* of the model transition Reynolds numbers but, from recent research at the Langley Research Center, the *trends* are also modified (e.g., effects on transition of bluntness, model configuration, unit Reynolds number, etc.). Another surprising result obtained in the Mach 3.5 Pilot Tunnel is the lack of any effect of wind tunnel noise on transition at the leading edge of a swept wing except when small trips are attached to the leading edge.

2. Advances in Quiet Nozzle and Tunnel Design

In order to apply nozzle wall transition data from these studies to the design of improved nozzles, it was essential to understand the instability mechanisms involved and to develop theoretical models that can be used for predictive purposes. The development of satisfactory models depended on the successful application by Malik of linear stability theory to the correct local mean flow conditions in nozzles and to test data on cones and flat plates in the Mach 3.5 Quiet Tunnel. The principal results of the nozzle investigation was to show that transition in the wall boundary layers of nozzles was caused by the Görtler instability mechanism in the concave curvature regions of the wall rather than Tollmien-Schlichting waves. We concluded that application of linear stability theory with the e^N method to the experimental data from nozzles for Mach numbers from 3 to 5 gave N-factors from about 6 to 10 for transition caused by Görtler vortices. Malik had shown previously that the use of $N \simeq 9$ to 11 predicted transition for several other different flows.

The use of $N \simeq 9$ for the Görtler instability then predicts quiet test cores in the new Mach 3.5 and Mach 6 axisymmetric-long pilot nozzles that are 3 to 4 times longer than observed in the previous test nozzles. These new nozzles utilize a region of radial flow which moves the wall inflection point far downstream and thereby delays the onset and amplification of the Görtler vortices. Recent data from the Mach 3.5 axisymmetric, long nozzle are in excellent agreement with the e^N theory ($N = 9.2$) for this type of advanced nozzle design.

A new Mach 8 nozzle has now been designed by these same methods. This nozzle is scheduled for installation and tests by 1991 in the Mach 8 Variable Density Tunnel at NASA Langley.

Modifications are now underway to an existing conical nozzle in one of the helium tunnels at NASA Langley to provide some quiet test flow for the purpose of calibrating the e^N method at Mach numbers from about 14 to 17. The advantage of the conical nozzle is that the Görtler instability is not expected to be present since the nozzle wall has no concave curvature.

3. New Large-Scale, High-Speed, Low-Disturbance Tunnel

The final engineering design of this tunnel has now been started. Existing high pressure air and vacuum systems at NASA Langley will be used. The tunnel is designed to accommodate nozzles of various lengths with Mach numbers ranging from 2 to 6.

A two-dimensional Mach 3.5 nozzle and a two-dimensional Mach 6 nozzle will be provided with the initial procurement. These nozzles are scaled-up by a factor of about three times larger than the corresponding pilot nozzles. The allowable roughness defects on the nozzle walls and models will also be a larger by the same physical scale-up factor. This eases the nozzle surface finish problems considerably and also allows for much improved duplication of model surface details such as suction slots, holes, steps, and other fabrication details related to laminar flow control techniques on real aircraft. This new tunnel is scheduled for operation in 1992.

4. Conclusions

The techniques and requirements for the design and fabrication of low-disturbance supersonic/hypersonic wind tunnels have been developed and tested. These new low-disturbance facilities will be essential for evaluating and controlling viscous flows for a variety of high-speed applications including the development and testing of laminar flow control techniques for the second-generation SST. Boundary-layer stability and transition prediction codes can be calibrate/validated only in these low-disturbance tunnels over the speed range for applications to supersonic and hypersonic aircraft of the Twenty-first Century.

TRANSITION IN HIGH-SPEED FREE SHEAR LAYERS

A. Demetriades

Montana State University
Bozeman, MT 59717

1. Nomenclature

The laminar free shear layers considered here are formed by combinations of the velocities U_1 and U_2 and momentum thicknesses θ_1 and θ_2 of two adjacent parallel flows (Figure 1). Thus, flows of common interest are symmetric wakes ($U_1 = U_2, \theta_1 = \theta_2$), asymmetric wakes ($U_1 = U_2, \theta_1 \neq \theta_2$), Chapman flows ($U_1 \neq U_2, \theta_1 = \theta_2 = 0$), separated boundary layers ($U_2 = \theta_2 = 0$) and partition flows or mixing layers ($U_1 \neq U_2, \theta_1 \neq \theta_2$). These flows have a wide variety of velocity profiles and thus, to the extent that the latter are important to stability, should have differing transition characteristics. It is also greatly significant that these flows undergo fundamental profile changes in the flow direction. Their initial profile is usually very different from their asymptotic or "fully-developed" profile – usually a Gaussian or a "Chapman" profile. The latter obtains from the classical viscous asymptote for $U_1 \neq U_2$ [1], and characterizes the only separate category ("Chapman flows") of unchanging profile in the streamwise direction.

2. Experimental Evidence

Transition in wakes ($U_1 = U_2, \theta_1 = \theta_2$) is typically shown on Figure 2, which relates the flow properties just outside the wake at transition (subscript e) and the distance X_T from the object to transition. Large scatter prevails because of differences in geometry (2-D and axisymmetric with attendant differences in the profiles), differences in diagnostics, differences in transition definition, inviscid gradients, etc. The data correlation, due to Zeiberg [2] and containing noisy wind tunnel and unit Reynolds number effects, says that X_T increases as Me increases and Re decreases. Figure 3 from [3] says that wake transition moves downstream as the object is heated to some power Q; Batt [4] has confirmed the opposite, i.e., a decrease of X_T if Q is negative (cooling).

"Pure" free-shear layers of the Chapman type ($U_1 \neq U_2, \theta_1 = \theta_2 = 0$) have been studied by Birch and Keyes [5] and Demetriades [6]. As in the case of wakes, experimentalists correlate $\mathrm{Re}_{XT} = U_1 X_T / \nu_1$ vs. M_1, except that now Re_{XT} is weighted by a function $f(\lambda)$ of $\lambda = (U_1 - U_2)/(U_1 + U_2)$, so that $\mathrm{Re}_{XT} f(\lambda) = g(M_1)$. This supposedly accounts for the "convective" effect. For the $f(\lambda)$ indicated by Threshold theory (see below), the data plot as in Figure 4. As with wakes, the data again show that X_T increases as M_1 increases, and as λ and Re decrease.

The largest volume of free layer transition data published concerns separated flows ($U_2 = \theta_2 = 0$) such as driven by adverse gradients; boundary layers shed off the trailing edges of bluff bodies belong to the same group because their profile adjusts to a zero-wall-shear boundary layer profile very quickly after separation (Figure 1). The asymptote of these flows is the Chapman profile. Figure 5 shows the data from References [7], [8], [11], [23], [24], [25], [26] in the form Re_{XT} vs. M_1; note that in this case $\lambda = 1$. These data are fairly consistent in showing an uninterrupted Re_{XT} increase from about 10,000 to about 1.3 million in the M_1 range from 0 to 8. At $M_1 = 8$ the point due to Hahn [7] is for a highly cooled fluid on the slow side and could rise to about 5 million if adjusted to adiabatic. In fact both Hahn [7] and Larson and Keating [8] show a decreased of x_T as the slow fluid cools down (Figure 6), in agreement with the results for wakes shown just above.

A flow of special interest to fluid-mixing technology is the partition flow ($U_1 \neq U_2, \theta \neq \theta_2$), Figure 1, where laminar boundary layers θ_1 and θ_2 grow independently on either side of the partition before they merge past the trailing edge. High-speed partition flows with substantiated measurements of transition have been reported by Brower [9] and recently by Demetriades [10]. In these measurements, λ ranged from 0.06 to 0.1, despite of which these data, shown on Figure 7, seem to agree with the separated $\lambda = 1$ boundary layers of Figure 5. It must be noted that low-λ partition flows are wake-like (Figure 1), and thus could just as readily fit the wake transition correlation, copied on Figure 7 from Figure 2.

Like the wake flows, when the "slow" side of the separated, Chapman and partition flows are heated, transition moves downstream [7], [8]. For the moment, the scarce data available show, roughly, that a factor of 2 increase in that temperature causes a factor of 2 increase in X_T. Another very obvious influence is that of the unit Reynolds number, as shown in Figure 8. However, note that if X_T "sticks" at some position for some reason (e.g., a finite number of unstable wavelengths needed to cause transition) then a unit Reynolds number such as shown on Figure 8 will result. Finally, Figure 9 shows an interesting insensitivity of Re_{XT} to tunnel noise [11]. At this point, this result is one-of-a-kind, but if true it should encourage more wind-tunnel measurements on free layer transition which, as the data presented above imply, are badly needed.

Although experiments such as those quoted here have so far revealed only trends, and located transition quantitatively by a factor of 5 at best, they have determined a mechanism of great value to the numerical analysis of mixing phenomena. In the classic picture, lateral spreading and mixing increase abruptly only after the appearance of turbulence (randomization). Roshko and his coworkers have supplied evidence [12] [13] that at low speeds this role is taken over by large-amplitude instabilities, e.g., vigorous lateral motions of entraining vortices. The high-speed data quoted above, however, especially the photographs of Roshko and Papamoschou [14], Brower [9], and this writer [6] show that such organized integral motions, although present, are greatly suppressed. To ensure that this conclusion also accounts for three-dimensional instabilities undetectable by photography, Brower [9] probed his partition flow with a hot film anemometer. His spectra show that a 120 KHZ oscillation dominates the region from 200 θ_1 to 400 θ_1, past the trailing edge, but thereafter the spectrum randomizes quickly to several hundred KHZ, and at 600 θ_1 the great majority of the energy is random.

3. Theoretical Rationale

The Stability-Transition Connection

From the experimentalist's viewpoint, it is important to mention the theoretical efforts made to explain the observations and anticipate the parameters controlling transition. Since all reports show the appearance of periodic phenomena ahead of transition, it is presumed that stability theory is most relevant. From among the rapidly multiplying reports of high-speed free layer stability calculations, the following are examples of this relevance.

First, there are examples of qualitative and often quantitative agreement between linear stability and *stability* experiments, although the latter are very few. Zhuang, et al. [15] have shown satisfactory agreement of the maximum amplification rate of a two-dimensional inviscid disturbance (normalized with the same at zero "convective" Mach number) with data from Papamoschou and Roshko [14], and Behrens and Ko [16] have produced fluctuation data in a hypersonic wake which support quite well amplification rate spectra calculated from linear stability theory (see Figure 10).

Second, there are computations from which the amplification-rate integral N, as found from linear stability theory, approaches the "accepted" values from boundary-layer transition studies (i.e., $N = 5-10$), when the integration stops where high-speed shear layer data indicate transition onset. Thus, Gasperas [17], using viscous compressible linear theory and a two-dimensional disturbance, finds $N = 6.6 - 7.9$ for the wake flows reported in [18] and [19].

Third, there is complete unanimity among stability calculations, most of which are not quoted here, that the parameters whose increase stabilizes the high-speed free layers are those whose increase also delays transition and vice versa. In general, as the Mach number is increased or the Reynolds number decreased or both, the amplified frequency band shrinks and the growth rates in that band are lowered; perhaps most telling is the extremely low rates predicted as soon as the convection Mach number exceeds unity [15]. Furthermore, there are now computations by Macaraeg, et al., showing, as per Figure 11, that for some profiles such as the Chapman flow of Figure 1, heating the "slower" fluid has a stabilizing effect on the shear layer and vice versa [20]. In the same reference it is demonstrated that the Prandtl number has a strong effect on the location and number of inflection points. Finally, the observed increased resistance to transition as λ is lowered is also fully borne out by the stability calculations [21].

In the view of this writer, a recurring deficiency in some stability calculations is the use of overly simplified laminar profiles, e.g., the hyperbolic tangent profile as representative of the far-field Chapman profile. This is done usually for convenience and after invoking some multiple of θ_1, as a measure of the asymptotic state. It should be borne in mind that for flows of practical interest the laminar profile undergoes fundamental changes between the solid object and the far field. The characteristic gage for asymptotic equilibration is not θ_1, but $\theta_1 \mathrm{Re}_{\theta 1}$ (the initial momentum thickness times the initial momentum Reynolds number). A review of laminar calculations for the flows of Figure 1 show that asymptotic profiles cannot be used as good approximations at distances smaller than 10-30 times $\theta_1 \mathrm{Re}_{\theta 1}$. It is much preferable that stability calculations are done for the laminar profile as it evolves along the flow direction, and N factors computed by integrating all the growth rates so found, as was done by Gasperas for wakes [17] and by Ragab for partition flows [22].

Semi-empirical Approaches: The Threshold Theory

Capable as it is of foretelling the destabilizing parameters and perhaps also the lateral spreading and development of organized structures, stability theory cannot predict the transition to turbulence without empirical inputs such as some set N value (the "e^N calibration"). Occasionally however, physical principles can be used to estimate the transition location or provide the factors which affect it. One such approach is the "Threshold Theory" of the 1970's [3], discussed here by way of example.

For any given flow, this approach first calculates the distribution of the turbulence Reynolds number $\mathrm{Re}_\Lambda = u'\Lambda/\nu$ assuming that the entire flow is turbulent and using the best information available on the distribution of the scale Λ and velocity fluctuation u' in the flow. It then dismisses as non-self-sustaining ("laminar") those portions of the flow where Re_Λ lies below a certain threshold (e.g., $\mathrm{Re}_{\Lambda 0} = 15$). For most flows, Re_Λ increases going downstream, so that a minimum "threshold distance" x_T is found for each flow (more accurately a "distance upstream of which that flow cannot be turbulent"). Thus, x_T or Re_{xT} can be found for each flow as a function of the same factors affecting Re in its turbulent counterpart, i.e., the geometry, Mach number, heat transfer, etc. The condition on x_T is of course one of necessity without one of sufficiency; nevertheless, the x_T values so found bear close and interesting similarities to experimental data on the transition distance. At the outset, this "Threshold Theory" reproduces quite well the most common known differences in transition among different flows (e.g., axisymmetric vs. 2-D jet, compressible vs. incompressible wake) and also gives transition Reynolds number close to those observed for such flows.

For two-dimensional, high-speed wake flows, the transition Reynolds number calculated with the threshold theory [3] is $\mathrm{Re}_{XT} = 10,000\sigma(\gamma-1)^2 M_\theta^4(1+Q)^2$ where σ and γ are the Prandtl number and specific heat ratio, respectively; Q the thermal power exchanged between the wake-shedding object and the flow; and Me($\simeq M_1$) the wake edge Mach number. This relation is plotted here in Figure 2 for $Q = 0$, and also for $Q \neq 0$ in Figure 3.

The application of the Threshold Theory to Chapman-type free shear layers [6] gives Re_{XT} as a function of γ, M_1, λ, the temperature-viscosity exponent k, and the ratio T_{02}/T_{01} of the two total temperatures:

$$\mathrm{Re}_{XT} = 5 \times 10^4 \frac{1}{r^2 G^2} \left(\frac{T_{DSL}}{T_1}\right)^{2(k+1)} \left(\frac{\lambda + 1}{\lambda}\right)^2$$

where, in addition $r = r(M_1), G = G(\lambda, M_1, T_{02}/T_{01})$ and the dividing streamline ratio $(TDSL/T_1) = \mathrm{fn}(\gamma, M_1, \lambda, T_{02}/T_{01})$. Computations with this formula are shown here in Figures 4 and 6. As is the case of wake flows, all major trends discernible in the experiments are also predicted correctly by this approach.

4. Summary and Conclusions

The high-speed free shear flows of interest exhibit a variety of profiles and most often a drastic profile change in the flow direction. In these flows, pretransitional instabilities and large scale structures are considerably suppressed. Experiments also show that transition to turbulence is much delayed as the Mach number is increased, and delayed further if the slower portion of the moving fluid is heated relative to the faster fluid. In flows with velocity jumps, the speed ratio will, if decreased, also delay transition. There is complete qualitative agreement between these observations and linear stability theory as applied to date on the free shear layer problem; a semi-empirical transition estimate based on the physics of turbulence is also in good qualitative agreement with the experiments. However, the data scatter is large and the reported experiments themselves are too few in number; it is crucial to find what effect wind-tunnel noise has on free layer transition, since this will determine the chances of enriching the data base. Partition-flow measurements are especially needed over a wide combination of Mach numbers and temperatures, as well as non-photographic data on laminar instabilities, and a close scrutiny of the reported unit Reynolds number effect. Theoretically, it is urged that linear stability calculations address the entire flow from the near field on, rather than depend on the asymptotic profiles.

References

[1] Chapman, D. R.: "Laminar Mixing of a Compressible Fluid," NACA TR 958, Washington, DC, 1950.

[2] Zeiberg, S.: "Transition Correlations for Hypersonic Wakes," AIAA J., Vol. 2, No. 3, March, 1964, p. 564.

[3] Demetriades, A.: "Transition to Turbulence in Two-Dimensional Wakes," AIAA J., Vol. 16, No. 6, June, 1978, p. 587.

[4] Batt, R. G.: "Experimental Investigation of Wakes behind Two-Dimensional Slender Bodies at Mach Number 6," Ph.D. Thesis, California Institute of Technology, Pasadena, CA, 1967.

[5] Birch, S. F. and Keyes, J. W.: "Transition in Compressible Free Shear Layers," J. of Spacecraft and Rockets, Vol. 9, No. 8, August, 1972, p. 623.

[6] Demetriades, A., Ortwerth, P. J., and Moeny, W. M.: "Growth and Transition Behavior of Arbitrary Laminar Free Shear Layers," AIAA J., Vol. 19, No. 9, September, 1981, p. 1091.

[7] Hahn, M.: "Experimental Investigation of a Separated Flow over a Cavity at Hypersonic Speed," AIAA J., Vol. 7, No. 6, June, 1969, p. 1092.

[8] Larson, H. K. and Keating, S. J., Jr.: "Transition Reynolds Numbers of Separated Flows at Supersonic Speeds," NASA TN D-349, Washington, DC, 1960.

[9] Brower, T. L.: "Experiments on the Free Shear Layer between Adjacent Supersonic Streams," M.S. Thesis, Montana State University, March, 1983.

[10] Demetriades, A.: "Transition to Turbulence in a Laminar Hypersonic Partition Flow," MSU/SWT IM 89-01, Montana State University, March, 1989.

[11] King, R., Creel, T., and Bushnell, D.: "Experimental Study of Free-Shear Layer Transition Above a Cavity at Mach 3.5," AIAA Paper No. 89-1813, AIAA, New York, 1989.

[12] Brown, G. L. and Roshko, A.: "On Density Effects and Large Structures in Turbulent Mixing Layers," JFM, Vol. 64, Part 4, 1974, p. 775.

[13] Konrad, J. H.: "An Experimental Investigation of Mixing in Two-Dimensional Turbulent Shear Flows with Applications to Diffusion-Limited Chemical Reactions," Ph.D. Thesis, California Institute of Technology, Pasadena, CA, 1976.

[14] Papamoschou, D. and Roshko, A.: "Observation of Supersonic Free Shear Layers," AIAA Paper No. 86-0162, AIAA, New York, 1986.

[15] Zhuang, M., Kubota, T., and Dimotakis, P. E.: "On the Instability of Inviscid, Compressible Free Shear Layers," AIAA Paper 88-3538, AIAA, New York, 1988.

[16] Behrens, W. and Ko, D. R. S.: "Experimental Stability Studies in Wakes of Two-Dimensional Slender Bodies at Hypersonic Speeds," AIAA J., Vol. 9, No. 5, May, 1971, p. 851.

[17] Gasperas, G.: "Transition Prediction for Two-Dimensional Viscous Supersonic Wakes," AIAA Paper No. 89-0284, AIAA, New York, 1989.

[18] Demetriades, A.: "Turbulence Measurements in a Two-Dimensional Supersonic Wake," Physics of Fluids, Vol. 13, No. 7, July, 1970, p. 1672.

[19] Demetriades, A.: "Heat Transfer Effects on Supersonic Wake Transition," Physics of Fluids, Vol. 13, No. 1, January, 1970, p. 204.

[20] Macaraeg, M. G., Streett, C. L., and Hussaini, M. Y.: "A Spectral Collocation Solution to the Compressible Stability Eigenvalue Problem," NASA TP No. 2858, NASA LaRC, 1988.

[21] Ragab, S. A. and Wu, J. L.: "Instabilities in the Free Shear Layer Formed by Two Supersonic Streams," AIAA Paper No. 88-0038, AIAA, New York, 1988.

[22] Ragab, S. A.: "Instabilities in the Wake/Mixing-Layer Region of a Splitter Plate Separating Two Supersonic Streams," AIAA Paper No. 88-3677, AIAA, New York, 1988.

[23] Crawford, D. H.: "Investigation of the Flows over a Spiked-Nose Hemisphere-Cylinder at a Mach Number of 6.8," NASA TN D-118, Washington, DC, 1959.

[24] Sato, H.: "Experimental Investigation on the Transition of Laminar Separated Layers," J. of the Phy. Soc. of Japan, Vol. 11, No. 6, June, 1956, p. 702.

[25] Reller, J. O., Jr. and Hamaker, F. M.: "An Experimental Investigation of the Base Pressure Characteristics of Non-Lifting Bodies of Revolution at Mach Numbers from 2.73 to 4.98," NACA TN 3393, Washington, DC, March 1955.

[26] Chapman, D. R., Kuehn, D. M., and Larson, H. K.: "Investigation of Separated Flows in Supersonic and Subsonic Streams with Emphasis on the Effect of Transition," NACA TR 1356, Washington, DC, 1958.

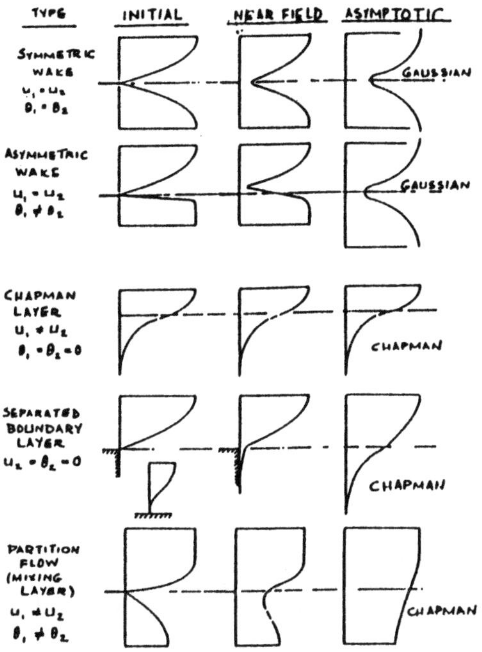

FIGURE 1. Typical free-shear-layer velocity profiles

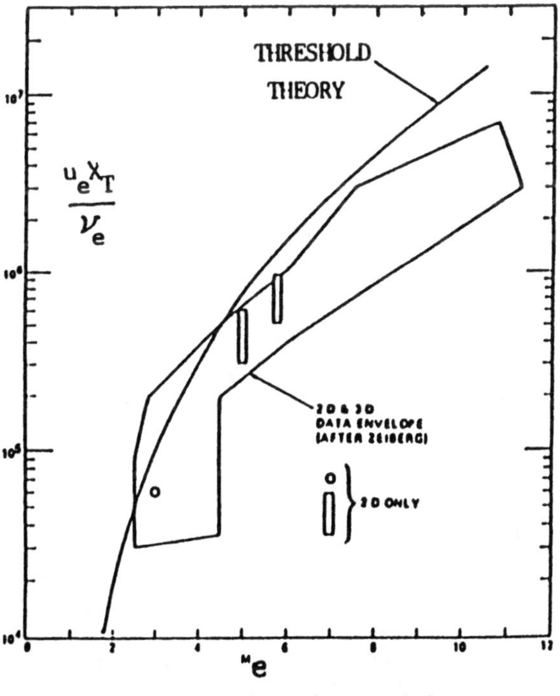

FIGURE 2. Correlation of transition to turbulence in high-speed wakes

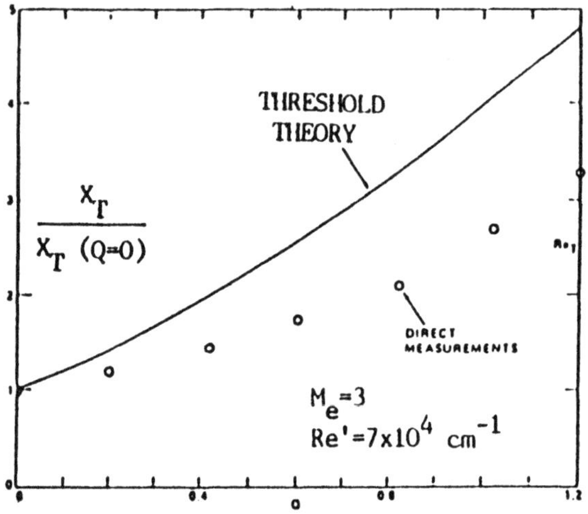

FIGURE 3. Test of the heat-transfer effect prediction of the Threshold theory, for wake flows. Data are from [19], and Q the non-dimensional power exchanged between object and fluid.

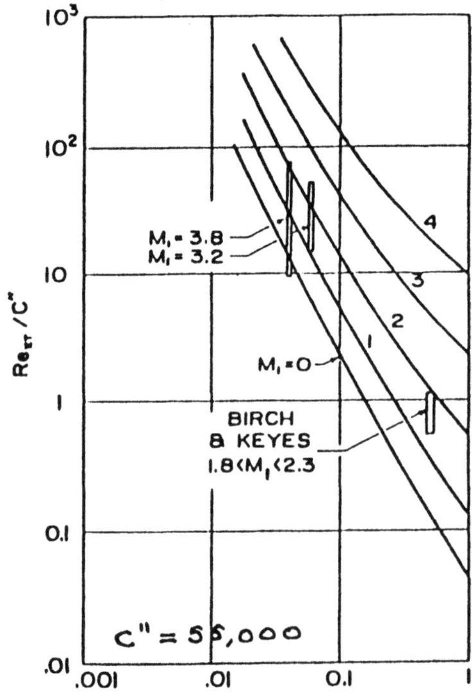

FIGURE 4. Transition in Chapman layers (slip-streams from crossing shock waves). The M=3.2,3.8 data are from [6] and the curves from the Threshold theory for such layers

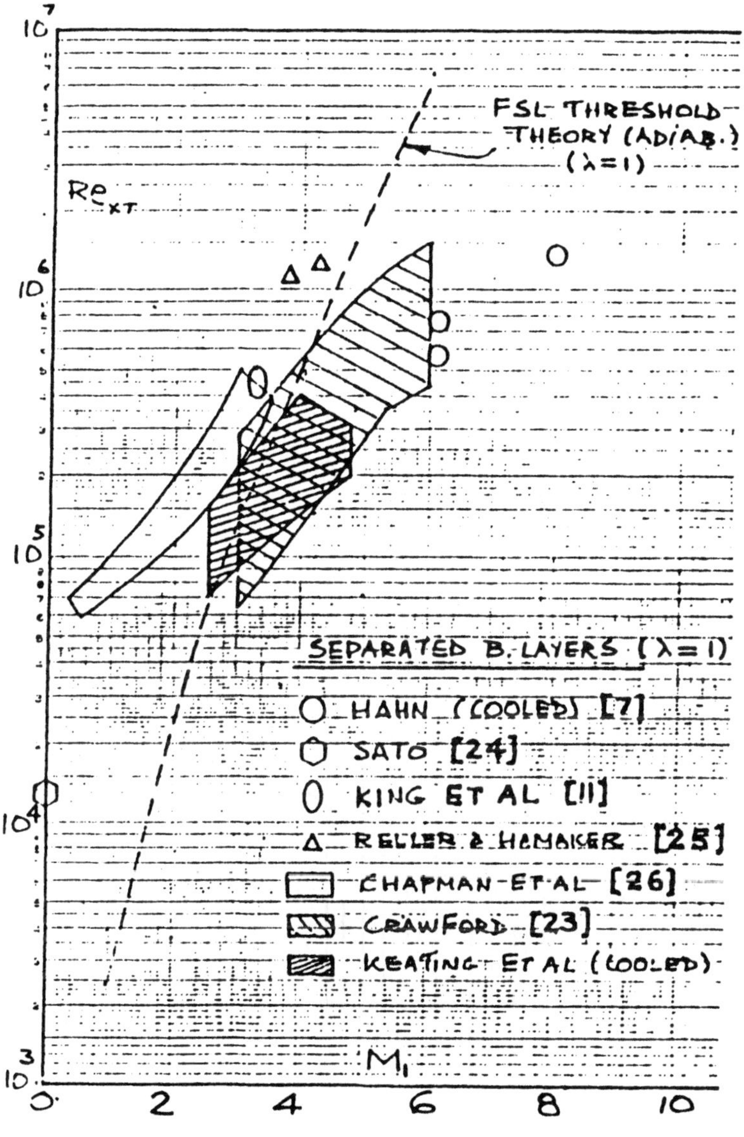

FIGURE 5. Transition in separated boundary layers, with x_T measured from separation; the dashed curve is the Threshold Theory adapted for Chapman flows but with x_T measured from the zero-thickness flow origin.

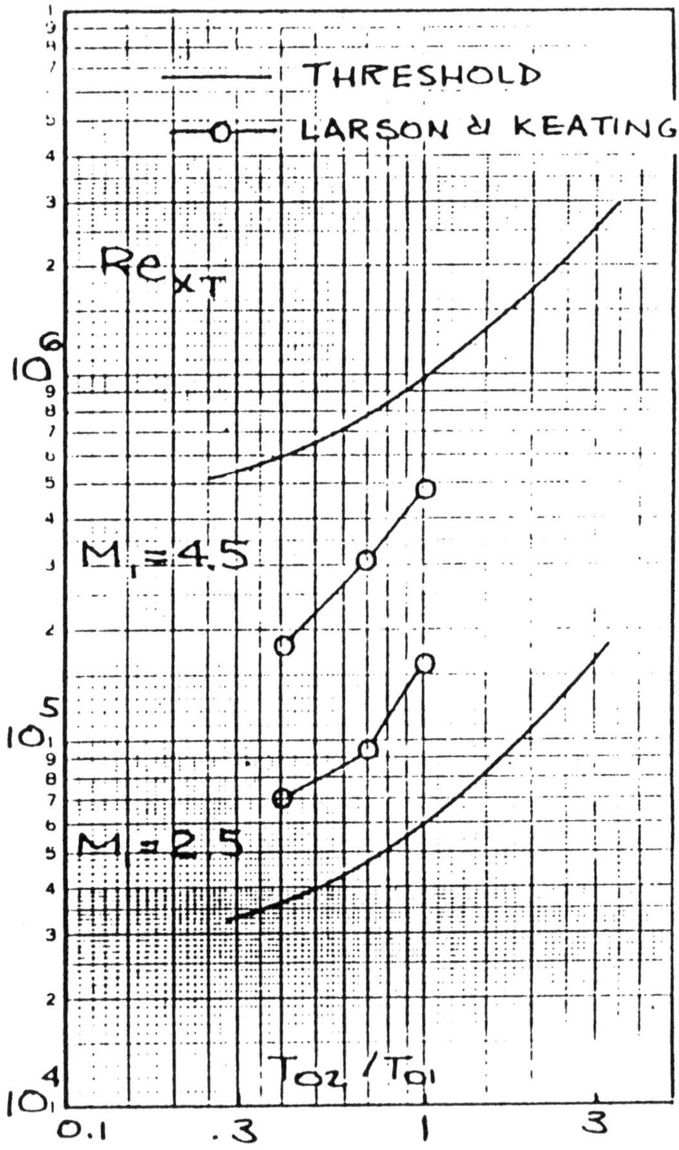

FIGURE 6.

Transition in separated boundary layers ($\lambda = 1$) when the "slow" stream total temperature T_{02} differs from that of the fast stream; curves are the Threshold theory for Chapman-type flows with x_T measured from the zero-thickness flow origin.

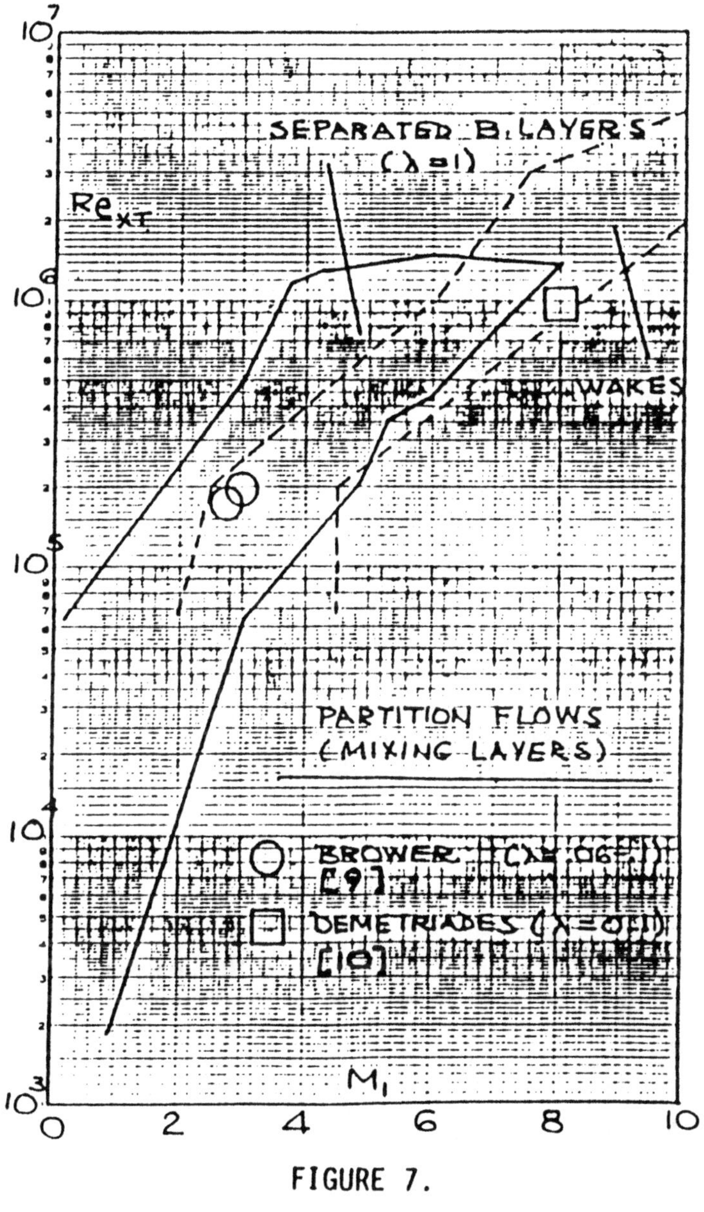

FIGURE 7.

Transition in partition flows (mixing layers) as signaled by randomness; x_T measured from the partition T.E.

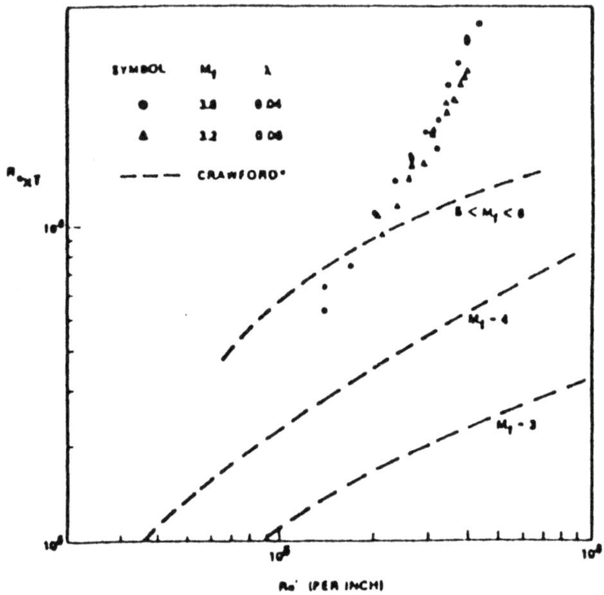

FIGURE 8. Unit Reynolds number effects on Re_{xT} for pure Chapman flows [6] (points are from crossing shocks) and separated boundary layers (dashed lines, from [23]).

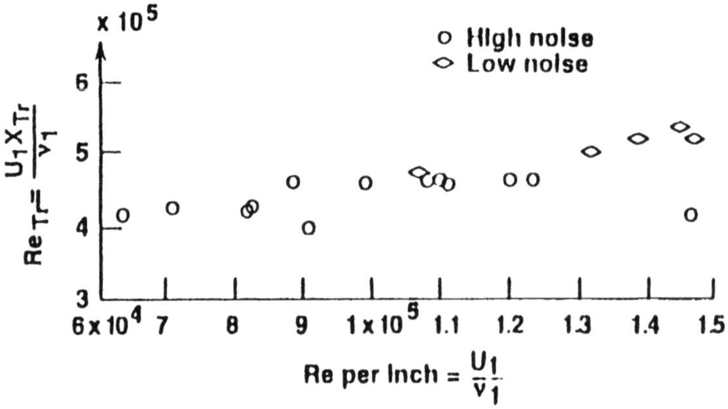

FIGURE 9. The findings of King et al [11] on the effect of a Mach 3.5 tunnel flow noise and unit Re effects on separated boundary layer transition.

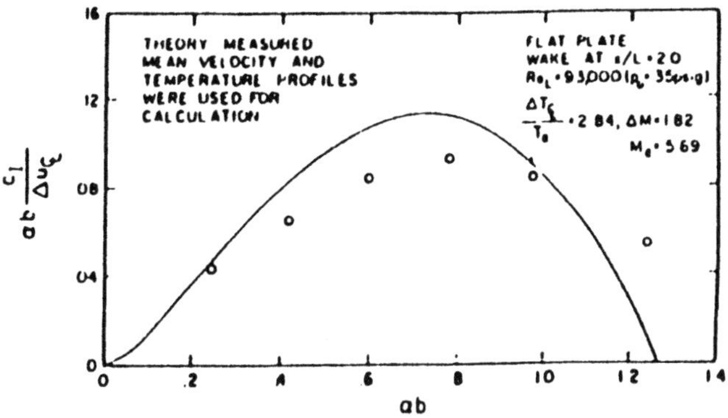

FIGURE 10. Comparison of amplification rate spectrum as computed and as measured for a hypersonic wake [16]

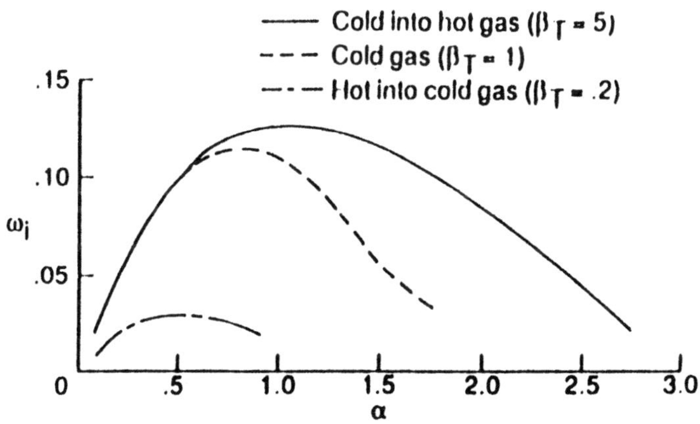

FIGURE 11. Dependence of the amplification-rate spectrum of tanhy shear layers on the temperature of the slow stream according to Macaraeg et al [20]. Slow stream is cooled when $\beta_T > 1$.

SOME COMPARISONS OF LINEAR STABILITY THEORY WITH EXPERIMENT AT SUPERSONIC AND HYPERSONIC SPEED

James M. Kendall

Jet Propulsion Laboratory
Pasadena, CA 91109

1. Introduction

The utility of stability theory in the prediction of transition at supersonic and hypersonic speeds has come to be widely accepted during the past few decades. The first compressible stability theory results to be obtained by direct numerical computation were produced during the late 1950s, and the work by Mack [1] and by others continues to evolve. Two principal results obtained early in the development were that for supersonic flow those waves (first mode) which are the counterpart of Tollmien-Schlichting waves in incompressible flow are more unstable when oblique than when two-dimensional, and that above a certain Mach number, there arises an acoustic mode (second mode) which can be highly amplified especially for a 2-D orientation. Supersonic and hypersonic stability experiments were reported by 1960 but were judged invalid for test of theory because the disturbance wave-fronts were of undetermined obliquity. The wave-front warpage problem was subsequently overcome by use of an improved disturbance-generator and by operating the tunnel such as to avoid eddy Mach-wave radiation from the wall boundary layers [2]. More recently, a point-source, pulse-like, generator was employed for study of a Mach 2 flat plate layer [3], an advantage being that waves of wide frequency and obliqueness ranges were excited simultaneously. In contrast with studies of controlled disturbances, measurements of waves initiated by wind tunnel sound have been obtained for both supersonic and hypersonic speeds. Because instability amplification does not usually account for the observed wave growth, the issues of forcing by the stream disturbances of the prevailing obliqueness spectrum and of linearity arise. In the subsequent sections, review is made of those experiments wherein comparison with theory is straightforward. It will be noted that experimental work to date pertains to zero-pressure-gradient flow, and that crossflow and Görtler instabilities have yet to be studied directly at high speeds.

2. Flat-Plate Experiments with Controlled Disturbances

The present author carried out flat-plate experiments [2] at Mach numbers of 4.5, 3.7, and 2.4 under the condition of laminar tunnel-wall boundary layers, whereby a low background fluctuation level prevailed. Disturbances were generated by an electrical glow-discharge of controlled frequency and amplitude using electrode pairs set within the plate surface at obliqueness angles of 0°, 30°, and 55°, the latter being near the angle expected from theory to be most amplified for first-mode waves. For Mach 4.5 the wave development was regular and uniform for combinations of wave angle and frequency where amplification resulted, as opposed to damping. Conversely, two-dimensional first-mode waves and oblique second-mode waves distorted during progression. Comparison with theory for amplified-wave cases is shown in Figure 1. The Mach 3.7 results verified the predicted absence of second-mode amplification, but the Mach 2.4 results were poor on account of glow-discharge non-uniformity at that tunnel condition. The Mach 2.0 point-source experiment [3] showed the most unstable wave angle to be near 60°, as predicted, but amplification rates were as large as twice the peak theoretical value [4].

3. Flat-Plate Experiments with Sound-Induced Waves at Supersonic Speed

The tunnel environment and the generation of waves by sound deserve mention. Laufer [5,6] conducted definitive studies on eddy Mach-wave radiation and noted its powerful influence on transition. The present author [7] showed by direct correlation that sound incident upon a flat plate initiated boundary layer waves and at Mach 4.5 incident sound produced a "footprint" exhibiting the obliquity expected of first-mode waves. The induced wave packets could be followed the full length of laminar run and into the turbulent layer. Packet growth commenced at the plate leading edge, in contrast to the theoretical result that damping should occur first. Mack [4] analyzed the problem of the response of a layer to a particular type of incident sound using solutions from compressible stability theory.

Results for Mach 4.5 are compared with the experiment in Figure 2 taken from Reference 7. The ordinate represents the amplitude of fluctuations within the layer at the vertical location of maximum value for several station along the plate and is normalized by the freestream fluctuation amplitude at similar frequency. The three dimensionless frequencies span much of the range of instability. The arrows denote theoretical neutral points, and curves to the left of these are theoretical results from forcing analysis, whereas curves to the right show predicted instability amplification. The message is that for this Mach number, forcing occurs only near the leading edge, and subsequent wave growth proceeds by instability. Unfortunately, the analysis could not be extended to other Mach numbers, and therefore this interpretation might not apply at other speeds.

The growth of "natural" waves was also measured for several other Mach numbers between 1.6 and 5.6. Comparison with stability theory was performed at Mach 2.2 and 4.5 as shown in Figures 3 and 4. For Mach 2.2, the peak growth rate and bandwidth both exceed prediction for wave angles near maximal instability. The area under the curve represents the growth of the total signal and is therefore much higher than prediction. Nonlinearity was probably not a cause as definite evidence of the onset of this was found at a more rearward station. The influence of forcing could not be assessed. For Mach 4.5, the comparison within the measured frequency band was more favorable in accordance with Figure 2. Demetriades [8] performed a similar experiment at Mach 3.0 and found amplification to conform with prediction fairly well at an intermediate station except for a substantial excess growth at low frequencies, but to differ at a more forward station and also at a more aft one. Nonlinearity seemed to be the cause of the latter discrepancy. Lysenko and Maslov [9] made no comparison with theory of the experimental growth rates obtained on a flat plate at Mach 2.0, 3.0, and 4.0, but qualitative comparison is offered here. For the adiabatic wall case, the Mach 2.0 results showed the maximum amplification rate and the frequency of this to be near predicted values, but the bandwidth exceeded prediction. The Mach 3 results extend in frequency to only half the theoretical upper neutral value, but within this range showed a peak value near that predicted. This is somewhat surprising because the Reynolds number of observation exceeded that at which transition usually occurs in conventional tunnels for this speed. For Mach 4.0, the peak amplification rates were commensurate with theory, but the unstable frequencies were approximately five-fold higher than predicted and theoretically-unstable frequencies were observed to damp. Thus, the results for this speed are difficult to understand.

4. Sound-Induced Waves on Cones at Hypersonic Speed

Three experimental studies of "natural" wave growth on sharp cones at zero angle-of-attack and at Mach numbers near eight have been reported [7,10,11]. The situation here differs from that of the supersonic flat-plate one on account of differences in the geometry, and because at this speed the second mode is far more unstable than the first. The results obtained by Stetson, et al. [11] typify those of the other two experiments and have been selected by Mack [12] for comparison with theory. As in Figure 5, the predicted center-frequency and bandwidth of second-mode waves is in good agreement with observation, but the predicted peak amplification exceeded observation by a substantial amount, especially at the higher Reynolds numbers. This difference might be due to lack of wave-front two-dimensionality in the experiments. The most striking difference between theory and experiment was that all experimenters observed a strongly-unstable band lying at frequencies above the second-mode one. It is not known whether this is a result of nonlinear second-mode oscillation or of an additional instability.

5. Experimental Research Needed

(1) Boundary layer receptivity to eddy Mach-wave radiation needs to be clarified, both experimentally and theoretically. The problem may be as important to supersonic aircraft design as it is to wind tunnel testing. For example, it is planned to maintain laminar layers over substantial areas of the wing of civil transport aircraft by use of suction, but the fuselage layer would be permitted to become turbulent and thus subject the wings to acoustic radiation. Theoretical work is needed also. Mack's forced response analysis [4] was applicable to only a limited Mach number range and also only to a region near the leading edge or tip of the test model. Thus, it cannot describe the laminar layer forcing that occurs, for example, when wind tunnel wall layer transition occurs sufficiently far downstream that the forwardmost radiation strikes the test model layer well behind the leading edge, as in the work by Chen, et al. [13].

(2) Theory and experiment are at widest variance at hypersonic speed. The three cone-flow experiments produced peak second-mode amplification rates substantially below prediction, and all showed substantial amplification at frequencies above the second-mode rate, where the theory predicts damping. The linearity, forcing, and wave-obliqueness questions could be sidestepped temporarily by studying the growth of intentionally-produced waves. There should be little difficulty in detecting weak waves in the presence of the much larger "natural" ones. A more difficult question would be whether the boundary layer possesses a zero-pressure-gradient profile over adequate length and time scales to respond in accordance with theory. The freestream $p\prime$ fluctuations buffet the test model layer and distort the mean profile. Thus, the low frequency signal components observable during stability measurements deserve analysis for profile distortion. Regarding the unexplained high-frequency amplification, others have suggested that the phase of those waves be compared with that of the second-mode ones. A constant relationship would signify nonlinearity, whereas independence would indicate a separate mode of oscillation.

(3) Investigation of the stability/transition relationship in zero-pressure-gradient layers will continue to be rewarding, but other transition problems await study. These include flows with longitudinal pressure gradient, with crossflow, and with surface concavity. Transition measurements, by themselves, are inadequate to ascertain what physical processes are important in a given situation. Probing the boundary layer with sensors possessing sufficient spatial and temporal resolution is essential for meaningful progress.

References

[1] Mack, L. M., "Review of Linear Compressible Stability Theory," ICASE/NASA Workshop on Stability of Time Dependent and Spatially Varying Flows, Hampton, VA, 1985 (published by Springer-Verlag).

[2] Kendall, J. M., "Supersonic Boundary Layer Stability Experiments," Proceedings of Transition Study Group Meeting, Vol. II, edited by W. D. McCauley, Aerospace Corp., San Bernardino, CA, 1967.

[3] Kosinov, A. D. and Maslov, A. A., "Development of Artificially Excited Disturbances in a Supersonic Boundary Layer," IUTAM Symposium on Laminar-Turbulent Transition, Novosibirsk, Springer-Verlag, 1984, pp. 601-606.

[4] Mack, L. M., "Linear Stability Theory and the Problem of Supersonic Boundary-Layer Transition," AIAA Journal, Vol. 13, No. 3, March 1975, pp. 278-289.

[5] Laufer, J., "Aerodynamic Noise in Supersonic Wind Tunnels," J. Aerospace Sciences, Vol. 28, No. 9, September 1961, pp. 685-692.

[6] Laufer, J., "Some Statistical Properties of the Pressure Field Radiated by a Turbulent Boundary Layer," Phys. Fluids, Vol. 7, No. 8, August 1964, pp. 1191-1197.

[7] Kendall, J. M., "Wind Tunnel Experiments Relating to Supersonic and Hypersonic Boundary-Transition," AIAA Journal, Vol. 13, No. 3, 1975, pp. 290-299.

[8] Demetriades, A., "Growth of Disturbances in a Laminar Boundary Layer at Mach 3," Phys. Fluids A, Vol. 1, No. 2, 1989, pp. 312-317.

[9] Lysenko, V. I. and Maslov, A. A., "The Effect of Cooling on the Supersonic Boundary Layer Stability and Transition," IUTAM Symposium on Laminar-Turbulent Transition, Novosibirsk, Springer-Verlag, 1984, pp. 495-502.

[10] Demetriades, A., "Laminar Boundary Layer Stability Measurements at Mach 7 including Wall Temperature Effects," AFOSR-TR-77-1311, 1977. See also: AIAA Paper 74-535, 1974.

[11] Stetson, K. F., Thompson, E. R., Donaldson, J. C., and Siler, L. G., "Laminar Boundary Layer Stability Experiments on a Cone at Mach 8. Part I. Sharp Cone," AIAA Paper No. 83-1761, 1983.

[12] Mack, L. M., "Stability of Axisymmetric Boundary Layers on Sharp Cones at Hypersonic Mach Numbers," AIAA Paper No. 87-1413, Honolulu, HI, June 1987.

[13] Chen, F.-J., Malik, M. R., and Beckwith, I. E., "Comparison of Boundary Layer Transition on a Cone and Flat Plate at Mach 3.5," AIAA Paper 88-0411.

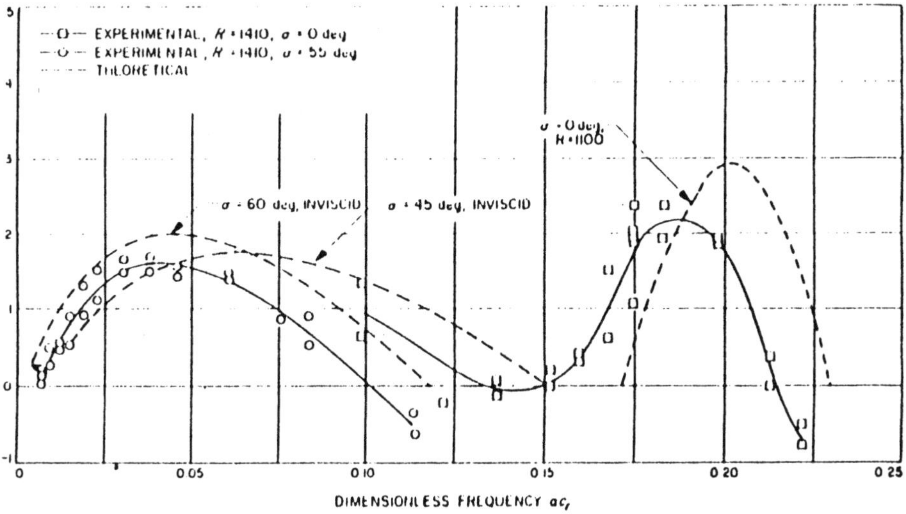

Fig. 1. Amplification Rate versus Frequency for Stimulated Waves, M = 4.5

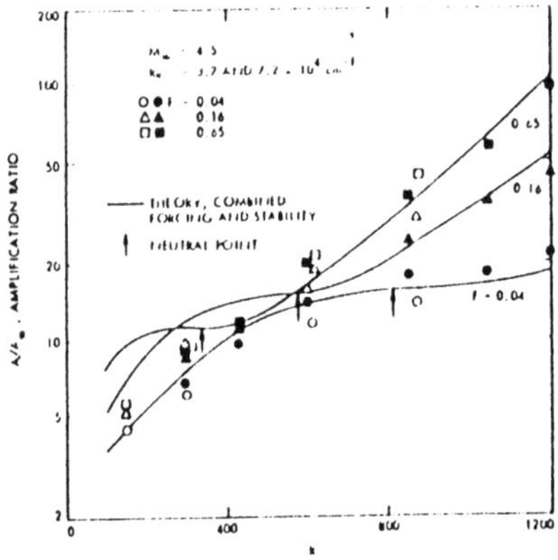

Fig. 2. Flat-Plate Fluctuation Growth due to Forcing and Instability. M = 4.5.

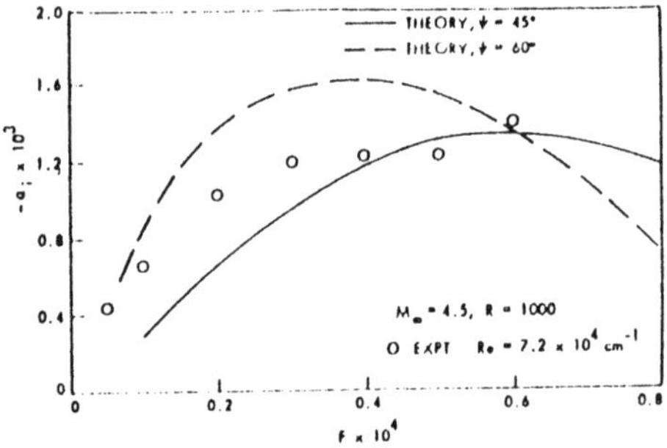

Fig. 4. Amplification Rate versus Frequency for Acoustically-Produced Waves. M = 4.5.

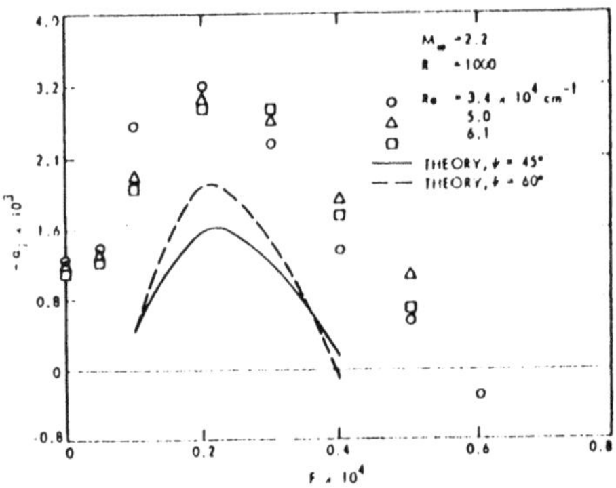

Fig. 3. Amplification Rate versus Frequency for Acoustically-Produced Waves. M = 2.2.

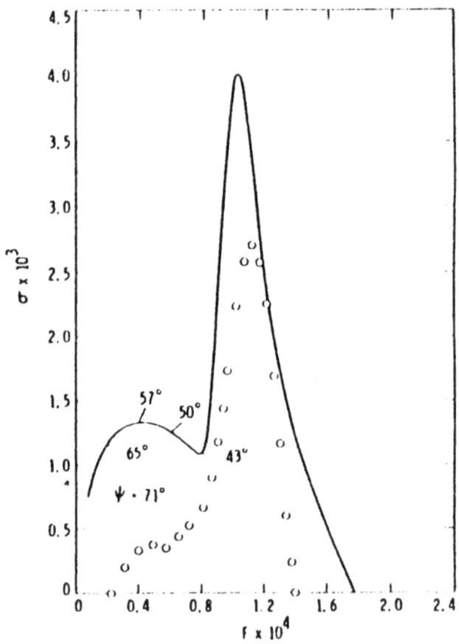

Fig. 5. Amplification Rate versus Frequency for Cone. M = 6.8. (from Ref. 12)

DOMINANCE OF "NOISE" ON BOUNDARY LAYER TRANSITION IN CONVENTIONAL WIND TUNNELS – A PLACE FOR THE "QUIET" BALLISTIC RANGE IN FUTURE STUDIES

Samuel R. Pate

Sverdrup Technology, Inc.
Tullahoma, TN

1. Introduction

It is the purpose of this paper to briefly review selected boundary layer transition experimental studies and some of the major contributions from 1905 to 1988 that have led to significant advances in boundary layer transition predictions. With this knowledge at hand, comments will be offered as to the type test facilities that could be used to make further significant advances in the future. Progress made in the understanding of laminar and turbulent boundary layer flows and boundary layer transitional flows is illustrated in Figure 1. The development process to date has covered approximately ninety years, beginning in 1905 with the definition of the boundary layer by Prandtl. For the purposes of discussion and evaluation, the development process has been divided into four major phases. As noted in Figure 1, the first three phases have lasted approximately thirty years each. To date, three phases have been completed. We are beginning the fourth phase. Whether or not a fifth phase will be necessary for boundary layer transition and boundary flow theory to become a mature science remains to be seen. The definition of a mature science in this paper is defined as the ability of the theoretical models to adequately predict the location of boundary layer transition relative to the design of ground test experiments and full-scale flight vehicles.

2. Phase I - The Beginning Era (1905-1930)

Phase I began in 1905 with Prandtl defining the existence of a boundary layer. A major advance occurred in the 1915 time period with the famous experiments of Prandtl on a sphere wherein the decrease in drag could be directly related to the boundary layer transitioning from laminar to turbulent flow and the amount of separated flow area decreasing.

A very major advance occurred around 1930, at the end of Phase I and the beginning of Phase II. At that time, the now famous Tollmien-Schlichting laminar boundary instability theory, commonly known as TS waves theory, was developed. The Tollmien-Schlichting instability theory, coupled with a theoretical laminar boundary layer profile, predicted that a laminar boundary layer could become unstable under certain test conditions and the instabilities amplified.

3. Phase II - The Basic Growth Era (1930-1960)

One of the most major advances made in transition research was the famous Schubaur-Skramstad experiments. These experiments, conducted in the mid-1940's, for the first time gave direct confirmation that Tollmien-Schlichting waves did in fact exist and were accurately predicted by the TS theory. Perhaps equally significant was the demonstration that theoretical studies and experimental studies were both required to successfully understand and verify the very complex incompressible laminar flow instability process. A third major contribution was made by Schubaur and Skramstad in the 1940's wherein their studies on subsonic transition on a flat plate revealed that the transition Reynolds number could be significantly adversely affected by the amount of freestream disturbances present in the test facility flow. These disturbances were directly related to velocity fluctuations, or turbulence, generated in the stilling chamber. Significant disturbances in subsonic wind tunnels were also identified as being generated by the fan system; whereas, these noise disturbances could be propagated back upstream and also have significant effect on the transition process. This was the first experimental study to show the dominating effect that wind tunnel flow field disturbances could have on the transition process and transition Reynolds number.

In the late 1950's and early 1960's, Spangler and Welles demonstrated that by varying the acoustical frequency, the maximum transition Reynolds numbers as obtained by Schubaur and Skramstad could be increased by about a factor of two. Once again, a significant effect of freestream disturbances on the boundary layer transition process was being demonstrated.

At the end of Phase II (1950's), significant advances in understanding the transition process were made by Michel, Emmons (the now famous Emmons Spots), Granville, and Smith. They developed concepts and contributed early theories and experimental data.

Significant contributions were also made by Van Driest during this time period in the development of laminar and turbulent boundary layer theoretical models and providing experimental data on the effect of surface roughness, wall cooling, and surface pressure gradients on the transition process at subsonic and supersonic Mach numbers.

During the 1950's and by the end of Phase II, it was recognized and accepted that the boundary layer transition process in subsonic wind tunnels was adversely affected by the presence of freestream disturbances. However, it was not generally recognized that significant disturbances existed in supersonic test facilities.

4. Phase III - The Advanced Era (1960-1990)

At the beginning of Phase III, during the late 1950's and early 1960's, Laufer (JPL) made significant contributions, to understanding and identifying flow disturbances present in supersonic wind tunnels. Utilizing the early 1950's research of Kouasznay, Laufer's compressible flow studies confirmed that there could be three primary sources of disturbances – velocity fluctuations, generally identified as turbulence, and generated in the stilling chamber; temperature fluctuations (entropy fluctuations), also stilling chamber generated; and radiated noise (sound waves) – generated by the turbulent boundary layer existing on the walls of conventional wind tunnels. Laufer's studies began to draw attention and concern to the fact that supersonic wind tunnel transition Reynolds numbers might be adversely affected by the presence of tunnel wall turbulent boundary layer generated aerodynamic noise.

In the mid-1960's, Pate (AEDC) developed a theory that transition Reynolds numbers in supersonic wind tunnels were dominated by radiated noise. His theory was developed based on a knowledge of Laufer's experiments and the observation that transition Reynolds numbers on flat plates increased in supersonic wind tunnels as the size of the wind tunnel increased and as the tunnel unit Reynolds numbers increased. Using flat plate transition data and known turbulent boundary layer characteristics for the tunnel wall turbulent boundary layers, Pate developed a transition aerodynamic noise dominance theory and supporting correlation. To verify the theory of radiated noise dominance, Pate conducted an extensive series of transition and noise measurement experiments on flat plates and sharp cones (1966-1970). These experiments were conducted in various wind tunnels ranging in size from one foot to sixteen feet test section heights over the supersonic Mach number range from 2.5 to 5.0. Additional experiments were conducted by Pate wherein the radiated noise field adjacent to the model could be controlled and the direct influence of radiated noise effects on boundary layer transition Reynolds numbers measured. These experiments confirmed that the transition process and transition Reynolds numbers on sharp flat plates and cones in supersonic conventional wind tunnels were dominated by radiated aerodynamic noise emanating from the turbulent boundary layer on the tunnel side walls. The initial correlation proposed by Pate was extended to include new experimental data (1978) and other existing data for sharp cones and plates ($M\infty = 3$ to 18). Data presented in Figure 2 illustrate how transition Reynolds numbers increase with increasing tunnel size and increasing unit Reynolds numbers as a direct result of the tunnel wall radiated noise decreasing.

During the late 1960's, Kendall (JPL) continued the experiments of Laufer wherein energy spectra generated by radiated noise emanating from turbulent boundary layers on conventional wind tunnel walls were measured. He was also able to control the boundary layer on the tunnel walls over a large Reynolds number range wherein the tunnel could be operated with laminar flow or turbulent flow on the tunnel walls. His data showed conclusively the large disturbance levels that existed from radiated noise when the flow was turbulent on the tunnel side walls. Kendall's experimental data, combined with Mack's (JPL) theoretical methods, made significant advances in understanding and predicting transition in the 1960's and 1970's.

In the mid-1970's, Donelson (AEDC) conducted experiments similar to Laufer's in the AEDC tunnels in which Pate conducted his experimental studies some years earlier. Donelson confirmed the same type disturbances identified by Laufer and Kendall. He also provided direct experimental evidence that as wind tunnel size increased and/or unit Reynolds numbers increased at supersonic speeds, radiated noise decreased.

An extensive program to measure transition Reynolds numbers on a sharp slender cone in transonic facilities and in flight was also begun in the mid-1970's. This program, led by Daugherty (AEDC), generated transition Reynolds numbers ($M\infty = 0.4$ to 1.8) data from a large number of U.S. and foreign transonic facilities and from flight experiments. The results of these extensive investigations confirmed that flow disturbances in transonic wind tunnels, generated primarily by wall porosity noise, had a dominating effect on the transition Reynolds numbers. Daugherty presented a correlation of transition Reynolds number as a function of the turbulence intensity measured in these wind tunnels. Also, interestingly, significant disturbance was measured in flight, and the flight data also correlated with the turbulence intensity parameter ($\overline{p}/q\infty$).

Harvey (NASA), in the 1970's, also successfully correlated transition data with the freestream disturbances parameter ($\overline{p}/q\infty$) over a Mach number range from $M\infty = 0$ to 4 and $M\infty = 4$ to Mach 23, once again providing strong evidence that flow field disturbance in conventional wind tunnels has a dominating effect on boundary layer transition on sharp cones and flat plates.

Early in the 1970's, NASA began an extensive program to develop a "quiet flow" experimental supersonic test facility wherein the boundary layer on the tunnel walls could be maintained laminar, thus eliminating radiated noise disturbances in the test section. Several concepts were evaluated in the early seventies, and an operational facility began in the late 1970's and early 1980's.

Beckwith and Chen (NASA), in the mid to late 1980's, successfully conducted transition studies on flat plates and cones in the NASA "quiet" tunnel at $M\infty = 3.5$. These transition data are presented in Figure 2. These new data were significantly higher than previous conventional wind tunnel data as shown in Figure 2. The new "quiet" tunnel transition data are also in good agreement with NASA's theoretical predictions as shown. The new NASA data are not only notably higher than previous wind tunnel data, they also do not exhibit a variation with unit Reynolds numbers. Pate's radiated noise dominance theory correlation and experiments had shown that the unit Reynolds numbers effect could be directly related to the radiated noise in the test section. The 1988 data presented by Beckwith and Chen supports the theory that in a "quiet" wind tunnel absent of radiated noise, transition Reynolds numbers on flat plates and sharp cones are significantly higher than conventional wind tunnel data and do not vary with change in unit Reynolds number. Also of significance is that the flat plate data were higher than cone data as predicted by the theory. These results are opposite earlier predictions that cone Ret values would be higher than flat plate Ret values. Pate's correlation (1970) also predicted that flat Ret values would approach, and at low tunnel disturbance levels, exceed cone Ret values.

Pate (1976) conducted additional transition studies on a sharp flat plate and sharp cone in a conventional AEDC hypersonic tunnel at $M\infty \sim 8$. These data confirmed that the flat plate Ret levels were higher than the cone values at $M\infty = 8$. Measurements by Harvey and Donelson have shown that the freestream radiated disturbance parameter $(\tilde{p}/q\infty)$ decreases significantly with increasing tunnel Mach number $(M\infty)$, increasing tunnel unit Reynolds number, and increasing tunnel size. The reversal in the ratio of cone to flat plate transition Reynolds numbers observed by Pate at $M\infty = 8$ suggests that the radiated aerodynamic noise levels at high Mach numbers may be reaching a "low" level relative to significantly affecting the transition process. At $M\infty \sim 10$ to 20, the freestream radiated disturbance parameter $(\tilde{p}/q\infty)$ is approximately 20% to 10% of the level existing at $M\infty = 3$. Future studies at $M\infty \gtrsim 10$ in a "quiet" tunnel are needed to verify this possibility.

Stetson (AFFDL) successfully conducted boundary layer stability measurements at $M\infty = 8$ in a conventional hypersonic wind tunnel (AEDC-B) in the mid-1980's. These measurements, made in the presence of freestream radiated noise disturbance, confirmed TS type disturbances as predicted by linear stability theory existed in the laminar boundary on a sharp cone.

In the mid-1960's to mid-1970's, transitioning experiments were conducted by Potter (AEDC) and by Sheets (NOL) on cone models in free flight ballistic ranges. These tests were conducted at supersonic and hypersonic speeds. They were designed to look at Mach number effects, wall temperature effects, and unit Reynolds number effects. These experiments allowed for the first transition measurements to be made on a slender cone in an "aerodynamic noise free" ground test environment at supersonic-hypersonic Mach numbers. Included in Figure 2 are ballistic range data and transition data reported by Potter and Sheets. The level and trends of the range data in Figure 2 can, to some degree, be related to a disturbance free environment in the ballistic range.

The range data presented in Figure 2 illustrate some rather significant trends: First, Sheets' data is at the same level as the NASA "quiet" tunnel cone data and exhibits a relatively small variation with unit Reynolds numbers. Second, at comparable unit Reynolds numbers, the ballistic range data of Potter is significantly lower than "noisy" wind tunnel data. Also, there is a definite strong trend of transition Reynolds number with unit Reynolds number in Potter's data. The cause for this trend to date is not known, but is surely not from the same generating source present in conventional wind tunnels. Other variables to note in comparing Sheets' data and Potter's data with the wind tunnel data are differences that existed in the test conditions (Mach numbers) and model geometries. Sheets' data were taken at a freestream Mach number of about 8, with a cone half angle of 5 degrees, nose tip radius of 0.001 inch, and local cone Mach number of about 7. Potter's data were taken at Mach number 5.0 ($M\infty = 4.3$), cone half angle of 10 degrees, and cone nose radius of 0.005 inch. Also, Sheets was very particular in keeping extreme sharpness on the cone models, nose bluntness of 0.001 inch or less. About a factor of five existed between the nose bluntnesses in the two experiments. It is very difficult to draw strong conclusions about the range data presented in Figure 2. Could the bluntness between the two models be a dominating difference with the amount of bluntness combined with extremely thin boundary layers on the cone be causing either an adverse or favorable effect? Also, what is the effect of the differences in Mach number? What are the effects of temperature ratio? Sheets' early experiments in this Mach number range showed that transition could be extremely sensitive to very small amounts of change in temperature ratios on the model.

As the end of Phase III approaches, there is wide-spread acceptance that freestream disturbances in subsonic, transonic, supersonic, and hypersonic conventional wind tunnels will dominate the transition process on sharp cones and flat plates. Therefore, it will be extremely difficult to develop experimental data from these facilities that can give adequate verification and confirmation of the transition theories being developed. "Quiet flow" facilities offer a possibility of generating boundary layer transition data free from dominating freestream disturbances. The NASA "quiet" tunnel that has been developed and successfully utilized by NASA at Langley is one such facility. The other existing facility is the ballistic range, which is free from flow field generated facility disturbances.

5. Phase IV (What's to Come)

Linear stability theory has continued to make significant contributions to understanding and predicting the transition process in Phases I - III. It is expected that additional major contributions will occur in Phase IV.

Numerical (Navier-Stokes) codes have also made significant contributions to understanding and predicting the transition process in Phase III. The contributions are expected to increase significantly during Phase IV.

The NASA "quiet" tunnels ($M\infty = 3$ to 18) offers a major opportunity to provide experimental data that will significantly advance our knowledge of the transition process and provide critical data for the development and verification of theoretical models and numerical codes.

The ballistic range also offers the potential for producing "disturbance free" transition data at conditions not available in the "quiet" tunnel.

There is a possibility that the conventional hypersonic tunnel ($M\infty \gtrsim$ 10 to 20) with its relatively low disturbance level can be a useful facility for additional transition studies.

6. Summary

It is evident that, in conventional facilities ($0 < M\infty \lesssim 8$), the freestream disturbances dominate the transition process on sharp flat plates and cones. To generate new and meaningful data in the future will require that boundary layer stability and transition data be generated in "disturbance free" environments. This leaves five possibilities: (1) exclusive use of a "quiet" tunnel as currently exists within NASA; (2) build new "quiet" tunnel in the future with increased tunnel size and extended test conditions to higher hypersonic Mach numbers; (3) use flight experiments, which in the past have been of limited success, particularly in development and confirmation of new transition theories; (4) use the ballistic range more extensively than it has been used in the past; and (5) utilize conventional hypersonic tunnels ($M\infty \gtrsim 10$) if it can be established that the relatively low disturbance levels do not dominate or significantly affect the transition process.

The ballistic range offers several desirable features: the ability to generate a wide range of Mach numbers, unit Reynolds numbers, and temperature ratios in a "noise free" test environment. Additional studies conducted in the ballistic range and the "quiet" tunnel on the effects of small amounts of nose bluntness under similar hypersonic high unit Reynolds number test conditions would be most valuable. Therefore, the ballistic range combined with the NASA "quiet" tunnel concept would appear to be the type facilities that look the most promising to generate experimental data in Phase IV that can offer guidance and confirmation to the development and acceptance of new and advanced boundary layer transition theories.

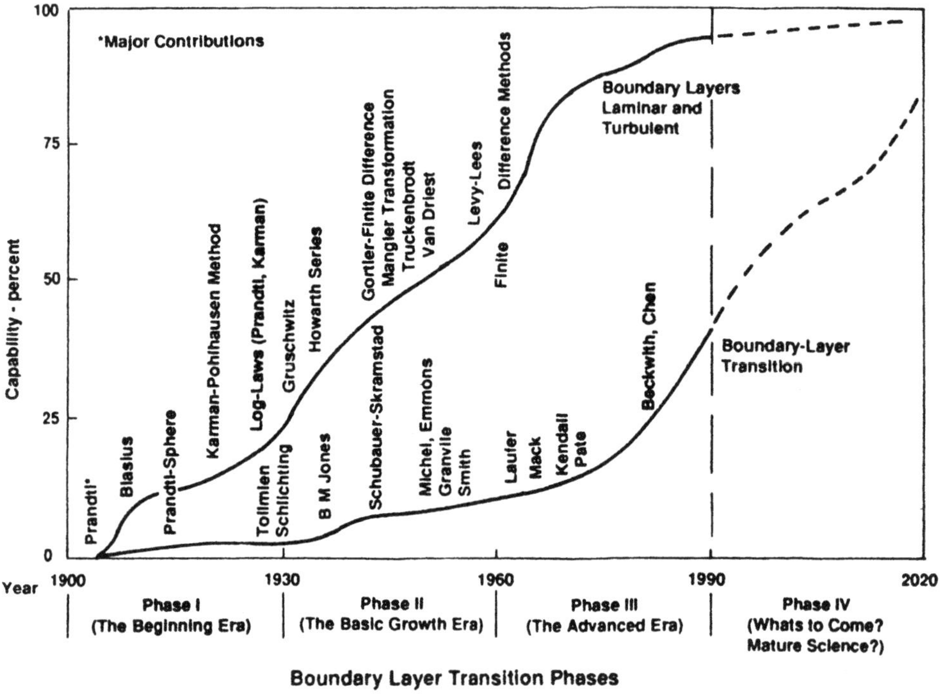

Figure 1. An attempt to picture the capability for analyzing and predicting boundary layers and laminar-turbulent transition. Chart considers only steady two-dimensional and axisymmetric, attached flow for Mach numbers less than 6. (From Smith, 1974 with estimates beyond 1974 by Pate)

Symbol	Model	M_∞	M_δ	T_w/T_{AW}	Nose Radius	Test Facility	Method of Measurement	Ref
□	Hollow CYL	3.0	3.0	~1.0	~ 0 in.	Conventional	Min. P_o' Probe	Pate
◨	Hollow CYL	3.0	3.0	~1.0	~ 0 in.	"Noisy"	Min. P_o' Probe	Pate
■	Hollow CYL	3.0	3.0	~1.0	~ 0 in.	Wind Tunnel	Min. P_o' Probe	Pate
□ (flagged)	Flat Plate	3.5	3.5	~1.0	~0.00005 in.	NASA "Quiet" Tunnel	Min. P_o' Probe	Beckwith, Chen
○	5-Deg	3.0	2.9	~1.0	≤ 0.003 in.	Conventional	Min. P_o' Probe	Pate
◑	Half	3.0	2.9	~1.0	≤ 0.003 in.	"Noisy" Tunnel	Min. P_o' Probe	Pate
○ (flagged)	Angle	3.5	3.4	~1.0	≤ 0.001 in.	NASA "Quiet" Tunnel	T_w Increase	Beckwith, Chen
●	Cone	7.5	6.9	~0.1	0.001 in.	Ballistic Range	Shadowgraph	Sheetz
— · · —	10-Deg	5.0	4.3	~0.2	0.005 in.	Ballistic Range	Shadowgraph	Potter

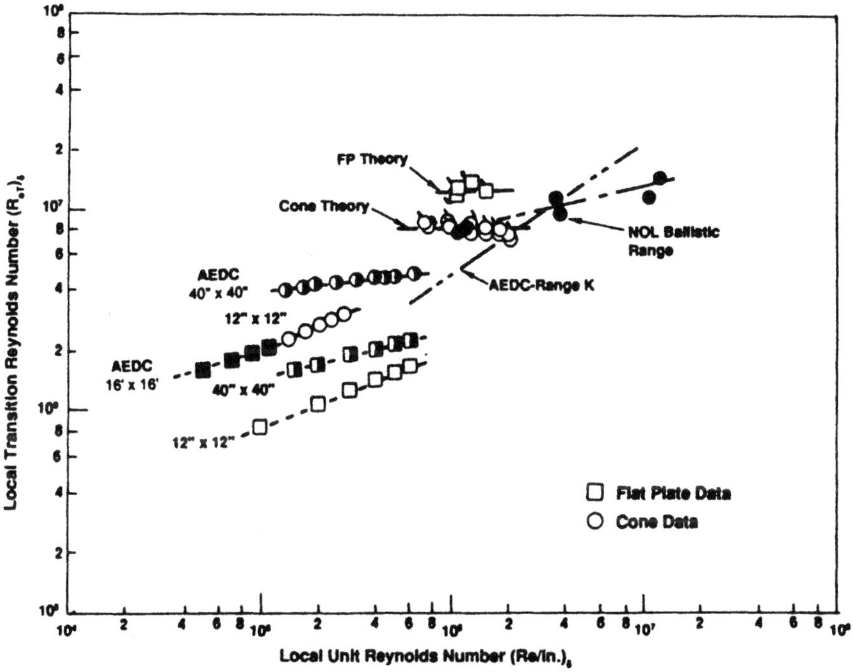

Figure 2. Comparisons of Transition On-Set Reynolds Numbers in "Noisy" Conventional Wind Tunnels, "Quiet" Wind Tunnel and "Quiet" Ballistic Ranges for "Sharp" Leading Edge/ Nose Flat Plate and Cone Models, M_δ = 2.9 to 6.9, Zero Angle-of-Attack

TRANSITION RESEARCH USING FLIGHT EXPERIMENTS

Eli Reshotko

Case Western Reserve University
Cleveland, OH 44106

The relationship between aerodynamic results obtained in wind tunnels and their counterparts in flight is one that continues to command the attention of the aerodynamics community (Ref. 1). This is especially true for a phenomenon that is as environmentally sensitive as boundary layer transition. It is well established for example for supersonic Mach numbers from about 1.5 to 8, acoustic radiation from the turbulent boundary layers developing on wind tunnel walls causes an early transition on test models that has no flight counterpart (see presentation by Sam Pate). Only the "quiet" tunnels described by Ivan Beckwith offer relief from this phenomenon in ground test facilities. Because of this, the transition community has looked to flight experiments as a means of obtaining proper transition information in an uncontaminated environment.

Flight transition experiments were performed in some profusion in the early to middle 1950's from the NACA Wallops Island facility using rocket propelled vehicles. The reports of the Rumsey-Lee group from NACA-Langley and the Disher group from NACA-Lewis document the extensive transition results obtained to Mach numbers of nearly four principally on axisymmetric geometries. These results were difficult to interpret, however, because of our minimal understanding at that time of stability fundamentals and our inability to determine what effects might have been due to model surface condition since the models were generally not recovered. Those flights, however, offered promise of length transition Reynolds numbers of 30-40 million and whetted our appetites for proper understanding of the factors affecting transition at supersonic speeds and the consequent opportunities for transition control. The need for closely controlled and interpretable flight experiments remains. Fortunately, we have the Dougherty and Fisher (Ref. 2) cone experiment as a fine example of what is possible in performing experiments of research quality in the flight environment. This experiment is all the more significant because of the extensive measurements reported by

Dougherty (Ref. 3) using the same cone model in most of the significant wind tunnels throughout the world.

The standards for research quality experiments on stability and transition are no different for flight studies than for wind tunnel experiments. These statements or "guidelines" as formulated by the U. S. Boundary Layer Transition Study Group (Ref. 4) are as follows:

1. Any effects specifically and only associated with test facility characteristics must be identified and if possible avoided.

2. Attention must be given to disturbances introduced by the model surface, model material, and internal structure. Experimental studies should include documentation of these various factors.

3. Details of coupling of disturbances of various kinds to the boundary layer must be understood theoretically and experimentally so that the sensitivity of the transition process to the flight environment might be determined.

4. Whenever possible, tests should involve more than one facility. Tests should have ranges of overlapping parameters, and whenever possible, experiments should have redundancy in transition measurements.

The implementation of these standards in flight experiments requires special attention to a number of factors.

1. The measurement of disturbance environment must be incorporated into the model design and in fact must be part of the model.

2. Attention has to be given to the maintenance and monitoring of test conditions such as Mach number, Reynolds number, angle of attack, yaw angle, and surface temperature for the duration of the measurement period.

3. Attention has to be given to the maintenance and monitoring of model surface conditions for each flight. This includes protection of the model surface before launch and recovery of the vehicle for inspection and reconditioning of the surface prior to the succeeding flight.

4. Because stability phenomena at supersonic and hypersonic speeds occur at frequencies of hundreds of kHz and even to MHz levels, there is a need for very high data sampling rates, especially when monitoring multiple channels. This poses special problems in data acquisition and data reduction. Reliable digital telemetering of data from the vehicle may also be necessary in order to minimize weight and volume of the data acquisition equipment.

Finally, the reader is referred to the paper of Dougherty and Fisher (Ref. 2) in order that the quality of that work might be better appreciated.

References

[1] "Boundary Layer Simulation and Control in Wind Tunnels," Report of the AGARD Fluid Dynamics Panel Working Group 09, AGARD Advisory Report No. 224, 1988.

[2] Dougherty, N. S., Jr.: "Boundary Layer Transition on a 10-degree Cone. Wind Tunnel/Flight Data Correlation," AIAA Paper No. 80-154, January 1980.

[3] Dougherty, N. S., Jr.: "Influence of Wind Tunnel Noise on the Location of Boundary-Layer Transition on a Slender Cone at Mach Numbers from 0.2 to 5.5. Vol. 1: Experimental Methods and Summary of Results. Vol. II: Tabulated and Plotted Data," AEDC-TR-78-44, March 1980.

[4] Reshotko, E.: "Boundary Layer Stability and Transition. Annual Review of Fluid Mechanics," Vol. 8, pp. 311-349, 1976.

HYPERSONIC TRANSITION TESTING IN WIND TUNNELS

Kenneth F. Stetson

Flight Dynamics Laboratory
Wright Research and Development Center

1. The Problem

Historically, the wind tunnel has been the major source of boundary-layer transition information. Often these wind tunnel data have become the primary data base used to develop transition correlations and to establish transition criteria for flight. During the late 1950's and the 1960's, the identification and understanding of wind tunnel freestream disturbances provided an explanation for wind tunnel transition Reynolds numbers being smaller than flight transition Reynolds numbers. A quiet wind tunnel (freestream disturbance amplitudes reduced to a small value) was proposed as a way of obtaining wind tunnel transition Reynolds numbers which would be comparable in magnitude to flight values. The purpose of this paper is to discuss the position that *one should not expect a transition Reynolds number obtained in any wind tunnel, conventional or quiet, to be directly relatable to flight.*

Only in a few isolated cases can one expect to duplicate hypersonic flight conditions in a wind tunnel. Furthermore, even though the configuration can be duplicated, it is usually of a relatively small scale. Thus, one must rely on similarity parameters and extrapolation procedures in order to use wind tunnel data for flight vehicle design and performance predictions. For wind tunnel transition experiments, in addition to similarity in terms of Mach number and Reynolds number, one must also be concerned with similarity of freestream environments and similarity of boundary-layer profiles.

An internal flow system, such as a wind tunnel, has a number of sources to generate velocity, temperature, and acoustical fluctuations not present within the atmosphere. It is well known that the freestream of a wind tunnel is a different environment than found in the atmosphere. One can reduce the amplitude of these disturbances, as in a quiet tunnel, but it is unrealistic to think of duplicating the atmospheric environment in a wind tunnel. The flight environment is mostly unknown and is probably time-dependent. So one must live with a wind tunnel environment which is different than flight.

One can duplicate a flight Mach number and Reynolds number, but generally in hypersonic wind tunnels it will not be possible to duplicate velocity or temperature. Therefore, it is not possible to maintain similarity of boundary-layer profiles between wind tunnel and flight. Since the boundary-layer stability characteristics are very sensitive to the profiles, differences in transition Reynolds number must be expected. Also, of lesser importance, the critical second mode disturbance frequencies of the disturbances that would be expected to cause transition will not be duplicated since the velocities will be different. Generally, the boundary-layer velocities will be larger in flight and result in larger disturbance frequencies.

The bottom line is that a hypersonic wind tunnel cannot duplicate the atmospheric environment or the boundary-layer profiles; therefore, there is no reason to expect the wind tunnel to duplicate the flight transition Reynolds numbers. A possible approach to obtain a solution to this dilemma is to take the same approach as is being used for other aspects of hypersonic aerodynamics – through a combination of analytical and experimental studies. To the extent possible, experiments should be conducted to define the instability phenomena, to compare with theory, to assist in the modelling of the instabilities for computation, to check the computational methods, and to evaluate the differences that occur because the environment and boundary-layer profiles have not been duplicated. The experimental requirements defined above require stability experiments, not transition experiments. When the only information obtained is the location of transition, it is impossible to know the disturbance mechanisms which caused the transition or any details of the transition phenomena.

2. Discussion

(a) Freestream Environment

The environment provides an extremely important initial condition for any boundary-layer transition problem. The environment provides the mechanism by which boundary-layer disturbance growth is generally initiated and establishes the initial disturbance amplitude at the onset of disturbance growth. Based upon the supposition that transition occurs when some boundary-layer disturbances have obtained the critical amplitude required for breakdown of the laminar flow, a change in the initial amplitude of the dominant disturbances changes the required period of growth to obtain the critical amplitude. Thus, a change in the environment will most likely change the transition Reynolds number.

Not all freestream disturbances are important to boundary-layer transition. Some freestream disturbances may have frequencies that do not correspond to unstable boundary-layer frequencies. Thus, these disturbances, upon entering the boundary layer, will be stable and attenuate. Other freestream disturbances may influence only slowly growing boundary-layer disturbances which do not grow large enough to affect transition. It is believed that the critical environmental disturbances are those disturbances of the same frequency as the boundary-layer disturbances responsible for transition. Therefore, it is important to identify the dominant boundary-layer disturbances and the amplitudes of the corresponding environmental disturbances at the same frequency.

Figure 1 is an example of boundary-layer disturbance growth on a flat plate at Mach 2, showing disturbance rms amplitude vs. frequency for several stations [1] (data obtained with a hot-wire). The frequency of the largest disturbances is decreasing with increasing distance along the plate, in response to the growing boundary layer. These disturbances are Tollmien-Schlichting disturbances and have wavelengths which are several times the boundary-layer thickness.

Figure 2 contains similar data obtained on a sharp cone at a Mach number of 8 [2]. The major disturbances in this case are second mode disturbances, the high frequency acoustical disturbances that are unique to high Mach number boundary layers. Second mode disturbances have a wavelength approximately twice the boundary-layer thickness. The point to be made from this comparison is that the major disturbances in a hypersonic boundary layer are of higher frequency than those of a subsonic/supersonic boundary layer. Thus, for a hypersonic boundary layer, one is concerned about a higher frequency portion of the freestream disturbance spectrum than for a supersonic boundary layer.

Figure 3 contains samples of the freestream disturbance spectra obtained in the Arnold Engineering Development Center Tunnel B at Mach 8. The trends of these data are similar to those of other wind tunnels, that of decreasing amplitude with increasing frequency until the signal becomes lost within the instrumentation noise.

Figure 4 identifies the frequencies, for several unit Reynolds numbers, at which the freestream disturbance signal becomes lost within the noise. The data band shown at each unit Reynolds number represents an uncertainty in the determination of that frequency. For some frequencies greater than the indicated values, there are presumably freestream disturbances still present, but they are of unknown amplitude since they are too small to be measured with a hot-wire anemometer. If one defines a quiet environment as one in which the disturbance amplitudes are too small to be measured, then those boundary-layer disturbances which have corresponding frequencies can be considered to be excited by a quiet environment. Thus, the typical wind tunnel situation is that the lower frequency Tollmien-Schlichting disturbances are excited by relatively large amplitude freestream disturbances and the higher frequency second mode disturbances are excited by small amplitude freestream disturbances. There are many hypersonic conditions where the second mode disturbance frequencies are high enough to correspond to the quiet portion of the freestream disturbance spectrum of a conventional wind tunnel. Since the second mode disturbances are expected to cause transition, these conventional wind tunnel situations represent quiet environment transition.

Figure 5 schematically illustrates the importance of the freestream disturbance spectrum in influencing transition in subsonic/supersonic and hypersonic wind tunnels. In subsonic/supersonic boundary layers, the major disturbances are of lower frequency and the corresponding freestream disturbances have large amplitudes. The large amplitude freestream disturbances produce wind tunnel transition Reynolds numbers which are considerably smaller than corresponding flight transition Reynolds numbers. If the wind tunnel freestream disturbance amplitudes are reduced, as in a quiet tunnel, the wind tunnel transition Reynolds numbers become more comparable to flight values. For hypersonic wind tunnels, the freestream disturbances at frequencies which correspond to boundary-layer second mode frequencies have much smaller amplitudes. In many cases, these amplitudes may be too small to be measured. Thus, both the hypersonic conventional wind tunnel and the hypersonic quiet tunnel may have "critical" freestream disturbances which are too small to be measured. Intuitively, it would be expected that the flight transition

Reynolds number would be larger than the quiet tunnel transition Reynolds number. Comparing transition Reynolds numbers does not provide an adequate solution, since they would be expected to be different and there is no way of knowing how different they should be. The basic question that needs to be answered is whether the transition phenomena are the same in all three cases (conventional wind tunnel, quiet tunnel, and flight). If the transition phenomena are the same and the different transition Reynolds number are only the result of a difference in the freestream disturbance amplitudes, then the problem is primarily a boundary condition problem. For this situation, compatibility of conventional wind tunnel, quiet wind tunnel, and flight transition Reynolds numbers becomes a matter of properly accounting for a boundary condition. All hypersonic boundary-layer stability data have been obtained in conventional wind tunnels. Hypersonic quiet tunnel stability data are needed to help answer these basic questions.

(b) Lack of Similarity of Boundary-Layer Profiles

The temperatures in the boundary layer of a hypersonic wind tunnel model will be less than the corresponding boundary-layer temperature of a flight vehicle. Therefore, it will not be possible to duplicate a flight boundary-layer profile in a hypersonic wind tunnel. The sensitivity of boundary-layer transition to changes in boundary-layer profiles is presently not well enough understood to evaluate this effect. However, stability calculations of Mack [3] suggest that the effects are significant. Mack made stability calculations corresponding to wind tunnel conditions and stagnation temperatures of 922°R and 1310°R. He noted that, "increasing the stagnation temperature has a considerable stabilizing influence at $M_e = 6.8$. The amplification rate is lowered at almost all frequencies and the unstable frequency band is narrowed by about 15%." A reduction in second mode amplification rates would be expected to increase the transition Reynolds number. If this is a consistent trend, then the large stagnation temperatures in flight should produce larger transition Reynolds numbers than found in wind tunnels, independent of the environmental effects. Also, the larger stagnation temperatures of a shock tunnel should produce larger transition Reynolds numbers than the long duration, conventional hypersonic wind tunnel. It is difficult to even speculate how transition in cold helium flow should relate to other situations.

(c) Lack of Second Mode Disturbances

There is a fundamental transition issue that applies to the interpretation of all hypersonic transition data and to the use of analytical prediction methods to predict hypersonic transition. It is generally assumed that boundary-layer disturbance growth will occur at all unstable frequencies. Such an assumption assumes that the environment is capable of exciting boundary-layer disturbance growth throughout the range of unstable frequencies. This is not necessarily the case for the high frequency second mode disturbances of a hypersonic boundary layer. There may be many hypersonic flow situations, both in ground test facilities and in flight, where the potentially most unstable boundary-layer disturbances may not be excited because the environment has not provided a suitable stimulus to excite them. If transition must wait for disturbances with a smaller growth rate to obtain the critical breakdown amplitude, then a delay in transition would be expected. Thus, some transition delay, due to a lack of environmental stimulus of the potentially most unstable disturbances, may be a common hypersonic occurrence. Figure 6 illustrates the potential problem. The major second mode disturbance frequencies are estimated for the boundary layer of a 7 degree half angle cone in a Mach 8 wind tunnel. For a freestream unit Reynolds number of 20 million, the most unstable second mode disturbances have frequencies greater than a megahertz. Available instrumentation cannot measure disturbances in this frequency range; however, intuitively it seems unlikely that the freestream would contain disturbance energy at such high frequencies. This problem is an additional possible complication in the interpretation and understanding of transition data which presently is not understood, but one should not ignore the situation.

3. Recommendations

It is recommended that the hypersonic transition problem be addressed though a combined theoretical, analytical, and experimental approach. On the experimental side, a plea is made for more boundary-layer stability data, in both conventional and quiet wind tunnels. Stability data, not transition data, are required to better understand transition phenomena, to evaluate stability theories, to evaluate analytical transition prediction methods, and to better interpret transition data.

References

[1] Laufer, J. and Vrebalovich, T., "Stability and Transition of a Supersonic Laminar Boundary Layer on an Insulated Flat Plate," *Jour. of Fluid Mechanics*, Vol. 9, Part 2, October 1960, pp. 257-299.

[2] Stetson, K. F., Thompson, E. R., Donaldson, J. C., and Siler, L. G., "Laminar Boundary Layer Stability Experiments on a Cone at Mach 8, Part 1: Sharp Cone," AIAA Paper No. 83-1761, July 1983.

[3] Mack, L. M., "Boundary-Layer Stability Analysis for Sharp Cones at Zero Angle-of-Attack," AFWAL-TR-86-3022, August 1986.

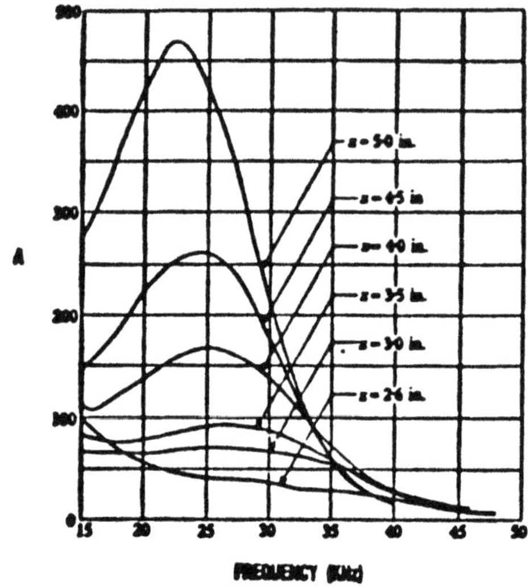

Fig. 1 Fluctuation Spectra (M_∞=2)

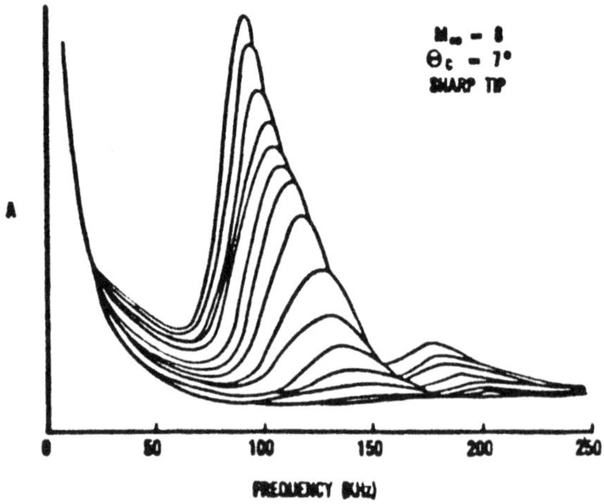

Fig. 2 Fluctuation Spectra (M_∞=8)

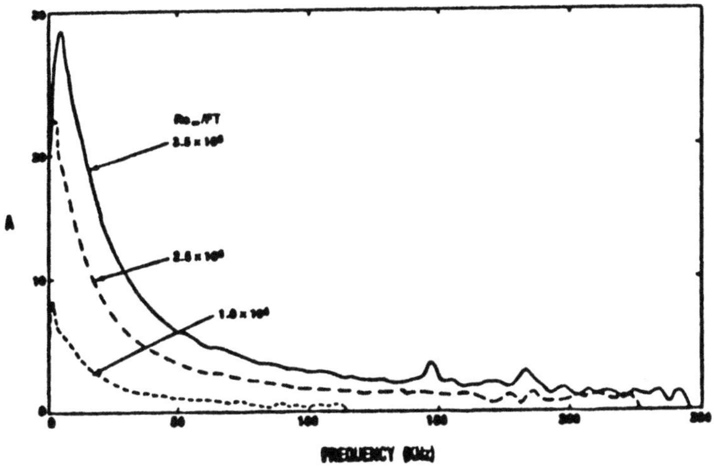

Fig. 3 Freestream Disturbance Spectra

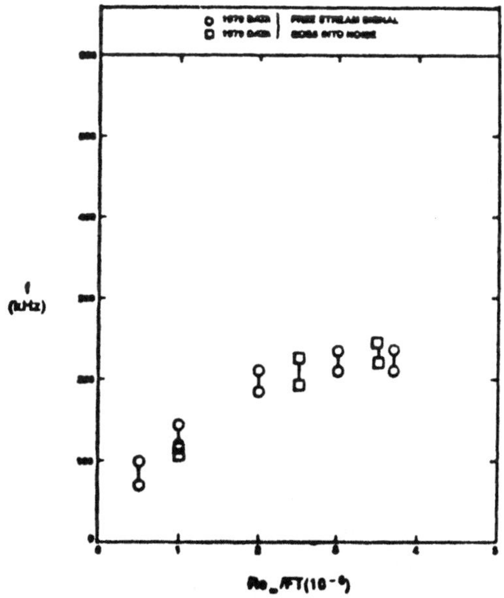

Fig. 4 Maximum Freestream
Disturbance Frequencies

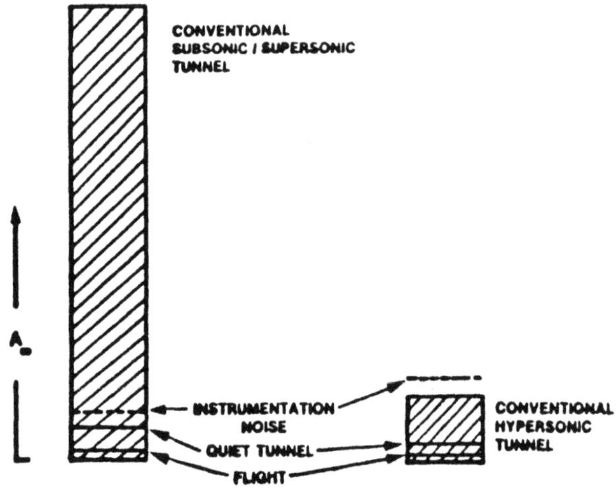

Fig. 5 Schematic of Freestream Disturbance Effects

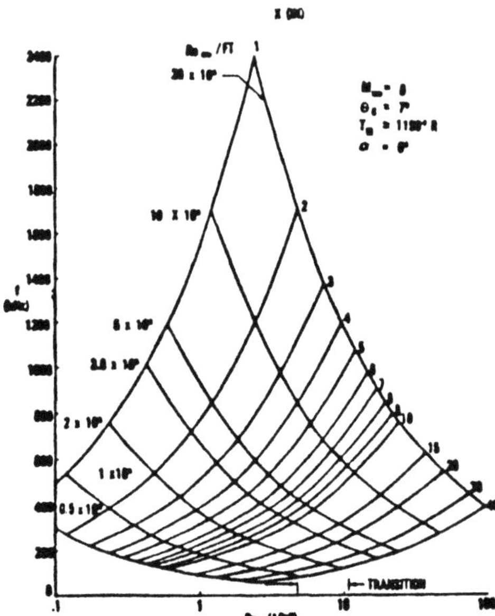

Fig. 6 Second Mode Frequencies

LOW-SPEED EXPERIMENTS

PANEL SUMMARY: LOW SPEED EXPERIMENTS

Percy J. Bobbitt

Distinguished Research Associate
NASA Langley Research Center
Hampton, VA 23665

The "Low Speed Experiments Panel" consisted of D. Arnal (ONERA/CERT), P. J. Bobbitt (NASA Langley), A. Braslow (NASA Langley consultant), T. C. Corke (Illinois Institute of Technology), M. Gaster (University of Cambridge), L. Maestrello (NASA Langley), and W. S. Saric (Arizona State University). Oral presentations were given by all except M. Gaster, who had to leave early, and T. C. Corke who was unable to attend due to his wife giving birth to their third child. Nevertheless, material provided by M. Gaster and T. C. Corke was presented by P. J. Bobbitt and M. Morkovin, respectively. All panel members provided written versions of their discussion except A. Braslow and W. S. Saric. However, some attempt will be made in this summary to give the gist of their remarks.

The chairman noted at the outset that the panel would consider both subsonic and transonic experimentation to be within their "jurisdiction." He also set forth a goal for his panel of establishing the state-of-the-art of low speed experimentation as well as providing suggestions for future directions and research. It was noted that the varied background, interests, and responsibilities of the panel members lent itself to the achievement of these objectives. Indeed, the presentations of the panelists ranged from very general or "broad-based" discussions (Braslow, Bobbitt, Saric, and Gaster) to the more detailed achievements of specific wind-tunnel-research programs (Arnal, Corke, and Maestrello). A summary of some of the more important points made by the panelists are summarized below.

The main thrust of A. Braslow's discussion was that while there is still much to be learned we cannot ignore the fact that a good engineering knowledge, gained from wind tunnel and flight experiments and improved boundary layer stability-codes, has already been established for wing design. Included in this "knowledge" is the understanding and data obtained from a number of recent complex flight research programs. Wind tunnel tests providing valuable data include the LFC and HLFC tests in the Langley 8-FT TPT of a seven foot chord model and a number of tests in the Langley Low Turbulence Pressure Tunnel (LTPT) of NLF airfoil concepts. It was noted that there have been many tests abroad of both varieties that have also contributed substantively. Braslow encouraged the continued emphasis on research aimed at a better understanding of transition physics but also of applied experiments such as those mentioned.

P. Bobbitt also took note of past progress but also detailed a number of concerns and problems pertaining to the basic understanding of transition physics and to the measurement of the important physical parameters in a transitional flow. He placed particular emphasis on the problems encountered in carrying out laminar flow research in a wind tunnel environment. A number of research opportunities (or needs) were cited and some shortfalls in current state-of-the-art instrumentation required for code validation pointed out.

W. Saric broke his discussion into two parts: stability experiments and future directions. With respect to the former he noted that "regardless of the goals of the experiment (control, 3-D, receptivity, nonlinear breakdown) the linear aspects of the problem must be correct." In addition, one must provide (measure) the initial conditions with accuracy for theory and computations. A number of pitfalls often seen in transition experiments were detailed and recipes provided for the best results. A list of the error sources discussed by Saric is given below.

<u>T-S Waves</u>

(i) Global Pressure Gradients

(ii) Local Pressure Gradients

(iii) Disturbance Source

(iv) 3-D Distortion and Freestream Effects

(v) Nonparallel Effects

(vi) Probe/Traverse/Model Problems

<u>C-F Vortices</u>
(without Disturbance Sources)

(i) Roughness

(ii) Freestream Turbulence

W. Saric completed his presentation by making forecasts of both near and far-term progress in the areas of receptivity, 3-D basic states, interaction/breakdown, roughness and transition control with most of the emphasis on the second, third, and fifth. The need for closely coupling theory and experiment was emphasized throughout his discussion.

T. C. Corke suggested that more experimentation should be directed toward understanding the first growth of 3-D disturbances. Particular attention was directed toward resonant interactions between the most amplified 2-D modes and 3-D modes. The 3-D modes in "real-life situations" are due, of course, to background disturbances; in controlled experiments they may be artificially generated. Three-dimensional modes with streamwise wave numbers near the subharmonic of the 2-D mode can satisfy the phase speed requirements required for resonance. As Corke points out "the implication of having two frequency interactions" relates to the transition-to-turbulence process itself. In a nonlinear interaction involving two frequencies, not harmonically related, "sum and difference interactions can rapidly fill the spectrum to give the energy saturation and broadband character associated with a turbulent state."

Finally, Corke further points out that 2-D/3-D interactions and the resultant spectral broadening provide a more realistic physical model of the transition process than the growth of a single mode. In order to explore this point, various methods of providing the required mode mix for boundary layers, jets, and wakes have been developed. In Corke's written paper, the three concepts are described and some sample results given.

D. Arnal's presentation also concerned the physics of transition but from a linear stability point-of-view. Low speed experiments and stability calculations (Mack code) for a swept wing in ONERA D airfoil were carried out to examine crossflow and Tollmien-Schlichting instabilities. Special emphasis was placed on the dependence of the amplification rate on angle of propagation. Leading edge contamination was studied using a wing fixed to the tunnel floor with a RA16SC1 airfoil section and the results agree with earlier investigations. A brief discussion was also given of the results from a series of thin-film gages distributed along and around the leading edge.

The only presentation to concentrate on the control of transition was L. Maestrello. He reviewed past steady (passive) and active techniques for transition and separation control as well as for turbulent drag reduction. Details of current research, both experimental and theoretical, on the use of a steady and unsteady heating to delay transition were given. The stabilizing mechanism in air is the cooling of the flow downstream of heat strip. Generally, the heat strip is located in a region of favorable pressure gradient near the leading edge of the airfoil or body. The success of these numerical and experimental activities are highly encouraging and militate for more "control" research but at higher Mach and Reynolds number (see Bobbitt paper).

M. Gaster argued for a better understanding of transition physics through the use of low speed tunnels and theory. He placed special emphasis on the excitation of Tollmien-Schlichting waves by free stream turbulence. Gaster also emphasized the need to understand various types of nonlinear interactions and the final stages of transition. Seven topics were listed by Gaster deserving of more attention. In addition to those already mentioned he highlighted transition control, surface quality, and instrumentation.

In general the audience's participation consisted of questions at the clarification type. The main exception was the discussion following the presentation of Morkovin (substituting for Corke) which continued after the formal presentations. This discussion centered on research pertaining to nonlinear interactions with some attempts made to relate the recent theoretical work of Herbert, DiPrima, Stuart, Smith, and many others to the "real" transition problem, the work of Corke and his colleagues, and to experimental research needs. One could conclude from this discussion that the linear, single-frequency problem was fairly well understood and the center of gravity of our theoretical and experimental research should be moving toward, if not centered in, the nonlinear regime. It was clear that the emphasis of some of the presentations on the need for a high order accuracy and detail in our experimental research was well received. Continued emphasis on improving our instruments and wind tunnels and our understanding of the data derived from these facilities (the receptivity problem) is needed. Finally I would recommend to the reader the additional information and research ideas detailed in the written papers of the "Low Speed Experiments Panel" members.

TRANSITION RESEARCH OPPORTUNITIES AT SUBSONIC AND TRANSONIC SPEEDS

Percy J. Bobbitt

Distinguished Research Associate
NASA Langley Research Center
Hampton, VA 23665

1. Introduction

Substantial progress has been made in a number of areas related to transition physics and applied transition technology in the past decade. This has been most noticeable in the subsonic and transonic speed regimes where improved stability theories and wind-tunnel facilities coupled with state-of-the-art boundary-layer and inviscid-flow codes have stimulated the design of a number of ambitious wind-tunnel and flight experiments. The data base and code validations obtained provide some of the resources required for the design of flight vehicles with large expanses of laminar flow.

Even with these recent advances, it is important to note that most experiments have either been of NLF or LFC concepts where longitudinal pressure gradient and/or steady suction control of transition have been employed. Spanwise pressure gradients have generally been small and in most cases ignored in the assessment. Heat, cold, and noise both steady and periodic have been employed to control transition (some with great success) but most experiments have been at low speed and low Reynolds numbers. Highly complex compressible stability codes have been formulated but they do not always give the same "answer" for the same input data. Recent CFD applications to transition simulation and stability theory improvements have provided a severe challenge for the experimentalists but only a few have accepted it. Our instrumentation and wind-tunnel facilities gradually improve, but not at the rate we would like. So transition research, like that in many other fields, seems to chase a moving target as it uncovers new problems as fast as it solves the old ones.

Most of the flow phenomena on an aerospace vehicle are affected in one way or another by the transition of the boundary-layer flow from laminar to turbulent. In the world of experimental aerodynamics, transition is often fixed in order that the data be more easily interpreted or to simulate, as best as one can, some high Reynolds number condition. When natural transition is permitted, its location is rarely measured. There are, nevertheless, a number of methods available to measure the transition point, or line, but they will not always agree.

A list of most of the flow phenomena influenced by transition is given in Figure 1. Also given in this figure are quantities which we have termed integrated phenomena. They determine the performance, stability, and to some extent, the weight of aerospace vehicles and consequently the efficiency and viability of a given aircraft concept. As indicated on the figure, all of the flow phenomena have an impact on the integrated phenomena. The pre-eminent importance of transition in fluid flows and aircraft design is abundantly clear to all.

There is much to do; the present discussion will point out a number of opportunities available to experimentalists in the subsonic and transonic flow regime. Included are needs to improve the quality of flow in our wind tunnels and the techniques used to measure it. Additional concerns treated in the paper are listed in Figure 2.

2. Wind-Tunnel Flow Quality

Most of us realize that the uncertainties in transition data obtained in wind-tunnel experiments can be quite large at high subsonic speeds due to the adverse effects of the environment, and designs based on these data will invariably be conservative. Flight tests have been carried out to examine both stability-theory and acoustic-environment concerns as well as the practical problems of powered suction surfaces (see Reference 1) on the more quiescent free-air environment. A recent paper (Reference 2) gives some results for the effects of the external environment on transition for several bodies of revolution. Large differences between free-air and wind tunnel transition are demonstrated.

Another concern pertaining to flow quality is the primary technique used to make the measurements. A number of reports over the past half dozen years (e.g., see Reference 3) have documented the fact that single-wire hot-wire anemometry techniques are greatly in error at high subsonic speeds. However, these are the type of data used to quantify flow quality and, consequently, are the basis for the empirical methods used to assess the effects of flow quality on transition. Furthermore, it was recently shown (Reference 4) that the single-wire technique may not even yield accurate results at low speed. The three-wire technique used in this investigation needs to be substantiated by other investigators across the speed range.

A three-wire probe, similar to those used in the aforementioned studies, is depicted in Figure 3. The method of calibrating such a probe and of extracting unsteady velocity, density, mass, temperature, and pressure data is documented in References 5 through 8. Figure 4 (from Reference 8) gives some example results along with data obtained using the more familiar two-wire and single-wire techniques. The differences are cause for concern and are generally attributed to the various assumptions required to extract data from the one- and two-wire probes. Fewer assumptions are required in implementing the three-wire technique.

So there is a great need for the flow quality in our tunnels to be remeasured and new assessments made of the effect of the wind-tunnel environment on transition using the best hot-wire and thin-film technology available. Basic experiments are needed where large speakers are used to introduce discrete tones as well as broad band noise over a wide range of intensities to the test section environment. Devices can also be designed and installed to vary the vorticity in the test-section and/or boundary layer flows (see References 9 and 10). Standard flat plate and cone models should be defined and used in these experiments. The AEDC transition cone (Reference 11) is a natural candidate for the latter.

3. Methods of Flow Quality Improvement

While it is important to understand the effect that wind-tunnel environments have on transition, it is also in our interest to try to improve these environments. Many tunnels over the past decade have been improved with the installation of anti-turbulence screens; several transonic tunnels have been, or soon will be, supplied with a downstream or diffuser choke. There are many other tunnel features which have been found over the years to be advantageous to the flow quality in the test section. A number of them are listed in Figure 5.

Figure 6 illustrates the large reduction in fluctuating pressure levels that can be obtained when a tunnel is choked at high subsonic speeds. The data shown were taken in the Langley 8-Foot Transonic Pressure Tunnel (TPT) and indicates an order of magnitude reduction in unsteady pressure level at $M_\infty = 0.95$.

Some idea of the beneficial effects of a series of screens (four) and a honeycomb in the settling chamber of a transonic tunnel (8-Ft. TPT) can be obtained from Figure 7. The fluctuating velocity data shown were obtained using a single-wire technique since only single-wire data were available for the tunnel without the screens and honeycomb as well as with the screens and honeycomb. While the magnitude of the data points are suspect, the large difference in levels is still felt to be a good indication of the improvement of flow quality due to the flow treatment.

Another item on the list of Figure 5 which has received some recent attention and deserves much more is the use of mufflers (or resonators) on the high-speed diffuser downstream of the test section and drive fans. Figure 8 depicts a proposed tunnel which has a diffuser muffler, an acoustically treated centrifugal fan and an "exponential" contraction. Predicted acoustic levels are shown in the figure for both two- and three-sided mufflers.

4. Unit Reynolds Number Effects and Flow Quality

One of the troublesome aspects of wind-tunnel testing is the masking effect the environment has on unit Reynolds number effects. Tunnels which control pressure level to vary Reynolds number typically use more or less power when the pressure level is increased or decreased, respectively. The increase or decrease in energy imported to the fan results in a higher or lower level of noise and vorticity in the test section. These changes in the environment affect the location of transition and in an unnatural way. Nevertheless, in the past they have often been attributed to "the unit Reynolds number effect."

A recent study, documented in Reference 2, has used both analytical and experimental data for the AEDC cone (Reference 5) to indicate how variations in unit Reynolds number can effect transition. Calculations and data are also given for an ogive/cylinder body of revolution in this reference. In the case of the AEDC cone, both flight and wind-tunnel data are available. Figure 9 shows flight data (F-15 flight tests) for the AEDC cone along with calculations from Reference 2. Local transition Reynolds number as a function of unit Reynolds number is given for Mach numbers of 0.837 and 1.543. Most noticeable in this figure is that at low unit Reynolds numbers there is a strong effect of unit Reynolds number while at high values, i.e., beyond $\frac{U_\infty}{\nu_\infty} \times 10^6$ of 1.5 to 2.0, there is almost no effect.

A look at the effect of a wind-tunnel environment on the variation of transition Reynolds number with unit Reynolds number is facilitated by Figure 10. Data and calculations are given for two wind-tunnel conditions and calculations for free-flight conditions at a Mach number of 0.8. It is again seen that there is little sensitivity of transition Reynolds number to changes in unit Reynolds number except of values of unit Reynolds number below 1.5×10^6.

Perhaps the most notable feature of Figure 10 is the change in level of transition Reynolds number with the change in the environment surrounding the cone. Finally, it should be noted that unit Reynolds number effects are quite different for an ogive/cylinder body than those just discussed for the cone (see Reference 2).

More research is required on unit Reynolds number effects for both simple and complex shapes. This type of experimentation might best be carried out in cryogenic pressure tunnels where Reynolds number can be changed without an associated increase or decrease in noise and vorticity levels. For conventional tunnels, the affect of environmental changes caused by changes in tunnel pressure level (and associated unit Reynolds number) must be accessed.

5. Spanwise Pressure Gradient

Spanwise pressure gradients just like chordwise pressure gradients can have favorable or unfavorable effects on boundary-layer stability. This means that an additional control can be exercised in the forward regions of a wing where one wants to maintain as large an area of laminar flow to as high a Reynolds number as possible. Reference 9 analyzes the effect of spanwise pressure gradient on both forward and aft swept wings and indicates that for normal elliptic or parabolic span loads the swept forward wing has a considerable advantage. Figure 11, taken from Reference 12, shows almost a 40 count drag reduction at a Reynolds number of 16×10^6 in going from an aft swept wing to an equivalent forward swept wing. This advantage is diminished as one goes to higher values of Reynolds number. It is important to note that too much tailoring of the spanwise pressure gradient can have an adverse effect on induced drag. Parametric experiments aimed at the exploration of spanwise pressure-gradient effects would be most useful.

Much of the laminar flow wing-design work is done with 2-D or quasi 2-D boundary layer methods coupled with 2-D or 3-D inviscid codes. The preferred combination, of course, is a 3-D boundary layer coupled with a 3-D inviscid code. Spanwise pressure gradients in the flow will then have a proper effect on the boundary layer profiles. The application of parallel stability codes to these profiles will compromise to some extent the utility of, and the flow physics contained in, the 3-D boundary-layer code. A non-parallel stability code is really what is required, but they are not in common usage. More applied code/experimental research is needed to explore the difference in parallel and non-parallel stability results.

6. Active Transition Control

Over the past half dozen years, there has been an increasing interest in the use of active means to control transition. Unsteady sucking and heating, as well as noise have been employed, either in experiments or theoretical simulations, with varying degrees of success. The experiments and simulations have all been at low speed and Reynolds number. Higher speed and Reynolds number experiments are needed to give us a few clues of the practical prospects for these techniques. Clearly diagnostic tests and theoretical studies should be integral parts of the research program.

A recent result from a transition control experiment (Reference 13) is given in Figure 12. It shows the output from a hot wire anemometer in the boundary layer of a flat-plate model. The control is affected by a heat strip located just downstream of the leading edge. The heat input is oscillatory.

Figure 12 shows the hot-wire output at different times for different levels of heat input starting with the plain, turbulent uncontrolled boundary layer. It is clear from the wire output for time step 6 that the boundary layer at the wire location has become laminar. The stabilizing effect that the control provides is thought to be the cooling of the boundary layer as it moves away from the heat strip.

Experiments like that of Reference 13 open up a whole new avenue of transition experimentation.

7. Stability Theory

Over the past decade, there has been a quantum leap forward in our ability to predict, on the basis of classical stability theory, the point at which the flow transitions from laminar to turbulent on a wing or body. Actually, transition occurs over a finite length which is dependent on the pressure distribution (chordwise and spanwise), condition of the surface (roughness and waviness), curvature of the surface, the external environment, the level of heat transfer at the surface, free-stream and local Mach and Reynolds numbers, whether the shear layer is attached or separated from the surface and if there is mass flow through the surface. There may be other factors but these are the major ones that past research has addressed.

There are several concerns related to the application of existing stability theory codes that need attention. One is the variance in N factors of transition determined by various stability codes. With the recent acquisition of high subsonic data in flight and in a wind tunnel there is an opportunity to define a number of test cases to examine these differences. Of course a concern relative to the N factors determined from the wind-tunnel data is the tunnel environment. Empirical curves formulated to aid in the evaluation of its affect on transition have been given by several investigators (see References 12, 14, and 15) but the accuracy of the results obtained using these curves depends on the hot-wire technique used. Furthermore, the empiricism only accounts for the RMS level of the disturbances when it is well known that the energy in the frequencies that are the most amplified during transition also plays a part in the initiation and rapidity of the transition process. As indicated in the section on flow quality, controlled experiments in this area are a priority need.

Empirical equations aimed at accounting for the effect of the environment on the amplification of the various types of disturbances yield lower values of N for a highly turbulent environment (such as, in a wind tunnel) than that for "free air." In other words, transition in a turbulent environment is predicted to occur at a lower level of wave amplification than in free air. Experimental data that back up this assumption is in short supply. It seems more likely that the amplification of waves <u>at transition</u> in a noisy environment will be about the same as in a quiet one. The quantities that should be affected the most (see Figure 13) are the location of the neutral point and the rate of growth of the amplitude of the unstable (two- and three-dimensional) waves. The neutral point (or points) may be extremely difficult to measure at high speed and only slightly less difficult to infer. Nevertheless, it should be attempted. Alternatively, it should be possible to measure the growth rate in controlled 2-D like (oblique waves may be generated) experiments to shed some light on the above conjecture. Navier-Stokes transition simulation with various prescribed disturbance (two-dimensional and three-dimensional) environments would also be enlightening.

The nonlinear interactions between the various instability modes have been the subject of much speculation but only a few analyses. Tollmien-Schlichting, Taylor-Görtler and crossflow disturbances can exist at the same location and the question of how one affects another is not trivial. High resolution experiments that can track the growth of several instability mechanisms to the point where turbulent bursts begin are required.

8. Transition Simulation Opportunities

The technology that holds the most promise for increasing our understanding of transition physics in the coming decade is that of transition simulation using the Navier-Stokes equations. Already the origination of the turbulent eddies (pin head vortices) and the interaction and breakdown of these flow features have been demonstrated (References 16, 17, and 18). Qualitatively these predictions agree well with both flow visualization and hot-wire measurements. However, a much more comprehensive measurement of transitional flow is required. As noted in the next section, continuous time varying measurements along a line or, better still, in a plane are needed to really validate simulation codes (Reference 19).

To date, the application of Navier-Stokes simulations to fluid flows have been restricted to channels, flat plates (temporal and spatial) and Taylor-Couette flow. The treatment of both swept and unswept leading edges and separation bubbles has begun. One would expect the simulation of transition under a shock interaction to soon "come under fire." Validation experiments to complement/validate these CFD results are a critical need.

9. Measurement/Instrument Needs

To keep pace with the advances in stability theory and Navier-Stokes simulations, more detailed, accurate, time-varying measurements are needed of transitional flows. Preferably they will be made using non-intrusive instruments, but accuracy may be a problem and many of the needed quantities cannot be measured (Reference 7).

Three-wire anemometry can provide state-variable data now unobtainable with state-of-the-art LDV systems but accuracy, in an absolute sense, is still to be defined. The development of laser-induced-flourescence (LIF) technology may help answer some of the accuracy questions and provide additional measurement capabilities.

As noted in Reference 19, one of the prime requirements in flow-field diagnostics is an instrument that can measure time varying data on a line or a plane. Point measurements in an unsteady flow can yield only limited understanding even when conditional-sampling, data-reduction techniques are used. Hot-wire rakes have been used on occasion but only one or two variables are measured or deduced. Again, LIF instrumentation may fill this void.

Measurements that are made simply to tell whether the flow is laminar or turbulent are of minimal utility in validating a transition code. One also wants to determine where transition started and where the flow becomes fully turbulent and if transition took place in a small separation bubble or in an attached boundary layer. The most amplified disturbance frequency (or frequencies) is also of interest. Reynolds stresses in the boundary layer following transition and as the boundary layer matures should also be determined. Thin film and laser instrumentation now allow most of these quantities and/or locations to be measured due to the high resolution and/or sensitivities available. This same resolution allows the measurement of direction of the most unstable frequencies but to date, no such measurement has been made. Note that this is important at transonic and supersonic speeds but not at subsonic speeds (Reference 14).

10. Momentum Thickness Transition Criteria

While most present-day "purest" disdain the use of criteria based on momentum-thickness Reynolds number to determine transition, the fact is that this approach was the preferred one for many years, particularly at low speed (e.g., see Reference 15). Experimental data and Navier-Stokes simulations that enables the determination of this quantity along with the required transition data at high speed are greatly needed to formulate the required empirical relationships. Clearly, for swept wings momentum thickness Reynolds numbers must be determined for both crossflow and streamwise boundary-layer profiles.

11. Induced Drag and Laminar Flow

A problem not yet addressed in the literature is how the induced drag of a wing is affected by the extent of laminar flow on its surface. While skin friction will probably go down as the extent of laminar flow is increased, it is also likely that the induced drag will go up. This reasoning is based on the assumption that a thick turbulent wake will not roll up as readily as a thin one. This in turn will have an affect on the onset flow (induced angle of attack) and the wing lift for a given incidence. Experiments and CFD/Navier-Stokes calculations should be carried out to explore these questions, perhaps, starting with the latter. Several experiments have been carried out that lend credence to the above hypothesis. One in which a saw tooth trailing edge was employed and a second where the trailing edge was corrugated. Both yielded a reduction in drag over that of a straight-trailing-edge wing and both provide a more turbulent wake.

12. Concluding Observations

A number of areas have been discussed and research needs delineated. They may be summarized as follows:

- Receptivity problem not well understood and further experimentation needed. Real and artificial (tunnel induced) effects must be sorted out.
- What constitutes good flow quality not well understood. Whole flow quality problem needs more attention. The amplification of disturbances (N factor) in various wind tunnel environments needs study.
- Unit Reynolds number effects in wind tunnels need additional study.
- The effect of spanwise pressure gradients on flow stability needs to be examined in a systematic manner to determine their effect on boundary layer stability and drag.
- Experiments that explore various means to actively control transition need vigorous pursuit.

- Navier-Stokes transition simulation holds tremendous promise for understanding transition physics and parallel experiments must be carried out.
- Further instrumentation advances needed for validation of transition simulations.
- Momentum thickness methods may be more valuable than realized for transition prediction and diagnostic experiments at high speed should be carried out to further explore them.
- Need to define/select standard flat plate and cone models.

Clearly there is a considerable amount of research to be done before the physics of transition can be predicted and understood to the level desired. The full potential of laminar flow technology will not be realized until most of this needed research is carried out.

References

[1] Natural Laminar Flow and Laminar Flow Control Symposium, NASA Conference Publication 2487, Hampton, VA, March 16-19, 1987.

[2] Goradia, S. H., Bobbitt, P. J., and Harvey, W. D.: Computational Results for the Effects of External Disturbances on Transition Location on Bodies of Revolution from Subsonic to Supersonic Speeds and Comparisons with Experimental Data. Aerospace Technology Conference and Exposition, Anaheim, CA, September 25-28, 1989, SAE Technical Paper Series 892381.

[3] Jones, G. S., Stainback, P. C., Harris, C. D., Brooks, C. W., and Clukey, S. J.: Flow Quality Measurements for the Langley 8-Foot Transonic Pressure Tunnel LFC Experiment. 27th Aerospace Sciences Meeting, Reno, NV, January 9-12, 1989, AIAA Paper No. 89-0150.

[4] Unpublished data from P. C. Stainback pertaining to hot-wire measurements in the NASA Langley Low Turbulence Pressure Tunnel.

[5] Stainback, P. C., Johnson, C. B., and Basnett, C. B.: Preliminary Measurements of Velocity, Density, and Total Temperature Fluctuations in Compressible Subsonic Flow. AIAA 21st Aerospace Sciences Meeting, January, 1983.

[6] Stainback, P. C.: A Review of Hot-Wire Anemometry in Transonic Flows. *ICIASF '85*, Stanford, CA, August, 1985, pp. 67-78.

[7] Bobbitt, Percy J.: Instrumentation Advances for Transonic Testing. Transonic Symposium: Theory, Application, and Experiment, NASA Langley Research Center, Hampton, VA, April 19-21, 1988, Jerome T. Foughner, Jr. (Compiler), NASA CP-3020, Vol. I, Part 2.

[8] Jones, G. S. and Stainback, P. C.: A New Look at Wind Tunnel Flow Quality for Transonic Flows. SAE 881452 Aerospace Technology Conference and Exposition, Anaheim, CA, October, 1988.

[9] Hall, A. A. and Hislop, G. S.: Experiments on the Transition of the Laminar Boundary Layer on a Flat Plate. ARC R & M 1943 (1938).

[10] Schubauer, G. B. and Skramstad, H. K.: Laminar Boundary Layer Oscillations and Stability of Laminar Flow. National Bureau of Standards Research Paper 1772. Reprint of a classified NACA Report first published in April, 1943, and later released as NACA War-time Report W-8; and J. Aero. Sci., Vol. 14, 69 (1947); see also NACA Rep. No. 909.

[11] Dougherty, N. S.: Influence of Wind Tunnel Noise on the Location of Boundary-Layer Transition on a Slender Cone at Mach Number from 0.2 to 5.5. AEDC TR-78-44, March, 1980.

[12] Goradia, S. H., Bobbitt, P. J., Morgan, H. L., Ferris, J. C., and Harvey, W. D.: Results of Correlations for Transition Location on a Clean-up Glove Installed on an F-14 Aircraft and Design Studies for a Laminar Glove for the X-29 Aircraft Accounting for Spanwise Pressure Gradient. Proceedings of a Symposium held at NASA Langley Research Center, NASA CP-3020, Vol. II, April 19-21, 1988.

[13] Maestrello, L. and Nagabushana, K. A., "Relaminarization of Turbulent Flow on a Flat Plate by Localized Surface Heating," AIAA 2nd Shear Flow Conference, Paper No. AIAA-89-0985, 1989.

[14] Mack, L. M.: On the Stability of the Boundary Layer on a Transonic Swept Wing. AIAA Paper No. 79-0264, AIAA 17th Aerospace Sciences Meeting, January, 1979.

[15] Granville, P. S.: The Calculations of Viscous Drag of Bodies of Revolution. Navy Department. The David Taylor Model Basin; Report No. 340 (1953).

[16] Biringen, S.: Final Stages of Transition in Plane Channel Flow," Journal of Fluid Mechanics, Vol. 148, 1984, pp. 413-442.

[17] Biringen and Caruso, M. J.: Symposium on "Research in Natural Laminar Flow and Laminar-Flow Control," March, 1987, NASA CP-2487, Part 2.

[18] Zang, T. A.: Numerical Computation of Transition to Turbulence. Instability and Transition Proceedings, NASA Langley Research Center, May 15 -June 9, 1989, Robert G. Voigt and M. Y. Hussaini (editors), Springer-Verlag.

[19] Bobbitt, P. J.: The Pros and Cons of Code Validation. AIAA 88-2535, Williamsburg, VA, June, 1988.

Nomenclatures

a_∞	speed of sound in free air
A	wave amplitude
A_0	wave amplitude at neutral stability point
c	airfoil chord
$\overline{c}$	mean aerodynamic chord of wing
C_D	drag coefficient, $= \frac{D}{\frac{1}{2}\rho_\infty v_\infty^2 \overline{c}}$
$C_{D,wake}$	total of skin friction, shock and form drags
C_L	lift coefficient, $= \frac{L}{\frac{1}{2}\rho_\infty U_\infty^2 \overline{c}}$

D	drag force
L	lift force
$\tilde{m}$	RMS value of fluctuating mass flow
m_∞	free stream mass flow, $= \rho_\infty U_\infty$
M_∞	free stream Mach number, $= \frac{U_\infty}{a_\infty}$
N	$ln\ A/A_0$, amplification factor
p_∞	free stream static pressure
$\tilde{p}$	RMS value of fluctuating pressure
R_∞	Reynolds number in free stream based on $\bar{c}$, $= \frac{U_\infty \bar{c}}{\nu_\infty}$
R	unit Reynolds number, $= \frac{U_\infty}{\nu_\infty}$
U_∞	free stream velocity
U_e	velocity at edge of boundary layer
$u(x,y,t)$	instantaneous velocity at a point x, y in the boundary layer
u'	x component fluctuating velocity
$\tilde{u}$	RMS value of fluctuating velocity
x	distance measured in streamwise direction
x_0	reference length, = 15 in.
y	distance measured normal to a surface
α	angle of attack
δ_t	turbulent boundary layer thickness at $x/x_0 = 1$ at time step 1
Λ_{LE}	leading sweep angle
ν_∞	kinematic viscosity of air in free stream
ρ_∞	free stream density
$\tilde{\rho}$	RMS value of fluctuating density

Transition Effects

- Flow Phenomena
  - Separated Flows (Massive, Vortices, Bubbles)
  - Shock Boundary Layer Interaction
  - Trailing Edge Interaction
  - Boundary Layer Thickness
  - Juncture Flows
  - Wing-tip Flows
  - Skin Friction
  - Vortex Breakdown
- Integrated Phenomena
  - Lift
  - Moment
  - Drag
  - Stability
  - Control Effectiveness
  - Buffet
  - Flutter
  - Pressures/Loads

Figure 1. Flow phenomena and integrated phenomena affected by transition.

- Flow Quality
- Methods of Flow Quality Improvement
- Unit Reynolds Number Effects
- Spanwise Pressure Gradients
- Active Transition Control
- Stability Theory
- Transition Simulation Opportunities
- Measurement/Instrumentation Needs
- Momentum Thickness Methods
- Induced Drag and Laminar Flow
- Concluding Observations

Figure 2. Topics discussed in present paper.

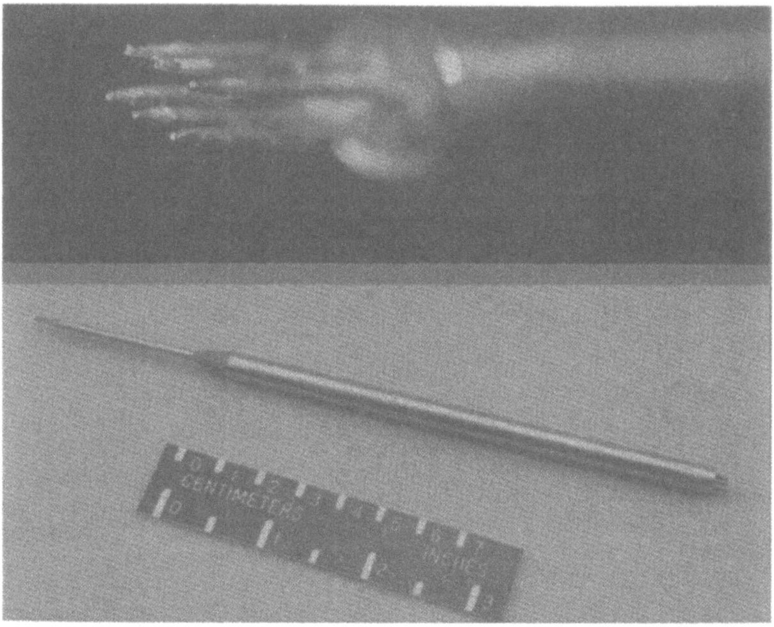

Figure 3. Enlarged photographs of a three wire hot-wire probe and probe head.

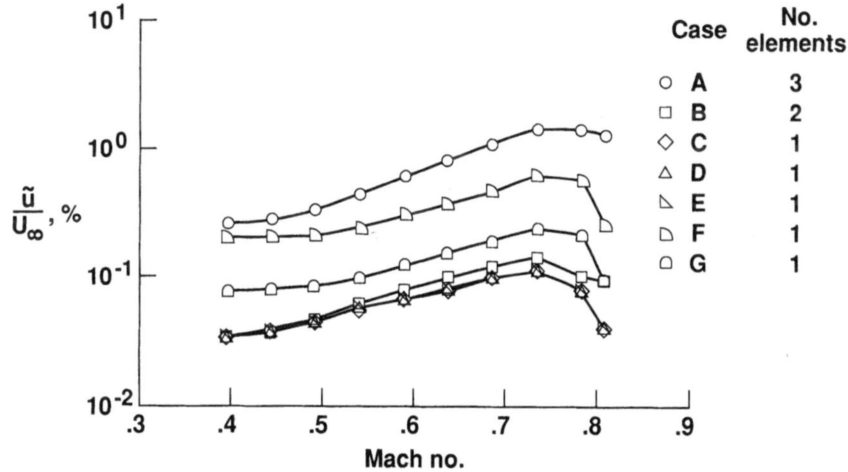

Figure 4. Comparison of single and multi-wire hot-wire-anemometer measurements of $\frac{\tilde{u}}{U_\infty}$, $\frac{\tilde{p}}{p_\infty}$, and $\frac{\tilde{m}}{m_\infty}$ in the Langley Research Center 8-ft Transonic Pressure Tunnel.

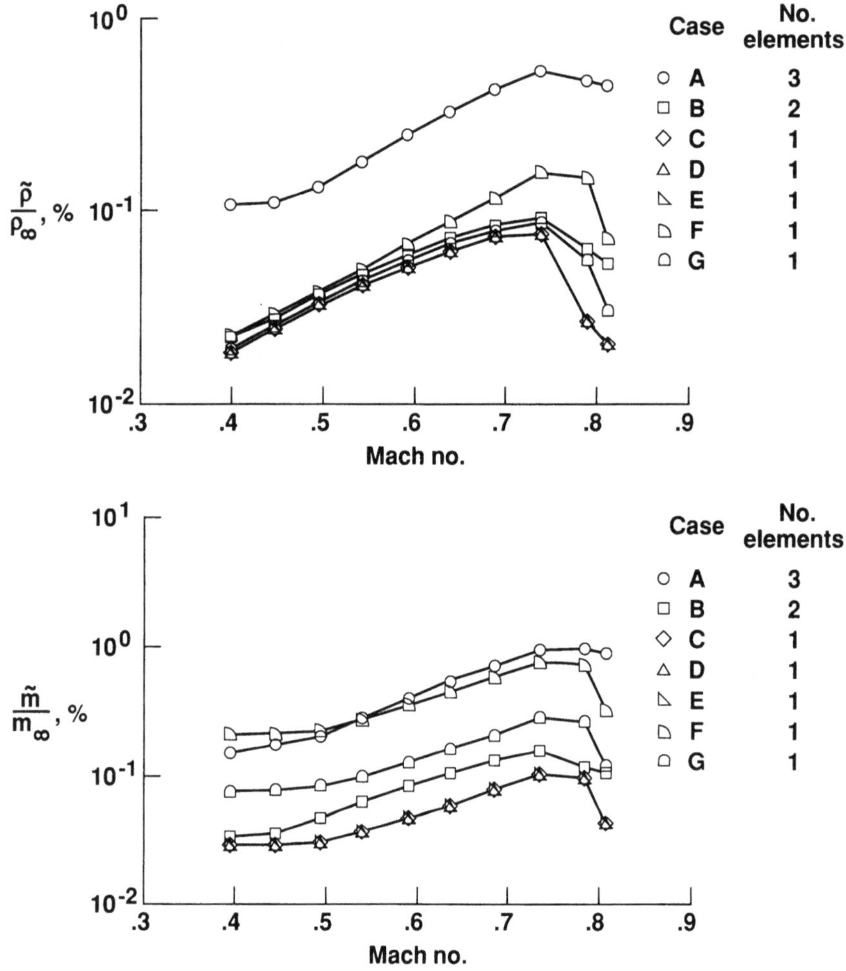

Figure 4. Concluded.

- **Honeycomb and Screens**
- **Large Contraction Ratio – Continuous Curvature**
- **High Speed Diffuser and Drive Fan Mufflers**
- **Contraction and/or Test Section Bleed**
- **Downstream Choke**
- **Exponential Horn Contraction for Open Return Tunnels**
- **Circular or Square Over the Whole Circuit**

Figure 5. Features of high-flow-quality tunnels.

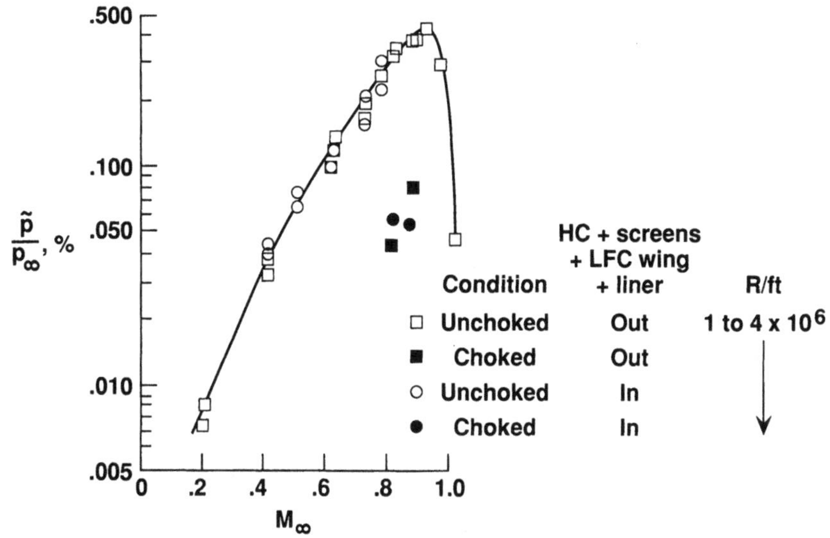

Figure 6. Variation of $\frac{\tilde{p}}{p_\infty}$ with M_∞ for both choked and unchoked high-speed-diffuser conditions and with and without the honeycomb, screens, Laminar-Flow-Control Wing and liner in the Langley Research Center 8-ft TPT.

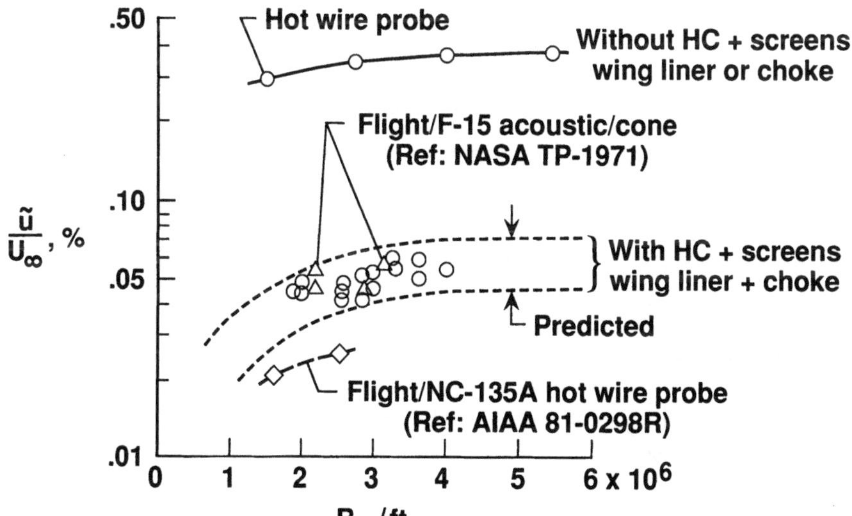

Figure 7. Single wire hot-wire-anemometer measurements of $\frac{\tilde{u}}{U_\infty}$ with and without honeycomb, screens, LFC wing, liner and choke in the 8-ft TPT compared to flight data.

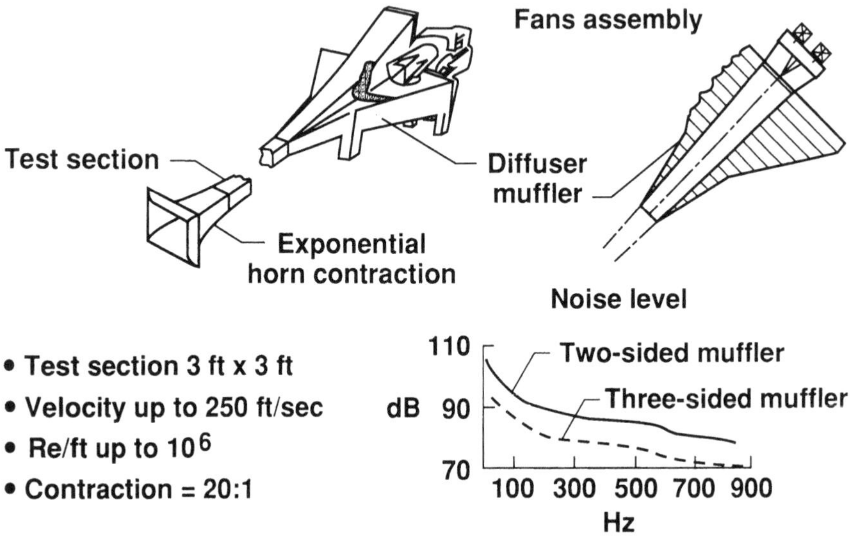

Figure 8. Features of a proposed high-flow-quality subsonic wind tunnel.

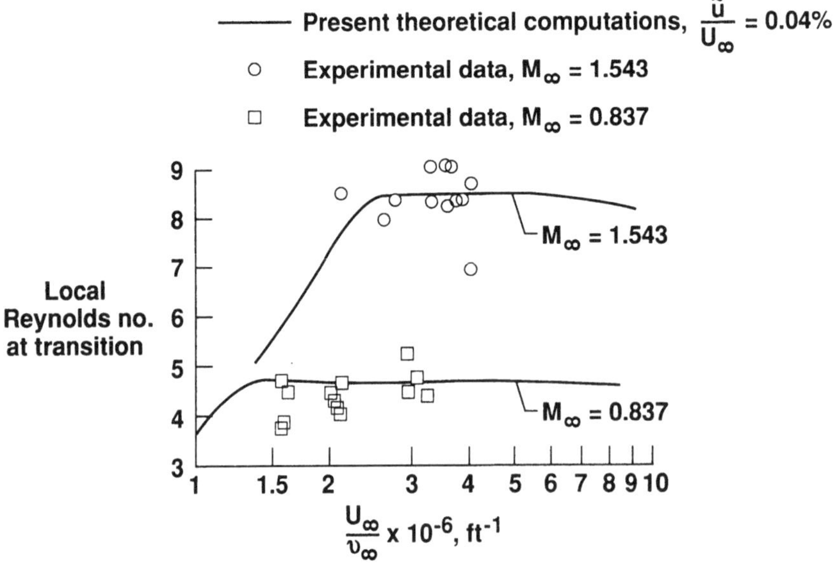

Figure 9. Prediction and experimental data for local transition Reynolds number versus unit Reynolds number for the AEDC transition cone for two Mach numbers and free-air condition.

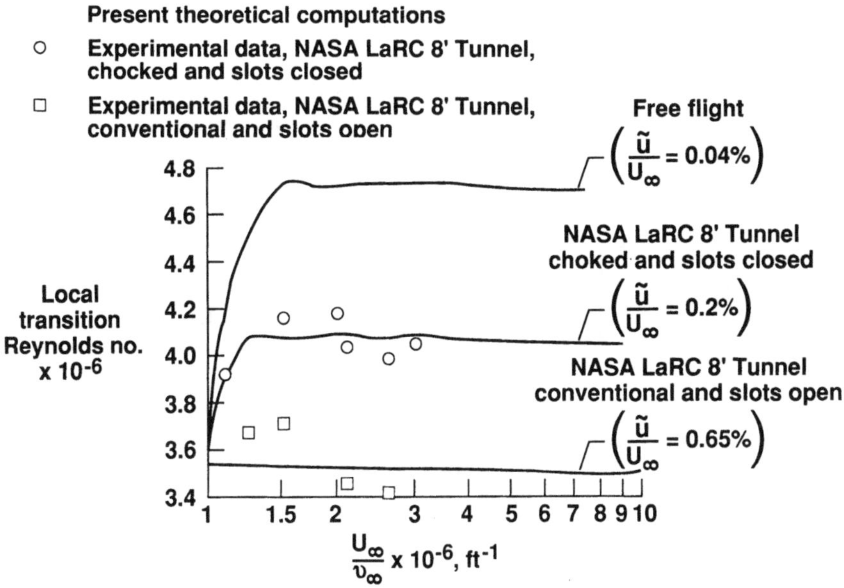

Figure 10. Prediction and experimental data for local transition Reynolds number versus unit Reynolds number for the AEDC cone at wind tunnel and flight conditions, $M_\infty = 0.837$.

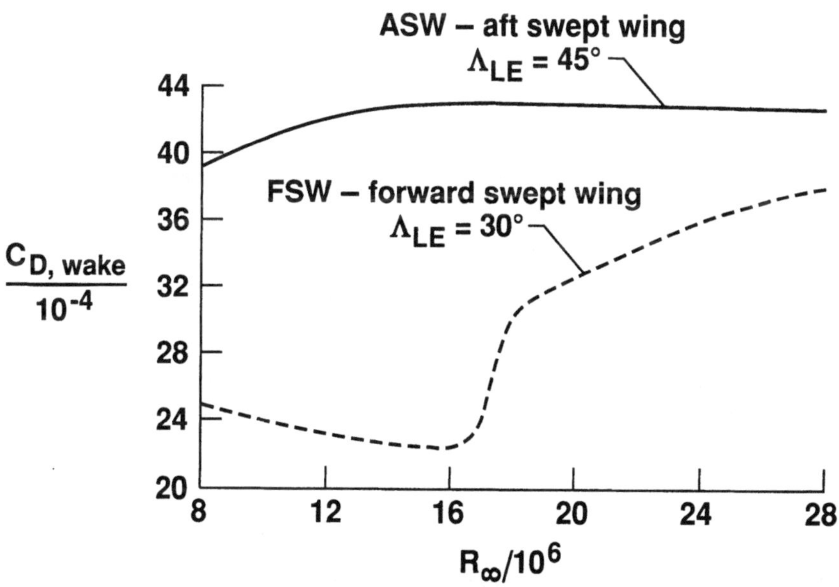

Figure 11. Wake drag versus chord Reynolds number for equivalent aft and forward swept wings.

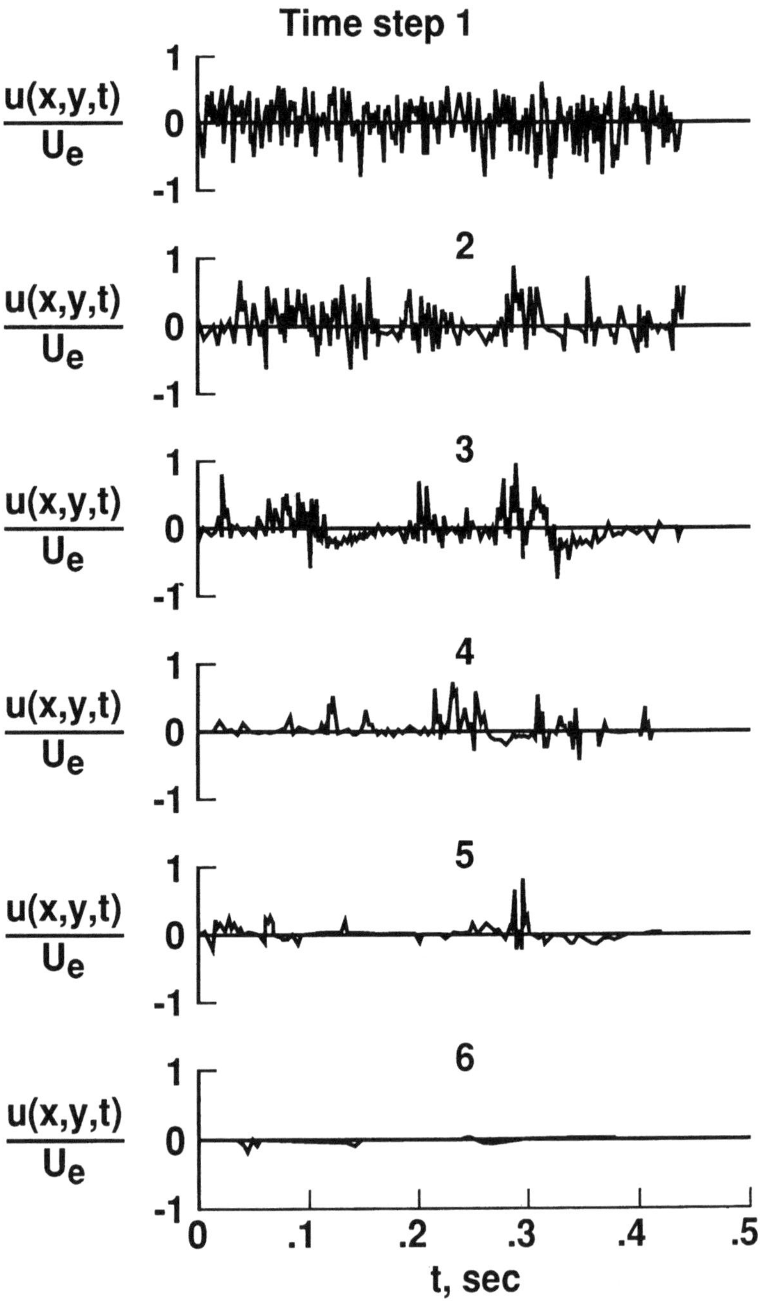

Figure 12. Longitudinal velocity time histories in the boundary layer of a flat plate at different times following the activation of an oscillatory heat input near the leading edge of the plate.

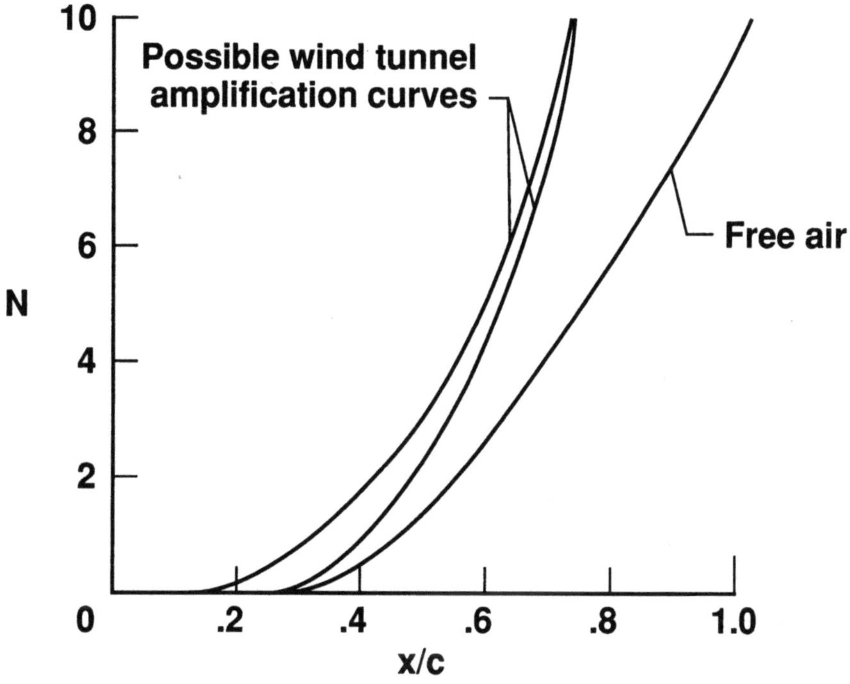

Figure 13. Possible wave amplification curves in free air and in a wind tunnel.

SOME TRANSITION PROBLEMS IN THREE-DIMENSIONAL FLOWS

D. Arnal

ONERA/CERT
Aerothermodynamics Department
2 avenue E. Belin, 31055 Toulouse
France

ABSTRACT

This paper deals with a brief description of transition problems arising on swept wings in incompressible flow. Two transition processes will be discussed: the cross-flow instability and the leading edge contamination.

1. Cross-flow Instability

In the absence of leading edge contamination (and of Görtler instability), it is well known that transition may appear through two mechanisms: streamwise instability (related to the properties of the streamwise profile u) and cross-flow instability (related to the properties of the cross-flow profile w).

A peculiar feature of the cross-flow instability is the appearance of streamwise, stationary vortices practically aligned with the free-stream velocity. Their wavelength is well predicted by the linear stability theory [1]. Travelling waves also exist; they are often more unstable than the zero frequency disturbances, and the measured frequency range agrees well with that predicted by the linearized theory.

In most of the fundamental studies devoted to this problem ([2] to [4] for instance), the external flow is accelerated all along the model, so that transition is mainly governed by cross-flow instability. In practical situations, however, the free-stream velocity exhibits a maximum, and there is a region around the point of minimum pressure where cross-flow and streamwise instabilities are mixed.

A series of experiments was performed on a symmetrical ONERA D airfoil, for $\phi = 60°$ and $\alpha_n = -4°$ (ϕ and α_n represent the angle of sweep and the normal angle of attack, respectively). In this case, the external velocity reaches its maximum at 50 percent of chord, see Figure 1a. Because detailed measurements of the instability waves are difficult to do, stability computations have been performed by using Mack's method [5], in which it is assumed that the amplification vector is normal to the leading edge. Let ψ_m be the most unstable direction with respect to the external streamline, for a given frequency. Figure 2 shows that ψ_m decreases progressively from the cross-flow direction to the streamwise one.

The problem becomes more complex for other experiments performed on an ONERA D airfoil equipped with a cambered leading edge. Figure 1b presents the free-stream velocity distribution corresponding to $\phi = 55°$ for several angles of attack, for a wind tunnel speed of $60ms^{-1}$. The spatial amplification rate α_i is plotted on Figure 3 as a function of ψ for a frequency equal to 3 KHz and for some stations located around the rather sharp peak of the external velocity. In this case, $\psi = 49°$ and $\alpha = -2°$. Up to 24.2 percent of chord, the most unstable direction ψ_m is close to the cross flow direction. But a short distance downstream, there is an abrupt change in ψ_m, which decreases up to -50°. At 28.6 percent of chord, the same frequency exhibits two directions of maximum instability (an absolute maximum close to -60° and a relative maximum around 80°). This leads to some questions concerning the application of the e^n method [6].

2. Leading Edge Contamination

The leading edge contamination occurs when a source of gross disturbances (wing-body junction for instance) produces turbulent structures which are not damped as they are convected along the leading edge; it is an example of "bypass" (Morkovin).

This problem was studied in the F1 wind tunnel at the ONERA Fauga-Mauzac Center, by using hot-films glued along the span of a swept wing. This wing was an RA16SC1 airfoil, the chord of which was constant an equal to 0.5 m. It was fixed on the wind tunnel floor [7].

As a first result, it was found that the well known criterion for leading edge contamination $\overline{R} = 245$, where $\overline{R}$ is a Reynolds number based on the characteristics of the leading edge (see [2], [8], [9], [10], for instance), was fairly well verified in the experiments. The spanwise evolution of the intermittency factor and the convection velocities of the turbulent spots were also measured.

At high angle of attack ($\alpha = 10°$), the hot films were placed on the upper and on the lower surfaces of the wing, 10 or 15 percent of chord of the attachment line, as it is indicated on Figure 4. An interesting result is that the turbulent spots generated by the leading edge contamination were observed on the lower surface (films 1, 3, 5, 7) but they were no longer visible on the upper surface (films 2, 4, 6, 8). This "relaminarization" was made possible by the fact that the pressure gradient on the upper surface is steeper than on the attachment line itself; however, further experiments are needed in order to see if this relaminarization is complete or not.

References

[1] Arnal, D. and Juillen, J. C.: "Three-dimensional transition studies at ONERA-CERT," AIAA Paper No. 87-1335, 1987.

[2] Poll, D. I. A.: "Transition description and prediction in three-dimensional flows," AGARD Report No. 709, 1984.

[3] Saric, W. S. and Yeates, L. G.: "Generation of cross-flow vortices in a three-dimensional flat-plate flow," IUTAM Symposium on Laminar-Turbulent Transition, Springer-Verlag, 1984.

[4] Bippes, H. and Nitschke-Kowsky, P.: "Experimental study of instability modes in a three-dimensional boundary layer," AIAA Paper No. 87-1336, 1987.

[5] Mack, L. M.: "Stability of three-dimensional boundary layers on swept wings at transonic speeds," IUTAM Symposium Transsonicum III, Springer-Verlag, 1988.

[6] Casalis, G.: Thesis, in preparation.

[7] Arnal, D. and Juillen, J. C.: "Etude de la transition et de la contamination de bord d'attaque sur ailes en fliche," AGARD/FDP Symposium, Cesme, Turkey, 1988.

[8] Pfenninger, W.: "Flow phenomena at the leading edge of swept wings," Recent Developments in Boundary Layer Research, AGARDograph 97, Part IV, 1965.

[9] Gaster, M.: "On the flow along swept leading edges," The Aeronautical Quarterly, Vol. XVIII, Part 2, 1967.

[10] Cumpsty, N. A. and Head, M. R.: "The calculation of three-dimensional boundary layers. Part II: attachment line flow on an infinite swept wing," The Aeronautical Quarterly, Vol. XVIII, Part 2, 1967.

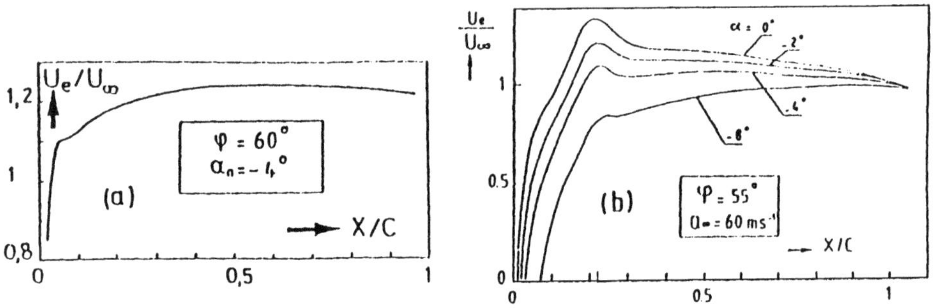

Figure 1 : Free-stream velocity distributions

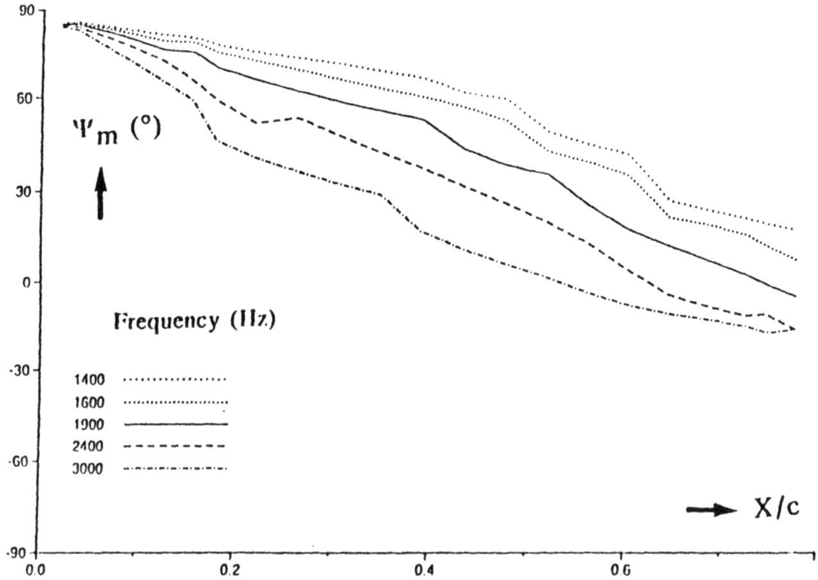

Figure 2 : Evolution of the most unstable direction for several frequencies

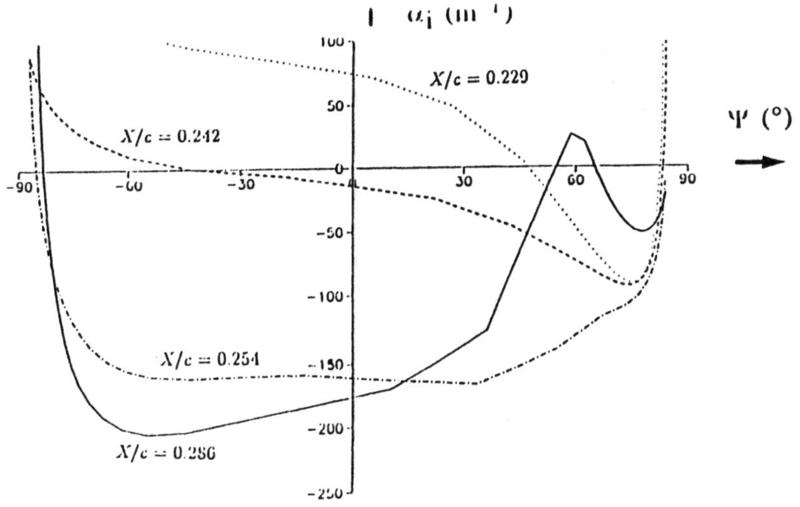

Figure 3 : Spatial amplification rates for f= 3 KHz, as a function of Ψ and of X/c

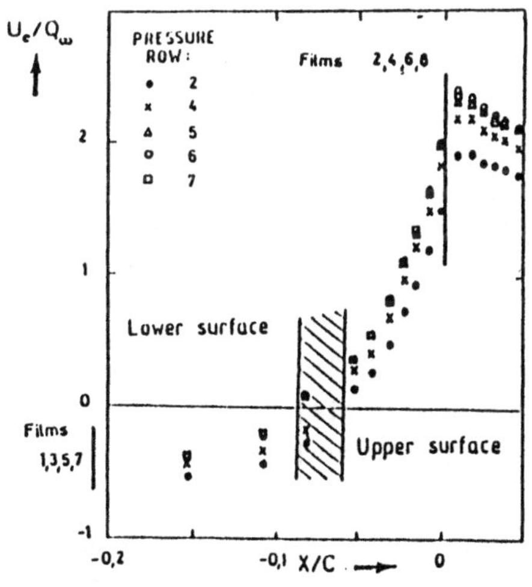

Figure 4 : Free-stream velocity distributions and hot-films positions for relaminarisation studies

THREE-DIMENSIONAL MODE RESONANCES IN BOUNDARY LAYERS, JETS AND WAKES

T. C. Corke

Fluid Dynamics Research Center
Mechanical and Aerospace Engineering Department
Illinois Institute of Technology
Chicago, IL 60616

1. Background

An important direction for experiments on instability and transition to turbulence should be towards understanding and promoting *mechanisms* which govern the *first growth of three dimensionality*. In this direction, the focus should be on *resonance* mechanisms, since these will require *lower initial amplitudes*, a factor which is consistent with "natural" conditions with weak stochastic disturbance input, which we are ultimately trying to predict. In the flow fields cited in the title, resonance interactions can occur between the most amplified two-dimensional modes (waves), predicted from linear theory, and three-dimensional modes. The latter results from three-dimensional background disturbances. In the boundary layer and wake the three-dimensional modes are initially damped and, from a linear stability point of view, are unimportant. In the thin shear layer of an axisymmetric jet, the three-dimensional (helical) modes have the same spatial amplification rate as the 2-D (axisymmetric) mode, and are thereby equally dominant. In all three cases, their combined presence leads to the nonlinear growth of the 3-D mode and transition to turbulence.

In resonance situations, the greatest focus should be on mode combinations that involve two different streamwise wave numbers ($\alpha = 2\pi f/C_r$). This is often dictated by the physics which can require the 3-D mode to have a streamwise wave number which is the near subharmonic of the 2-D mode to satisfy phase speed requirements (Craik (1971) mechanism). The implication of having two frequency interactions is however more deep, and relates to the transition to turbulence process. In a nonlinear interaction involving two frequencies, not harmonically related, the resulting sum and difference interactions can rapidly fill the spectrum to give the energy saturation and broadband character associated with a turbulent state.

Even with two initially harmonically related modes, slight detuning produced by weak background disturbances can produce broadband sum and difference interactions that achieve the same end result. This is fundamentally different from the forms of transition involving a single mode, such as fundamental mode transition in boundary layers (K-type; Klebanoff et al., 1962). The resonant interaction between *two* modes allows an efficient mechanisms for intercomponent energy transfer and early spectral broadening not present in a single (fundamental) mode breakdown. Further, the process of transition can be explained through a *secondary instability* of basic modes *rather than* through higher order mechanisms often attributed to higher mode generation.

To reinforce this aspect, it is important to realize that such resonant interactions between fundamental 2-D and 3-D modes produce sum and difference modes which are *three-dimensional.* For an interaction between modes with harmonic streamwise wave numbers, $(\alpha_1, 0)$ and $(\alpha_1/2, \pm\beta)$, sum and difference interactions result in the following 3-D modes, $(3\alpha_1/2, \pm\beta)$, $(5\alpha_1/2, \pm\beta)$, $(\alpha_1, \pm 2\beta)$, $(0, \pm 2\beta)$. In a "detuned" situation with interacting modes with non-harmonic streamwise wave numbers, $(\alpha_2, 0)$ and $(\alpha_1, \pm\beta)$, such interacted modes as $(\alpha_1, \pm 2\beta)$, $(\alpha_2 - \alpha_1/2, \pm\beta)$, $(3\alpha_1/2 - \alpha_2, \pm 3\beta)$, $(5\alpha_1/2 - 2\alpha_2, \pm 3\beta)$, etc. produce a much more rich distribution of 3-D spanwise wave numbers. Such detuned conditions leads to a lower energy saturation limit and more rapid spectral filling, more consistent with observations in naturally forced conditions, than with "tuned" fundamental/subharmonic conditions. This makes studying these important.

2. Experimental Approach

How do we perform *controlled experiments* with 3-D mode inputs? In past experiments on transition, generally only two-dimensional modes were directly forced. Any three-dimensional interaction were the result of indirect stochastic input from background disturbances. The difficulties in this approach is that 3-D disturbance content is essentially undefined, both in amplitude and wave numbers. Further, these vary between different facilities and are difficult to accurately document. With random input to three dimensional modes, there are also potential problems in achieving spatial stationarity of 3-D features so that phase averaged statistics or flow reconstructions are difficult or impossible. Lastly, there is a lack of control to select for study less *amplified* spanwise wave number conditions.

To overcome these problems, our approach (Corke and Mangano, 1987, 1989, Mangano, 1987, Dal-Ferro, 1987, Corke, 1989, in boundary layers; Kusek et al., 1989, in jets; and Krull, 1989, in wakes) has been to seed 3-D oblique modes along with 2-D modes. The particular amplitudes and wave number content of these were such to produce a resonant growth of the 3-D mode. Sample schematics of how this is done in the three cited flow fields is shown in Figures 1 through 3. In each of these, the method for seeding disturbances is one which is most suited to the particular frequencies (streamwise and spanwise), required initial amplitudes and receptivity to our controlled input, which was dictated by each particular flow situation.

In the boundary layer, the relatively lower frequencies of unstable plane TS modes (< 100 Hz) allowed the use of heating wires, elevated at the height of the critical layer. These were operated in a time-periodic fashion to produce a periodic v-component through a local change in viscosity. The mechanisms for resonance were those postulated by Craik (1971) and Herbert (1983, 1988).

At comparable velocities in wake of a thin airfoil, the frequencies of the plane sinuous or varicose modes are a factor of 6 to 7 times higher (< 600 Hz) than in the boundary layer. These frequencies preclude the use of periodic heating because of its limited frequencies response. An alternate approach shown in Figure 2 uses separately controlled piezoelectrically active film elements on the top and bottom surfaces. These are operated to introduce a periodic v-component (normal to the surface) to feed unstable modes. The piezoelectric film is wrapped around the airfoil in a continuous sheet so that no surface imperfections are produced which can alter the wake thickness and thereby affect the wake instability characteristics.

In the jet, the thin shear layer at the exit is *highly* receptive to acoustic disturbances. Therefore in this case, we utilized miniature speakers which were placed around the jet circumference (Figure 3). Although these face downstream, we expect that acoustic streaming directs sound to the point of receptivity at the exit edge. As with the other two cases, this method of introducing the controlled disturbance input is designed to have a *negligible passive effect* on the flow properties. To verify this, we documented the Strouhal number versus Reynolds number dependence of the dominant instability modes (fundamental axisymmetric and $m = \pm 1$ helical) without and with the speaker array. A sample comparison taken from Kusek et al. for this jet is shown in Figure 4, where HMG refers to the *h*elical *m*ode *g*enerating speaker array. Since the Strouhal number, St_θ, dependence scales with the local momentum thickness of the exit shear layer, this measure gives a sensitive indication of changes which are due to the passive presence of the speaker array on the exit face of the jet.

In all these techniques, there remains a commonality in approach. All of them introduce controlled three-dimensional (oblique) time-periodic waves through a spanwise phase shift between neighbor elements (heating segments, piezoelectric elements, or speakers). The phase shift patterns produce oblique wave pairs of equal angle but opposite sign. (See schematic at bottom of Figure 1). The wave angle and thereby spanwise wave number is proportional to the phase shift between elements. Simultaneous with these is added a two-dimensional disturbance (zero phase shift in span). The amplitudes, streamwise and spanwise wave numbers, and initial phase difference between 2-D and 3-D inputs are all individually controllable. This is important because the resonance mechanisms appropriate to these flows is sensitive to all these parameters.

The time periodic disturbances are clocked by a computer which controls the experiment and acquires the data time series. Because the 3-D modes are seeded directly, there is a high degree of phase locking (high linear coherence) in 3-D space. This allows multiple spatial point statistics to be computed from time series which are not simultaneously sampled from multiple sensors. That is, a single sensor sampling at different points in space and time are related by the same phase clock. This is a very convenient feature which eliminates the need for complex and potentially flow altering multiple sensor arrays, or interfering upstream sensors.

Another common feature of these approaches to produce 3-D modes is that they are limited in their forcing amplitude. Because they are designed to act on a nonlinear resonance mechanism, they are not required to "force" but rather to "seed" these modes. Because of the 3-D phase coherence, only amplitudes on the order of the background disturbance are required. This is a very important point since a common feature of earlier experiments on transition in boundary layers *per se* (Klebanoff and Tidstrom, 1959; Klebanoff et al., 1962; Williams et al., 1984) have used large initial TS forcing amplitudes (greater than 1 percent) to accelerate transition. This is a practice not limited to experiments, but also common to numerical simulations (Zang and Hussaini, 1985). Although these map out a valid phenomena, their relation to the more general *and basic* low initial disturbance path to transition in natural boundary layers may be tenuous. We then advocate the use of low initial amplitudes which are commensurate with natural transition conditions. Focusing on resonance mechanisms offers this approach.

3. Time-Series Analysis

The emphasis on nonlinear two frequency interactions that we propose requires the use of different types of frequency analysis. In particular, linear coherence is an indication of linear phase locking that exists for a mode at two points in space. For two points spaced in the flow direction, a high coherence would indicate linear phase locking required for a matching of phase velocities between resonant modes. In order to document the nonlinear phase locking that exists to produce sum and difference modes, one needs to calculate the normalized cross-bispectrum, or cross-bicoherence (CBC). A more detailed discussion of this with applications to jets and boundary layers is given by Corke, 1987. A sample result taken from Corke (1989) for the nonfundamental/subharmonic resonance growth of 3-D modes in a boundary layer, is given in Figure 5.

The cross-bicoherence (CBC) is defined as

$$\beta_{ijk}(\alpha_1, \alpha_2) = \frac{|\overline{R}_{ijk}(\alpha_1, \alpha_2)|^2}{< |\overline{u}_i(\alpha_1)|^2 |\overline{u}_j(\alpha_2)|^2 >< |\overline{u}_k(\alpha_3)|^2 >}$$

where

$$\overline{R}_{ijk}(\alpha_1, \alpha_2) - < \overline{u}_i(\alpha_1)\overline{u}_j(\alpha_2)\overline{u}_k^*(\alpha_3) >$$

is the cross-bispectrum, with i, j, and k equal to either i or j, referring to one of two mean removed time series, u, with wave numbers α_1, α_2, and α_3 respectively. The notation (| |) is the modulus of the component, (–) denotes the transformed function in the wave number domain and <> the ensemble average. These wave numbers are related to each other such that $\alpha_1 + \alpha_2 + \alpha_3 = 0$. In the experiment from which Figure 5 was derived, the time series were the streamwise velocity fluctuations, acquired at a particular spatial location in the boundary layer, $u(x, y, z, t)$, and at a fixed upstream reference. For the case when i, j, and k are the same, that is a single time series, $\beta_{ijk}(\alpha_1, \alpha_2)$ will be symmetric with respect to α_1 and α_2, and is to be referred to as the auto-bicoherence (ABC).

In the above equation, the calculation of the quantity

$$< |\overline{u}_i(\alpha_1)|^2 |\overline{u}_j(\alpha_2)|^2 >$$

requires considerable computer time and memory storage. Therefore, it was convenient (and common) to replace it by

$$< |\overline{u}_i(\alpha_1)| >^2 < |\overline{u}_j(\alpha_2)|^2 > .$$

Justification for this substitution is given by Corke et al. (1990).

As with the linear coherence, the value of the CBC varies between 0 and 1. High values of the CBC indicate a high degree of triple phase locking

$$\theta_i(\alpha_1) + \theta_j(\alpha_2) = \theta_k(\alpha_3)$$

from realization to realization. Here $\theta_i(\alpha_1)$ is the instantaneous circular phase angle of $\overline{u}_i(\alpha_1)$. The maximum level of the CBC estimate is however dependent on the signal-to-noise ratio of the data series. When no such phase-locking exists the CBC has a value of zero.

The upper frequency limit of the cross-bicoherence is set by the Nyquist criterion, such that the sum of any two frequencies, f_1 and f_2, cannot exceed half the sampling frequency. The region of validity takes the form of the bold outlined right triangles (top and bottom) shown in Figure 5. Other regions outside of this, not exceeding the Nyquist limit, are omitted because of symmetry. The magnitude of the CBC is drawn as constant level contours, in this case with values of 0.35, 0.65, and 0.85.

By our convention, the frequencies for the upstream reference, designated by subscript 1, are read from the abscissa. The frequencies read on the ordinate correspond to the downstream position and are designated by subscript 2. The ordinate frequencies are either added (top) to or subtracted (bottom) from those read on the abscissa to give the third frequency which is read as the intercept of a -45 degree line to the abscissa. The third frequency also corresponds to the downstream position (subscript 2). The downstream coordinate is indicated in the middle of the figure. For references, above each of the CBC estimates appears the power spectra (in dB) of u-time series from the upstream fixed reference position and from the downstream position. These are shown as respective solid and dashed curves. They have been plotted so that the autospectra of the downstream time series has been shifted downward 20dB with respect to the reference autospectra. The full dynamic range of both is 50dB. The dotted curve is the linear coherence between the upstream reference and downstream time series. Its ordinate value ranges from 0 to 1.

Figure 5 documents the triple phase locking that existed in the sum and difference interactions produced by initial modes consisting of a 2-D TS wave at a frequency of 35.8 Hz, and pairs of oblique waves of equal but opposite wave angles at a frequency of 16.0 Hz. As a result of the detuned fundamental/subharmonic interaction, the 3-D mode at 16 Hz produces by summing interaction its harmonic mode (32 Hz), to which it is linearly phase locked. This is indicated by the high linear coherence (dotted line) at these two frequencies. These two modes mediate energy transfer to other sum and difference modes, as is evident by the vertical alignment of peaks in the CBC at 16 and 32 Hz. Such interactions result in the multiple discrete peaks seen in the spectra at the downstream position (dashed line). The arrows depict how, for example, a mode at 12.2 Hz can interact with the 16 Hz mode to produce a mode in the low frequency range (3.8 Hz), as well as interact with the 32 Hz mode to produce a mode in the higher frequency range (44.2 Hz).

By this approach, one can systematically follow the origin of secondary modes which occur as a result of sum and difference interactions of primary instabilities. By following the development of these in space, one can further infer the direction of energy transfer between modes. This, for example, can give insight to the origin of broadband spectral content associated with transition to turbulence. In conjunction with experiments involving multiple mode interactions, it forms a powerful diagnostic tool.

References

[1] Corke, T., 1987. Measurements of Resonant Phase Locking in Unstable Axisymmetric Jets and Boundary Layers. In "Nonlinear Wave Interactions in Fluids," eds. Miksad, Akylas, and Herbert, Applied Mech. Div. of ASME, New York, pp. 37-65.

[2] Corke, T., 1989. Resonant Three-Dimensional Modes in Tran-

sitioning Boundary Layers – Structure and Control. AIAA 2nd Shear Flow Conference, Paper AIAA-89-1001.

[3] Corke, T. and Mangano, R., 1987. Transition of a Boundary Layer: Controlled Fundamental-Subharmonic Interactions. Turbulent Management and Relaminarization IUTAM Symposium Bangalore, India, Springer-Verlag, pp. 199-213.

[4] Corke, T. and Mangano, R., 1989. Resonant Growth of Three-Dimensional Modes in Transitioning Blasius Boundary Layers. To appear in *J. Fluid Mech.*

[5] Corke, T. C., Shakib, F., and Nagib, H. M., 1990. Mode Selection and Resonant Phase Locking in Unstable Axisymmetric Jets. To appear in *J. Fluid Mech.*

[6] Craik, A. D. D., 1971. Nonlinear Resonant Instability in Boundary Layers. *J. Fluid Mech., 50*, p. 393.

[7] Dal-Ferro, P., 1987. Later Stages of Transition in Three-Dimensional Seeded Boundary Layers. M.S. Thesis, Illinois Institute of Technology.

[8] Herbert, Th., 1983. Secondary Instability of Plane Channel Flow. *Phys. Fluids, 26*, p. 871.

[9] Herbert, Th., 1983. Subharmonic Three-Dimensional Disturbances in Unstable Plane Shear Flows. AIAA Paper 83-1759.

[10] Herbert, Th., 1988. Secondary Instability of Boundary Layers. *Ann. Rev. Fluid Mech., 20*, p. 487.

[11] Klebanoff, P. S., Tidstrom, K. D., 1959. Evaluation of Amplified Waves Leading to Transition in a Boundary Layer with Zero Pressure Gradient. NASA TND-195.

[12] Klebanoff, P. S., Tidstrom, K. D., and Sargent, L. M., 1962. The Three-Dimensional Nature of Boundary-Layer Instability. *J. Fluid Mech. 12*, p. 112.

[13] Krull, J., 1989. Resonant Growth of Three-Dimensional Modes in Far Wakes. M.S. Thesis, Illinois Institute of Technology.

[14] Kusek, S., Corke, T., and Reisenthel, P., 1989. Control of Two- and Three-Dimensional Modes in the Initial Region of an Axisymmetric Jet. AIAA 2nd Shear Flow Conference, Paper AIAA-89-0968.

[15] Mangano, R., 1987. Transition of a Boundary Layer: Controlled Fundamental-Subharmonic Interactions. M.S. Thesis, Illinois Inst. of Technology.

[16] Williams, D., Fasel, H., and Hama, F. R., 1984. Experimental Determination of the Three-Dimensional Vorticity Field in the Boundary-Layer Transition Process. *J. Fluid Mech., 149*, p. 179.

[17] Zang, T. A. and Hussaini, M. Y., 1985. Numerical Experiments on the Stability of Controlled Shear Layers. Paper AIAA-85-1698.

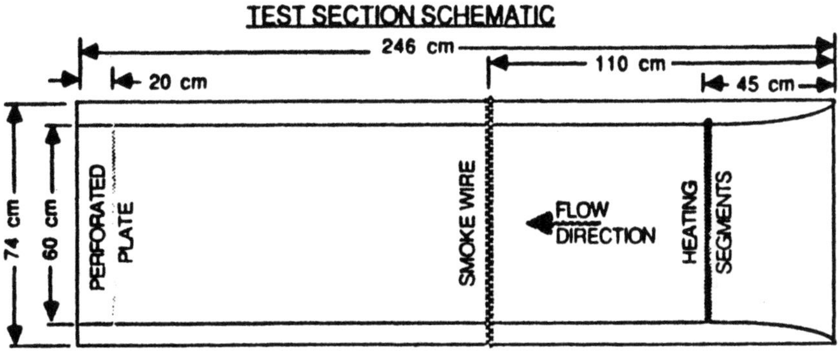

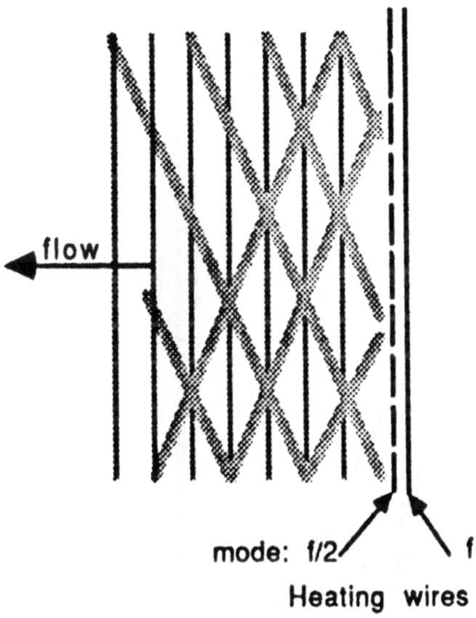

Figure 1. Schematic of Heating Wire Arrangment for Introducing Combined TS and Oblique Wave Modes in Boundary Layers.

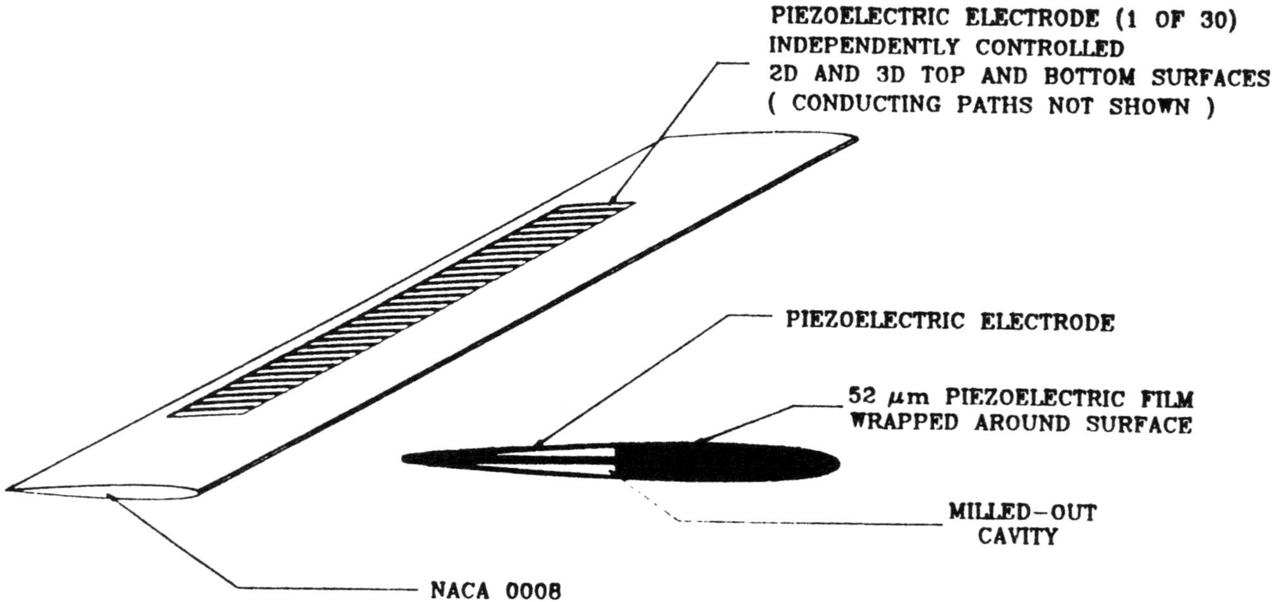

Figure 2. Schematic of Piezoelectric Element Array for Introducing Combined 2- and 3-D Modes in the Far Wake of an Airfoil.

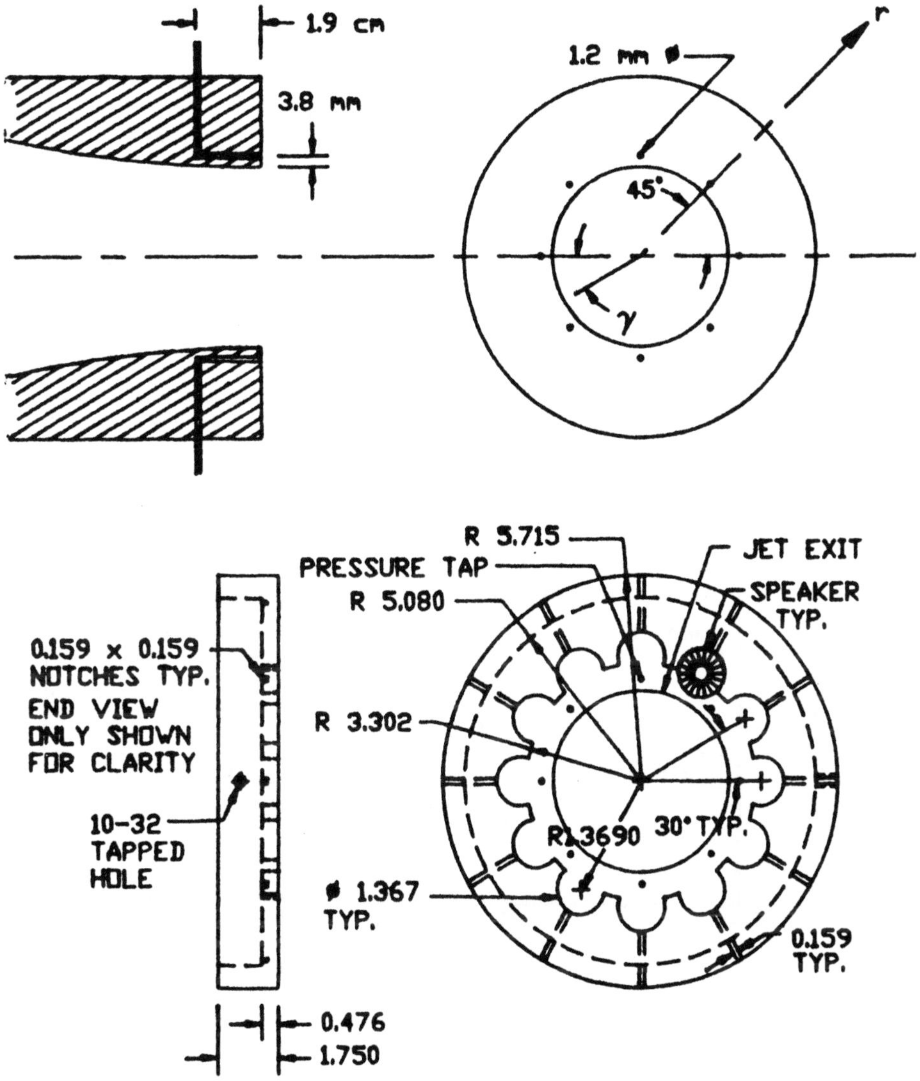

Figure 3. Schematic of Azimuthal Array of Miniature Speakers for Introducing Combined Axisymmetric and ±m Helical Modes in a Circular Jet.

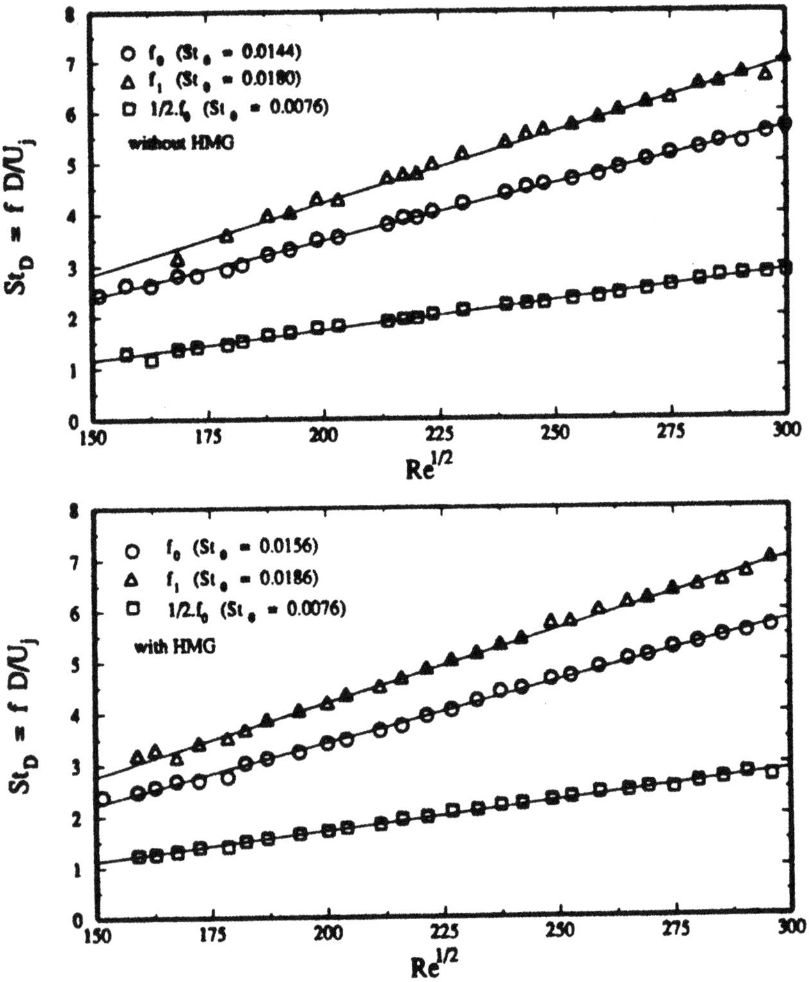

Figure 4. Strouhal Number Dependence of Dominant Natural Instability Modes Without (top) and With (bottom) Passive Speaker Array.

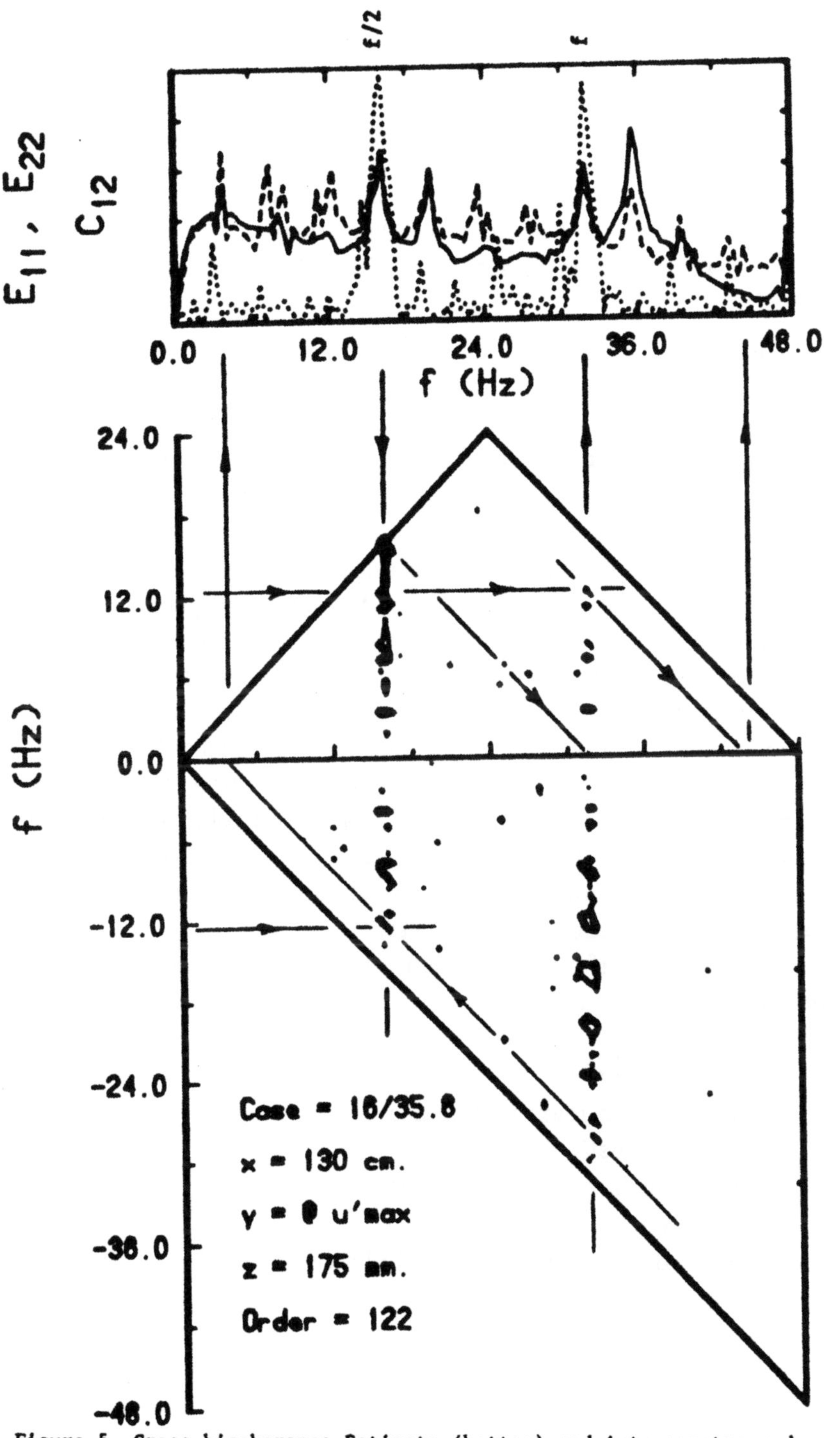

Figure 5. Cross-bicoherence Estimate (bottom) and Auto-spectra and Linear Coherence (top) for Detuned Resonance Case in Boundary Layer Transition.

THE ROLE OF THE LOW-SPEED WIND TUNNEL IN TRANSITION RESEARCH

M. Gaster

Department of Engineering
University of Cambridge
Trumpington Street, Cambridge, CB2 1PZ, U.K.

We need to understand much more about the physical processes that take place during the transition from laminar to turbulent flow if we are going to be in a stronger position to predict the event. The aim of much of the present transition research is directed to the prediction of transition on the wings of modern transonic aircraft. It is therefore desirable to consider the stability of the types of boundary layers that arise on real swept wings in compressible flow. Such problems are complicated and are therefore difficult to study, especially through experiment. It is much easier to investigate the various phases of the transition process at low speed. I do not believe that the physics are greatly different at low speed and so we have the opportunity of experimentally investigating some of the fundamental aspects of the process without the added complexity of working in a high-speed tunnel. The low-speed wind tunnel has a vital role to play in such research. Not only is it much easier to do the work in a low-speed tunnel, but in general turbulence levels are lower and the costs of operating the facilities are much less.

We need to perform experiments to get better insight into such aspect of transition as:

A. RECEPTIVITY

- i) Free-stream turbulence
- ii) Acoustic excitation
- iii) Surface panel vibration
- iv) Particles in the fluid (ice)
- v) Roughness/sound/turbulence

B. TYPES OF NONLINEAR INTERACTION

i) Periodic wavetrains

ii) Combinations of 2- and 3-dimensional periodic waves

iii) Wave Packets
Random wave trains

iv) i), ii) & iii) with small additions of random noise excitation

C. DEVELOPMENT OF IRREGULAR CHAOTIC MOTION

Later stage of evolution prior to breakdown. Study of this phase in terms of dynamical system theory.

D. FINAL BREAKDOWN STAGE

Formation of embryo turbulent spots.
The criteria for breakdown are really not known.

In almost all the topics listed, we can only expect to make progress if we can first understand the fundamental physics of the flow process. Only then will it be possible to make quantitative predictions and to control events to our advantage. At present, one of the best tools that we have for predicting transition is the e^n method. Surely if we can also incorporate the influence of the environment into the scheme by taking due account of the 'Receptivity' and also provide a good criterion for breakdown to turbulence we will be able to do even better. This is an important topic worthy of study. It may well be that the nonlinear stage of evolution involving various complex modal interactions leading to chaos is of interest to the scientist, but of little immediate value in solving the current practical engineering problems posed by laminar flow aircraft. Nevertheless, I still believe that such studies need to be carried out if in the long term we want a full understanding of transition.

Low-speed tunnels generally exist in universities equipped to do such long term research projects at low cost. There are many establishments looking for support in order to study these fundamental aspects of transition. The low-speed wind tunnel is ideal for this type of research where it is often necessary to suppress the natural system noise so that one can inject suitable artificial forms of excitation. This is relatively easily done in a good low-speed tunnel, but is very difficult indeed to do in a high speed flow. There are also a number of more applied topics that can also profitably be investigated in low-speed tunnels.

E. CONTROL

i) Passive - suction or cooling
ii) Active control
iii) Leading edge contamination and its control

F. SURFACE QUALITY

Surface wavyness

We have no theoretical basis for defining the acceptable limits on wavyness, steps, or roughness.

G. INSTRUMENTATION

Development of various means of detecting transition in the laboratory and in flight.

The topic of most importance at present seems to be A (i), the response of the boundary layer to free-stream turbulence. There have been numerous attempts to understand the way that Tollmien-Schlichting waves are excited by turbulence, but the matter is still obscure. The process will only be properly understood when the experiments are carried out in conjunction with a theoretical model of the process. Then measurements to test the analytical model can be made. A proper understanding implies that the amplitude of the excited waves can be predicted in a specified environment. Once the problem is understood in the above sense, we can extend or modify the analytical model to cope with the additional complication of compressibility.

TRANSITION DELAY AND RELAMINARIZATION OF TURBULENT FLOW

Lucio Maestrello

NASA Langley Research Center
Hampton, VA 23665-5225

1. Introduction

In recent years, the application of boundary layer control has been investigated for low drag or high lift. The first deals with laminar flow control and turbulence manipulation whereas the second deals with separation control. Recent experiments have demonstrated that laminar flow can be obtained by contouring the surface to maintain a favorable pressure gradient over the surface. In addition, improvement in lift and stall characteristics have been demonstrated for an airfoil by local excitation at the leading edge.

Different types of flow control devices or methods have been investigated. They are: shaping the contour as in the case of natural laminar flow control, suction, injection, surface cooling or heating, surface oscillations, vibrating ribbons, riblets, large eddy breakup devices (LEBU) and additives.

At NASA Langley Research Center much progress has been made in natural laminar flow control as well as in transition delay and turbulent control. Since 1978 riblets and LEBU's have been investigated for viscous drag reduction. Recently, localized surface heating has been use to affect the relaminarization of turbulent flow and transition delay. A description of some of these techniques with emphasis on the latter will be given.

2. Transition Control

Several techniques have been used in the past to delay transition both in air and water. One such technique is uniform surface heating. Recently, localized heating rather than uniform heating has been deemed as an effective technique. In air, the natural cooling of the flow downstream of the heating element becomes the stabilizing mechanism. The fact that cooling can stabilize the laminar boundary layer subjected to two-dimensional small disturbances in supersonic flows was shown by Lees [1]. Dunn and Lin [2] found that a moderate reduction in wall temperature is sufficient to increase the critical Reynolds number of all three-dimensional disturbances. However, they concluded that no matter how low the surface temperature may be there is always some unstable three-dimensional disturbance at sufficiently large Reynolds number. McCroskey [3] found experimentally that the addition of heat to an incompressible boundary layer over an insulated flat plate by means of a line heat source (local) along the leading edge delayed the onset of transition to turbulence. He further showed that the velocity and temperature profiles resemble that of an equivalent compressible boundary layer with density variation related to temperature on an insulated flat plate. His work shows the effect of Mach number and heat addition upon the stability and transition of the boundary layer. This effect is opposite to that of uniform surface heating to the boundary layer, a technique successfully used for drag reduction, in water. Reshotko [4], Reshhotko et al. [5], and Strazisar et al. [6] have shown that in water the disturbance amplitude and bandwidth decreases as the surface heating increases. The difference in the effect for water and air can be explained as follows: for water the viscosity decreases and the temperature increases, whereas for air the effect of temperature on viscosity is opposite.

Another technique is by the application of broad band disturbance wave cancellation using wave superposition principle. As shown by Thomas [7], disturbances in a laminar boundary layer can be reduced by superimposing (active) controlled disturbances of proper phase and amplitude which lead to delayed transition and to a increase in the critical Reynolds number. A new approach to boundary layer control is by active surface heating. Liepmann and Nosenchuck [8] in their experimental study used a localized active surface heating technique to attenuate naturally-occurring Tollmien-Schlichting waves in water. This technique was later analyzed theoretically by Maestrello and Ting [9] by the method of matched asymptotic expansions as a triple deck problem. Using wave cancellation Pupator and Saric [10] have shown experimentally, in air, successful control of random two-dimensional disturbances. Recent numerical and experimental work on transition and its control are reported in ref. [11-19].

The experiments by Maestrello and Nagabushana [20] conducted at NASA Langley Research Center demonstrate that localized heating, in air, can effectively delay transition along a flat plate and reduce viscous drag. In the experimental setting, the local heat flux at the nose of the leading edge of the flat plate in the proximity of the stagnation point strongly couples with the flow field causing a profound delay in the transition location. As a result, the rate of natural cooling becomes the major cause of flow stabilization as the temperature downstream of the heating element adjusts to its ambient value.

The process of transition delay observed from a fixed hot-wire probe placed in a turbulent region of the boundary layer goes through several stages as the heating is turned on: first, from turbulent to randomly intermittent; second, periodically intermittent to sparsely intermittent; then into a single event; and finally, into laminar flow. The new profile adjusts to a lower Reynolds number (based on the momentum thickness) than that of the previous turbulent state, as a result of shift in transition location. Figure 1 shows the drag reduction for each stage of transition delay at a fixed location, as the heating is increased until the flow becomes completely laminarized. Although the experiments are conducted in low speed flow, it is a natural feature in high speed flow to observe stability as a result of cooling in the flow direction. For this reason, hypersonic flows remain laminar and a hot jet remains quiet.

The amount of power input to the heating element depends also on the receptivity of the boundary layer to the external interference. In this experiment, the major interference is the noise originating in the diffuser-fan region located downstream of the test section. The noise propagates upstream and the mechanism that couple with the flow is via the pressure gradient. Therefore, higher power to the heater was required to reduce the viscous drag than the power gain due to the downward shift in transition location.

3. Turbulence Control

Turbulence control can be divided into turbulence manipulation and relaminarization. Relaminarization [21] is a process by which an initially turbulent flow is rendered effectively laminar, whereas in turbulence manipulation the flow structure is changed using devices such as riblets, LEBU's etc.

Among nonplanar surface approaches, riblets are the best established means of turbulent viscous drag reduction [22, 23]. These studies suggest that the viscous effect, which induces drag reduction due to slow, quiescent flow in the riblet valleys, outweighs the drag increase due to the increase in wetted area. Large eddy breakup devices (LEBU) have been shown, in a number of careful low-speed experiments, to effect large skin friction reduction over long distances downstream. It has been shown that LEBU's modify large-eddy structure of the turbulent boundary layer inhibiting Reynolds stress production near the wall [24, 25] far downstream of the device.

In another experiment at Langley the feasibility of relaminarizing turbulent flow over a flat plate with an upstream concave-convex curvature, in air, has been demonstrated using localized heating on the concave part of the curved surface [26]. The heat source is placed in a region of favorable pressure gradient, since the flow in this region is highly receptive to external excitation. Before the application of control, the flow upstream and downstream of curvature is turbulent. With the application of control, the flow downstream of curvature becomes relaminarized, with velocity profile closely following a Blasius profile with lower Reynolds number based on momentum thickness than that of previous turbulent state. Relaminarization is due to the temperature gradient downstream of the heat source. The rate of cooling downstream of the heater acts as stabilizing mechanism, as seen in transition delay experiment.

The flow over the curved surface has been investigated by thermal mapping using infrared technique. The centrifugal forces induced by curvature effect exhibits a strong instability producing a system of three-dimensional counter-rotating vortices with their axes in the streamwise direction. These vortices are commonly referred to as Görtler vortices or Taylor Görtler vortices [27, 28]. Observing Görtler vortices [26] over the curved surface downstream of heating element, a definite increase in the vortex wave length can be seen, as the heating at the source is increased, Figure 2. As the resulting cooling increases, these vortices become weaker in strength, leading to a stable flow and the flow downstream of the curvature becomes fully laminarized.

An important engineering application of the Görtler vortex study is the flow through a wind tunnel contraction. Controlling and dissipating these vortices are essential for achieving disturbance-free flow in test section. Beyond a certain distance further downstream of the curved surface, the flow goes back to its original turbulent state, following the transition process.

In these experiments the Reynolds number is not appreciably high but the possibility of the extension of this mechanism to turbulence control at moderate Reynolds number over a complex geometry is encouraging.

Another application of this technique which has a practical interest is the flow along a fuselage type of body where the mean flow is three-dimensional. As a part of future work on flow control, a model has been built with a heating element placed on an elliptic nose to test the effect of three dimensionality and pressure gradient. In addition, numerical work is in progress to study the effect of temperature gradient on the stability of boundary layers in supersonic flow due to the interaction of viscous and inviscid modes using the complete Navier-Stokes solver. Also, a study to minimize the total drag on a wedge and on a cone is being conducted using Navier-Stokes solver combined with Newton's method to optimize both inviscid and viscous drag at hypersonic flow where delay transition is due to natural heating.

References

[1] Lees, L., "The Stability of the Laminar Boundary Layer in a Compressible Fluid," NACA-TR-876, 1947.

[2] Dunn, D. W. and Lin, C. C., "On the Stability of the Laminar Boundary Layer and a Compressible Fluid," Journal of Aeronaut. Sci., Vol. 22, 1955.

[3] McCroskey, W. J., "Experimental Investigation of Boundary Layer Transition on a Flat Plate with a Point Heat Source at the Leading Edge," Princeton University Report No. 623, 1962.

[4] Reshotko, E., "Heated Boundary Layers," 12th Symposium on Naval Hydrodynamics, 1978.

[5] Reshotko, E., Gran, R. L., and Tzou, K. T. S., "Re-establishment of Laminar Flow on a Heated Submerged Vehicle," DARPA Report by Dynamics Technology Inc., Torrance, CA, 1979.

[6] Strazisar, A. J., Reshotko, E., and Prahl, J. M., "Experimental Study of the Stability of Heated Laminar Boundary Layers in Water," J. Fluid Mech., Vol. 83, pp. 225-247, 1977.

[7] Thomas, A. S. W., "The Control of Boundary Layer Transition Using a Wave Superposition Principle," J. Fluid Mech., Vol. 137, pp. 233-240, 1983.

[8] Liepmann, H. W. and Nosenchuck, D. M., "Active Control of Laminar Turbulent Transition," J. Fluid Mech., Vol. 118, pp. 201-204, 1982.

[9] Maestrello, L. and Ting, L. "Analysis of Active Control by Surface Heating," AIAA Journal, Vol. 23, pp. 1038-1045, 1985.

[10] Pupator, P. T. and Saric, W. S., "Control of Random Disturbances in a Boundary Layer," AIAA 2nd Shear Flow Conference Paper No. AIAA-89-1007, 1989.

[11] Kazakov, A. V., Kogan, M. N., and Kuparev, V. A., "Increase in Stability of a Subsonic Boundary Layer Upon Heating of the Surface Near the Leading Edge of a Streamlined Body," Sov. Phys. Dokl. 37(7), pp. 539-540, 1985.

[12] Bayliss, A., Maestrello, L., Parikh, P., and Turkel, E., "Numerical Simulation of Boundary-Layer Excitation by Surface Heating/Cooling," AIAA Journal, Vol. 24, pp. 1095-1101, 1986.

[13] Maestrello, L., Badavi, F. F., and Noonan, K. W. "Control of the Boundary Layer Separation about an Airfoil by Active Surface Heating," AIAA 88-3546, 1988.

[14] Ladd, D. M. and Hendricks, E. W., "Active Control of Two-Dimensional Instability Waves on an Axisymmetric Body," Experiments in Fluids, Vol. 6, pp. 69-70, 1988.

[15] Struminski, V. V., Dovgal, A. V., Lebedev, V. B., Levchenko, B. YA., Timofeev, V. A., and Formichev, V. M., "Theoretical and Experimental Investigation of the Stability of the Boundary Layer to Unsteady Heating on the Surface," Reprint No. 3-87, Inst. of Theor. and Appl. Mech., Academy of Science, USSR, 1987. (Russian Edition).

[16] Breuer, K. S., Haritonidis, J. H., and Landahl, M. T., "The Control of Transient Disturbances in a Flat Plate Boundary through Active Wall Motion," Phys. Fluids, pp. 574-582, 1989.

[17] Bushnell, D. M., "Application and Suggested Directions of Transition Research," 4th Symposium on Numerical and Physical Aspects of Aerodynamics, 1989.

[18] Blackwelder, R. F., "Some Ideas on the Control of Near-Wall Eddies," AIAA Paper No. 89-1009, 1989.

[19] Beckwith, I. E., Creel, R. C., Jr., Chen, Fan-Jeng, and Kendall, J. M., "Free Stream Noise and Transition Measurements on Cone in a Mach 3.5 Pilot Low Disturbance Tunnel," NASA-TP-2180, 1987.

[20] Maestrello, L. and Nagabushana, K. A., "Relaminarization of Turbulent Flow on a Flat Plate by Localized Surface Heating," AIAA 2nd Shear Flow Conference, Paper No. AIAA-89-0985, 1989.

[21] Narashima, R. and Sreenivasan, K. R., "Relaminarization of Fluid Flows," Adv. Appl. Mech. Vol. 19, 1979.

[22] Walsh, M. J., "Turbulent Boundary Layer Drag Reduction Using Riblets," AIAA paper No. 82-0169, 1982.

[23] Wilkinson, S. P., Anders, J. B., Lazos, B. S., and Bushnell, D. M., "Turbulent Drag Reduction Research at NASA Langley: Progress and Plans," Int. J. Heat and Fluid Flow, Vol. 9, pp. 266-277, 1988.

[24] Corke, T. C., Guezennec, Y., and Nagib, H. M., "Modifications of Drag of Turbulent Boundary Layers Resulting from Manipulation of Large-Scale Structures," Symposium on Viscous Drag Reduction, 1979.

[25] Anders, J. B., Hefner, J. N., and Bushnell, D. M., "Performance of Large-eddy Breakup Devices at Post-Transitional Reynolds Numbers," AIAA Paper No. 84-0345, 1984.

[26] El-Hady, N. M. and Verma, A. K., "Growth of Görtler Vortices in Compressible Boundary Layers along Curved Surfaces," AIAA Paper No. 81-1278, 1981.

[27] Swearingen, J. D. and Blackwelder, R. F., "The Growth and Breakdown of Streamwise Vortices in the Presence of a Wall," J. Fluid Mech., Vol. 182, pp. 255-290, 1987.

Figure 1. Wall Law - Transition Delay

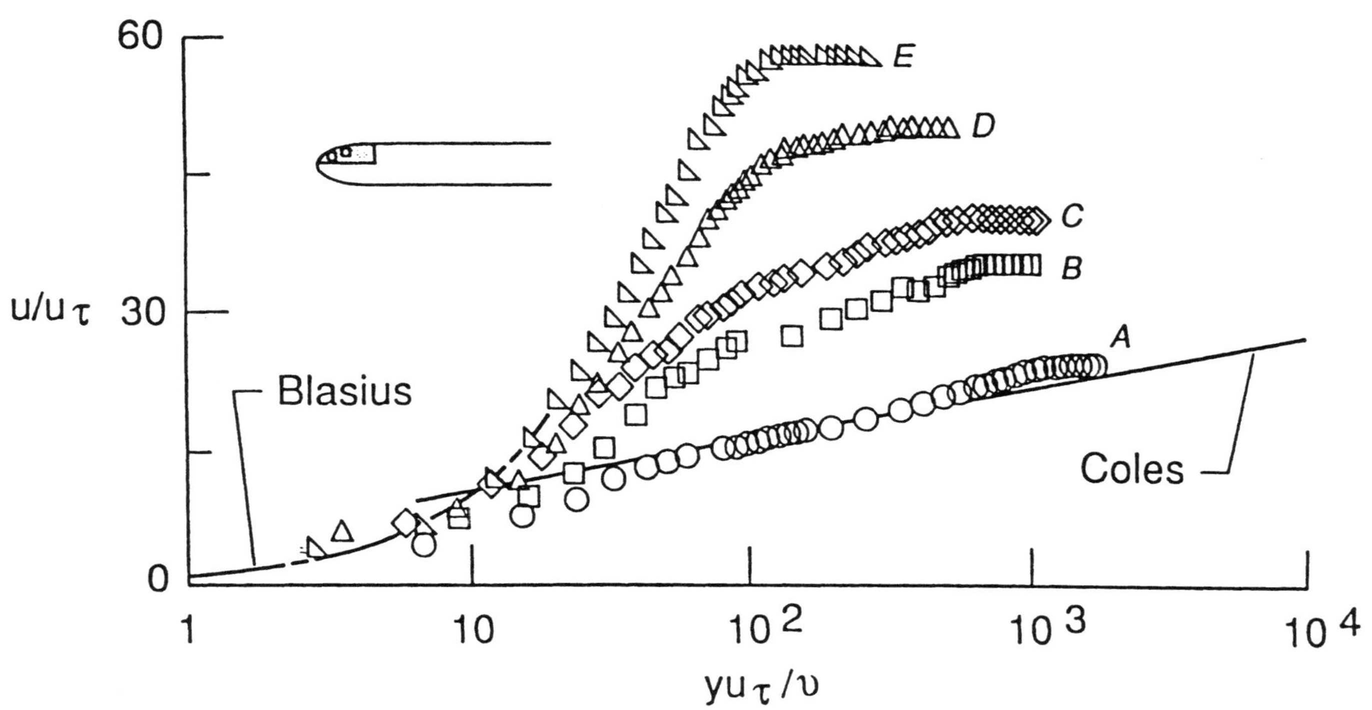

Figure 2.

SURFACE TEMPERATURE OVER THE CONCAVE/CONVEX SURFACE

Turbulent Boundary Layer | **Laminarized Boundary Layer**

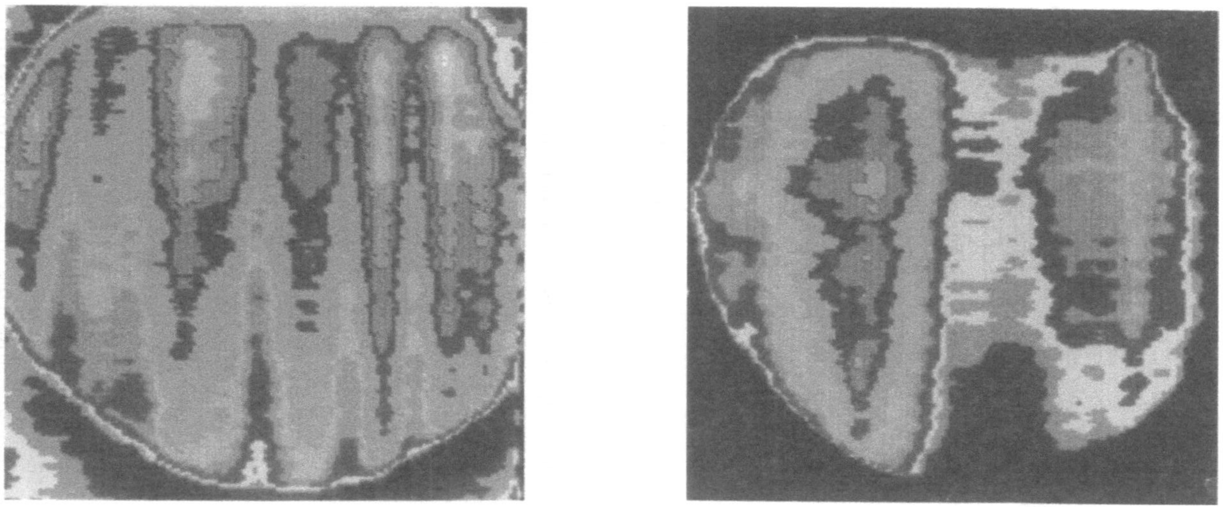

LOW-SPEED EXPERIMENTS: REQUIREMENTS FOR STABILITY MEASUREMENTS

William S. Saric[1]

Arizona State University
Tempe, AZ 85287-6106

ABSTRACT

The difficulties of conducting low-speed stability and transition experiments are discussed. The sources of experimental error are introduced and certain remedies are proposed. The result is a list of requirements that must be satisfied in order to be able to successfully interpret a stability experiment.

1. Introduction

The basic idea of stability is that very small effects produce large changes in the basic state. When transition to turbulence proceeds through loss of stability, the process critically depends on these small effects. Thus, unlike many of the situations in turbulent boundary layers, measurements of stability characteristics require a special sensitivity to environmental conditions.

The present paper addresses some of the sources of error that may occur due to this special sensitivity. These errors then affect the measurements during a boundary-layer stability experiment. The emphasis is strictly on low-speed experiments within the guidelines of the workshop.

The objective of this paper can be stated as follows. Regardless of whether the experimental objectives are transition control, three dimensionality, secondary instabilities, nonlinear breakdown, or receptivity, the overriding rules of conducting a stability experiment are: (1) the linear problem must be correct and (2) initial conditions must be provided for theory and computations. If one can show the comparison with linear theory for the particular flow, then the basic state is probably what it is advertised to be. Full documentation of physical properties, background disturbances, initial amplitudes, and

[1]This work was supported by AFOSR Contract No. F49620-85-C-0089 and by NASA Langley Grant NAG 1-805.

spatial variations must be provided to the analyst. Although these seem like simple requirements, there are experiments presently being conducted that ignore these precepts. In the sections that follow, examples are discussed that illustrate the difficulty of establishing (1) and (2). However, in contrast to the role of unknown receptivity problems, all of the examples are real, correctable effects. The format for the discussion is a problem statement followed by a solution.

The first example we consider is the incompressible, isothermal flow over a flat plate with zero pressure gradient. This is a Tollmien-Schlichting (T-S) dominated Blasius boundary layer. The streamwise, normal, and spanwise coordinates are x, y, and z, respectively. The basic-state velocity vector is (u, v, w) and the disturbance-state velocity is (u', v', w'). The example considered in Section 12 is a 3-D basic state with accelerating pressure gradient. This is a crossflow (C-F) dominated swept-wing flow.

2. Global Pressure Gradients

Problem: Whether one uses a blunted flat plate or a sharp flat plate at negative angle of attack, a pressure gradient is present at the leading edge and a finite recovery distance is required. If the pressure gradient is zero after this recovery distance, the boundary layer is Blasius but referenced to a different chordwise location, x. Thus, there is a *virtual* leading edge from which the measurements and the Reynolds number must be referenced. If this is unaccounted for, it is very easy to have 20-30% errors in x (and 10-15% errors in R). For example, in order for theory to agree with the linear part of the well-known nonlinear work of Klebanoff et al. [1] one must apply a correction to x (Herbert [2]).

Solution: Measure $u = u(y)$ within the boundary layer and calculate the displacement thickness, δ^*. Then, the virtual leading edge is $x_v = (\delta^*/1.72)^2 U_0/\nu$. The boundary-layer Reynolds number, R, used for stability calculations, is given by $R = (U_0 x_v/\nu)^{\frac{1}{2}}$. Because of traverse effects and tunnel side-wall blockage, x_ν may actually change with different chordwise measurements. Therefore, δ^* should be measured (and R calculated) at each location. With modern, computer-controlled experiments this is no problem.

3. Local Pressure Gradients

Problem: It is difficult to accurately measure small changes in the pressure coefficient, C_p. Thus, the flow may not be locally Blasius and the stability characteristics may be quite different. For example, a decrease in C_p of approximately 0.01 over 100 mm corresponds to Falkner-Skan pressure-gradient parameter, β, of approximately +0.1. For $\beta = +0.1$, the minimum critical Reynolds number (based on $\mathrm{Re}_x^{\frac{1}{2}}$) is changed by a factor of 3!

Figure 1 is the neutral stability curve from Herbert and Bertolotti [3] that compares nonparallel theory with experiments [4], [5]. One possible explanation of why the measured minimum critical Reynolds number, $\min R_{\mathrm{crit}}$, is around 230 as compared to the value of 300 given by the theory is the extreme sensitivity of $\min R_{\mathrm{crit}}$ to the smallest of pressure gradients (in this case adverse). Other plausible explanations for this discrepancy between theory and experiment are presented in Section 4.

A weak adverse pressure gradient can also explain why instabilities are measured at dimensionless frequencies above 250 contrary to theory. The range of unstable frequencies could increase dramatically if the measurements were made in the weak adverse pressure-gradient region of the recovery zone of the leading edge. Because of the low Reynolds number needed, the measurements of $\min R_{\mathrm{crit}}$ (Ross et al. [5]) were conducted close to the leading edge and close to the disturbance source (also see Section 4).

Solution: With extreme care one may be able to measure $\Delta C_p \approx 0.3\%$. This may not be enough. However, one could measure $u = u(y)$ and calculate the shape factor, $H = \delta^*/\theta$. Here, $\beta = +0.1$ corresponds to a $\Delta H = 7\%$. The measurement of changes in the shape factor is more reliable than ΔC_p. From the ideas of Section 2, one should already have $u(y)$ at each chord location and hence the pressure gradient can be verified.

One should avoid measurements of $\min R_{\mathrm{crit}}$. Not only is this a very difficult measurement to interpret, but changes in $\min R_{\mathrm{crit}}$ have very little to do with transition.

One should also avoid "flat" plates made of *rolled* material. Klebanoff [private communication] measured a periodic pressure distribution on the original Schubauer and Skramstad [4] flat plate and had a new plate made in 1971 for his own work.

4. Too Close to Disturbance Source

Problem: A disturbance source such as an air jet, or heated wire, or vibrating ribbon, locally creates a disturbance that is not a T-S wave but has a T-S wave as one of the modes in the distribution. A relaxation distance is required to attenuate the more stable modes so that the least stable (the T-S wave) remains. If measurements are made within this relaxation distance, some strong stabilizing effects may be measured.

Problem: Ribbons and wires have wakes that cause a kink in the basic-state velocity profile. This kinked profile can be subject to a different set of instabilities (inviscid) that could be *subcritical* to T-S waves. This could also account for Ross et al. [5] only seeing Branch II neutral points at higher frequencies and lower Reynolds numbers than linear theory. There is a relaxation distance for the basic-state profile to "heal" from the effects of the disturbance source.

Problem: When measuring close to the disturbance source one cannot take advantage of the unstable growth of the disturbance in order to measure reasonably-sized disturbances. Therefore, there is a tendency to increase the amplitude into what may be the nonlinear range and hence cause subcritical behavior. In a recent nonlinear, nonparallel analysis, Sen and Verma [6] show that the high amplitudes of Ross et al. [5] near Branch I may have caused subcritical nonlinearities. This represents one more plausible explanation of the discrepancy between theory and experiment. It would be nice to have some additional experimental evidence since, at least to max u' rms amplitudes of 1.5% U_0, this author has measured linear, 2-D waves near a vibrating ribbon as shown in Figure 2.

Solution: One should determine the relaxation distance downstream of the disturbance source. This would depend on the type of source used, but it should be in the range of about 10 boundary-layer thicknesses. This can be verified by first comparing the $u' = u'(y)$ distribution with linear theory to the accuracy shown in Figure 2. Local growth rates should also be compared as a function of input amplitude. These comparisons should be documented if it is required to measure close to the disturbance source.

5. Too Far from the Disturbance Source

Problem: One would like to carry out stability measurements

over a wide range of Reynolds numbers while keeping the disturbance source fixed. Unfortunately, a typical disturbance source has a finite span and Mack [7] showed that the domain of dependence of a finite-span disturbance source propagates from each end toward center span at an angle of approximately 12 degrees. Outside of this triangular domain, the disturbance amplitude is different from linear theory. This is analogous to boundary-condition domain of dependence in hyperbolic systems. Ross et al. [5] had a vibrating ribbon span of only 250 mm.

Solution: Just as one is limited in the useful chord of the model due to sidewall contamination, the distance downstream of the disturbance is similarly limited. If b is the span of the disturbance source and Δx is the distance in x from the source, then the centerline measurements should be made such that $\Delta x/b < 2$. For off-centerline measurements, this value is obviously smaller.

6. Impure Disturbance Input to the Boundary Layer

Problem: One wishes to study nonlinear wave interactions and the nonlinearities of the disturbance source impose different initial conditions on the nonlinear components. For example, if one wishes to study the nonlinear interaction of waves with two frequencies f_1 and f_2, when the disturbance source, such as a vibrating ribbon, is oscillated at too high an amplitude, the disturbance source will input $2f_1, 2f_2, f_1 - f_2, 2f_1 - f_2$, etc. into the boundary layer.

Problem: One wishes to invoke active wave cancellation into the boundary layer through a disturbance source. In this case a feedback signal is processed by the computer and relayed to a disturbance source. However, the D/A converter is a low-pass filter, the vibrating ribbon is a low-pass filter having a typical linear-oscillator response, and the boundary layer is a band-pass filter/amplifier having its unique response curve. Thus the boundary-layer response is much different than the original input.

Solution: First, one must always directly measure the disturbance-source response and the boundary-layer response in order to establish the initial conditions (see Saric and Reynolds [8]). A tailored boundary-layer response can be obtained using inverse Fourier techniques (see Pupator and Saric [9]).

7. Three-Dimensional Distortion of the Basic State

Problem: Streamwise vortical structures in the basic state produce a weak spanwise periodicity that is strongly susceptible to secondary instabilities. These spanwise variations have been carefully documented by Klebanoff et al. [1], Nishioka et al. [10], and Anders and Blackwelder [11]. They strongly influence the type of breakdown to transition that is observed (Saric and Thomas [12] and Singer et al. [13]).

Solution: The source of these spanwise variations is probably from immeasurable disturbances in the freestream and is perhaps uncontrollable (see Section 8). The next best thing, however, is to always do a spanwise traverse of both the basic state and the disturbance state in the early stages of growth. Provide the appropriate 3-D amplitudes for comparisons with theory and computation.

8. Freestream Disturbance Effects

Problem: Unknown receptivity issues such as the roles of freestream turbulence and sound in creating T-S waves and 3-D structures inhibit the understanding and control of transition. It is certainly clear that a naked statement of u' rms fluctuations is not enough to describe a particular wind-tunnel environment.

Solution: Until we really understand the receptivity mechanisms, it is important to document the freestream disturbance environment as completely as possible. In addition to u' rms, one should quote: (1) band pass and spectrum for all measurements, (2) u', v', w' rms at different positions, (3) spatial correlation measurements of all components to separate turbulence from sound, and (4) transition Reynolds number at different unit Reynolds numbers (see Saric et al. [14]).

9. Nonparallel Effects

Problem: Both analysis [3] and Navier-Stokes computations [15] have shown that if one follows the maximum of the $u'(y)$ distribution, nonparallel effects are not important and the Orr-Sommerfeld equation is a very good approximation. However, if one instead follows $\eta = y(U_0/\nu x)^{\frac{1}{2}}$ = constant, Figure 9 of Saric and Nayfeh [16] (confirmed by Herbert and Bertolotti [3]) shows that very strong nonparallel effects can be observed. The difficulty is that we must attempt to describe the spatial stability of a 2-D basic state. Simple concepts from 1-D basic states may not always be adequate.

Solution: Ideally one should measure disturbance energy. However, in thin laminar boundary layers, accurate resolution of the velocity component normal to the wall, v', is not within reach. The next best thing is to measure $A(x) = \int_0^\infty |u'|^2 dy$. This is close to energy and it integrates out the y-dependence. In Sections 2 and 3, a strong case is made for accurately measuring $u(y)$ in order to locally determine x_v, H, and R, so that $u'(y)$ is always available. With no extra work, one can document this less ambiguous stability criterion that is more applicable to 2-D flows (see Reynolds and Saric [17] and Saric and Reed [18] for examples).

10. Probe and Traverse Interference

Problem: The traverse mechanism may be too large or too close to the hot wire. A multi-wire rake may have too much local blockage. What could happen in these cases is that the pressure field around the probe support, although unseen in a basic-state measurement, can strongly influence T-S wave amplitude.

Solution: Fix a very small hot wire to the plate at a location where $u/U_0 = 0.3$. Establish the amplitude of a controlled T-S wave as measured by the fixed wire. Move a traverse-mounted hot wire to the same location and see if the T-S amplitude has changed. This is the most sensitive and the only means for determining whether one has eliminated traverse and probe-support interference problems. One should be sensitive to the fact that traverses and probe supports that are quite suitable for turbulent boundary layers may not be suitable for laminar stability work.

11. Smoke-Wire Effects

Problem: Most smoke wires used for flow visualization have diameters in the range of 50-80 μm. When used in stability experiments they are placed near the critical layer. The wake of the smoke wire causes a kink in the basic-state profile which alters its stability characteristics. As a result, if one examined the amplitude growth of a T-S wave in the stream direction, one would observe an almost step-like increase in amplitude downstream of the smoke wire. The amplitude could easily change by a factor of 3 due to the smoke wire.

Thus, in contrast to its universal use in turbulent boundary layers, special care must be exercised with laminar stability.

Solution: If one wishes to visualize the wave patterns at different amplitudes, the measurements with the hot wire should always be done with the smoke wire in the flow. In the same way, visualization should always be accompanied by direct measurements. Do not believe any measurements taken within 15δ of the smoke wire.

12. Crossflow Roughness

Problem: In contrast to the Blasius flow problem, it is not clear at this time whether roughness can be completely eliminated from the crossflow-dominated instability. The stationary vortex structure and the "saw-tooth" transition pattern are strongly influenced by roughness (Müller and Bippes [19], Arnal and Juillen [20], Saric et al. [21], Dagenhart et al. [22]). The crossflow instability is amplified near the attachment line where the boundary layer is still very small so that even roughness of the order of 10 μm can influence the stationary structure. In spite of the fact that the travelling waves are calculated to be more unstable than the stationary waves, the strong initial conditions established for the stationary structure could cause a different mix of stability modes.

Solution: At the present time, one can only caution the experimentalist to exercise care in surface preparation. Moreover, the surface finish should be documented. Presently, it is not being done.

13. Conclusions

Boundary-layer stability experiments are extremely sensitive and require a very conservative approach for measurements. A list of precautions is given which, as time will demonstrate, is not complete. As we begin to learn more about receptivity and as we work more closely with the Navier-Stokes simulations, a better quantification of the role of initial conditions will be established. Thus, the prospects of control of transition will be enhanced.

Acknowledgements

Throughout the course of conducting boundary-layer stability experiments, I have learned a great deal from my colleagues who freely discussed their own ideas and techniques with me and offered suggestions to my own work. Many of the ideas in this paper came from these discussions. The principal contributors were James Kendall, Philip Klebanoff, Victor Levchenko, Hasan Nagib, Mark Morkovin, Werner Pfenninger, and Gregory Reynolds to whom I would like to express my thanks.

References

[1] Klebanoff, P. S., Tidstrom, K. D., and Sargent, L. M., "The Three-Dimensional Nature of Boundary-Layer Instability." *J. Fluid Mech. 12*, 1962, pp. 1-34.

[2] Herbert, T., "Three-Dimensional Phenomena in the Transitional Flat-Plate Boundary Layer." *AIAA Paper No. 85-0489*, January 1985.

[3] Herbert, T. and Bertolotti, F. P., "Stability Analysis of Nonparallel Boundary Layers." *Bull. Amer. Phys. Soc. 32*, 1987, p. 2079.

[4] Schubauer, G. B. and Skramstad, H. K., "Laminar Boundary-Layer Oscillations and Transition on a Flat Plate." *J. Res. Nat'l. Bur. Stand. 38*, 1947, pp. 251-283.

[5] Ross, J. A., Barnes, F. H., Burns, J. G., and Ross, M. A. S., "The Flat Plate Boundary Layer, Part 3. Comparison of Theory with Experiment." *J. Fluid Mech. 43*, 1970, pp. 819-832.

[6] Sen, P. K. and Verma, R., "On the Non-Linear Stability of Boundary-Layer Flow Over a Flat Plate Considering Non-Parallel Effects." To be published 1990.

[7] Mack, L. M., "Line Sources of Instability Waves in a Blasius Boundary Layer." *AIAA Paper No. 84-0168*, January 1984.

[8] Saric, W. S. and Reynolds, G. A., "Experiments on the Nonlinear Stability of Waves in a Boundary Layer." In *Laminar-Turbulent Transition*, Eds.: R. Eppler and H. Fasel, Springer-Verlag, 1980.

[9] Pupator, P. and Saric, W. S., "Control of Random Disturbances in a Boundary Layer." *AIAA Paper No. 89-1007*, March 1989.

[10] Nishioka, M., Iida, S., and Ichikawa, Y., "An Experimental Investigation of the Stability of Plane Poiseuille Flow." *J. Fluid Mech. 72*, 1975, pp. 731-751.

[11] Anders, J. B. and Blackwelder, R. F., "Longitudinal Vortices in a Transitioning Boundary Layer." In *Laminar-Turbulent Transition*, Eds.: R. Eppler and H. Fasel, Springer-Verlag, 1980.

[12] Saric, W. S. and Thomas, A. S. W., "Experiments on the Subharmonic Route to Transition." In *Turbulence and Chaotic Phenomena in Fluids*, Ed.: T. Tatsumi, North-Holland, 1984.

[13] Singer, B., Reed, H. L., and Ferziger, J., "The Effects of Streamwise Vortices on Transition in the Plane Channel." To appear in *Physics of Fluids A*, December 1989.

[14] Saric, W. S., Takagi, S., and Mousseux, M. C., "Fundamental Requirements for Freestream Turbulence Measurements." *AIAA Paper No. 88-0054*, January 1988.

[15] Bertolotti, F. P., Herbert, T., and Spalart, P. R., "On the Stability of the Blasius Boundary Layer." *J. Fluid Mech.*, 1989, in press.

[16] Saric, W. S. and Nayfeh, A. H., "Nonparallel Stability of Boundary Layers with Suction." In *Symposium on Laminar-Turbulent Transition*, AGARD Conference Proceedings No. 224, May 1977.

[17] Reynolds, G. A. and Saric, W. S., "Experiments on the Stability of Flat-Plate Boundary Layer with Suction." *AIAA Journal*, Vol. 24, No. 2, 1986, pp. 202-207.

[18] Saric, W. S. and Reed, H. L., "Effect of Suction and Mass Injection on Boundary-Layer Transition." *AIAA Journal*, Vol. 24, No. 3, 1986, pp. 383-389.

[19] Müller, B. and Bippes, H., "Experimental Study of Instability Modes in a Three-Dimensional Boundary Layer." In *Fluid Dynamics of Three-Dimensional Turbulent Shear Flows and Transition.* AGARD Conference Proceedings No. 438, October 1988.

[20] Arnal, D. and Juillen, J. C., "Etude de la Transition et de la Contamination de bord D'Attaque Sur Ailes en Fleche." In *Fluid Dynamics of Three-Dimensional Turbulent Shear Flows and Transition.* AGARD Conference Proceedings No. 438, October 1988.

[21] Saric, W. S., Dagenhart, J. R. and Mousseux, M. C., "Experiments in Swept-Wing Transition." *Numerical and Physical Aspects of Aerodynamic Flows*, Vol. 4, Ed.: T. Cebeci, Springer-Verlag, 1989.

[22] Dagenhart, J. R., Saric, W. S., Hoos, J. A., Mousseux, M. C., "Experiments on Swept-Wing Boundary Layers." In *Laminar-Turbulent Transition*, Eds.: R. Michel and D. Arnal, Springer-Verlag, 1990.

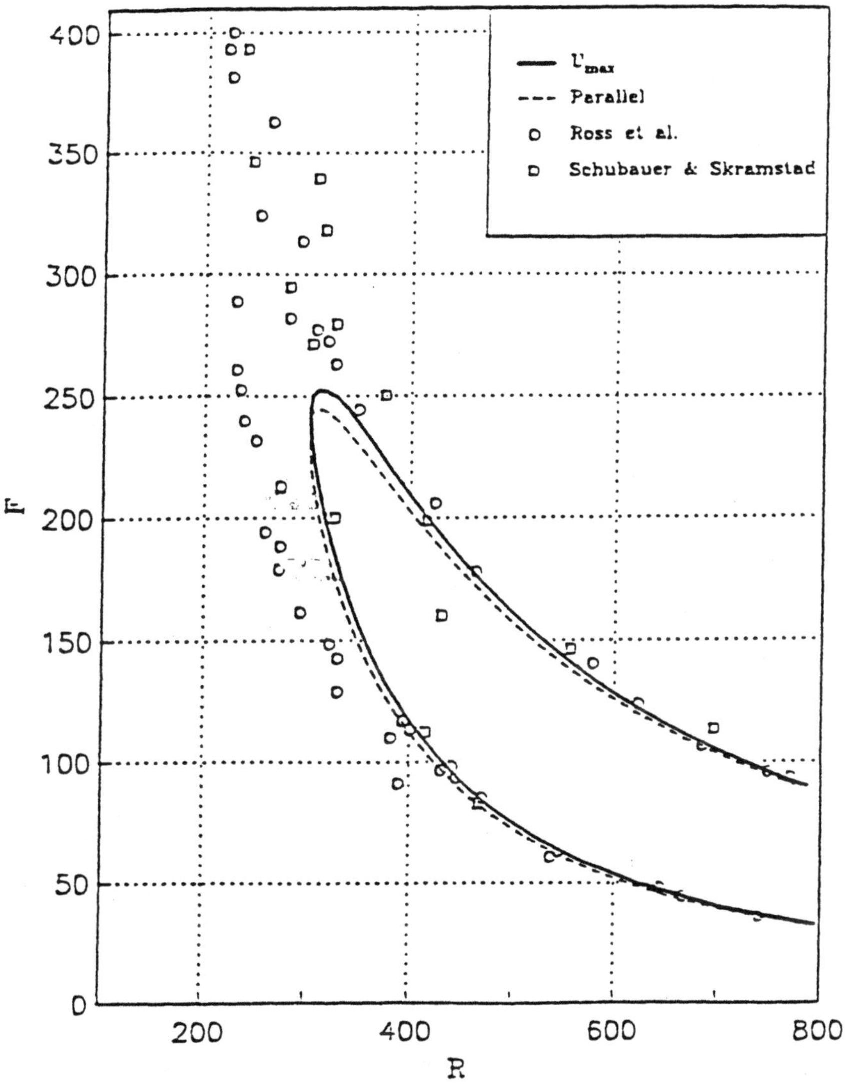

Figure 1. Neutral stability comparisons of experiments [4], [5] with theories [3], [15] for parallel (dashed line) and nonparallel (solid line) calculations. Here, $F = 2\pi f\nu/U_o^2 \times 10^6$ and $R = (U_o x/\nu)^{1/2}$. Figure courtesy of Thorwald Herbert.

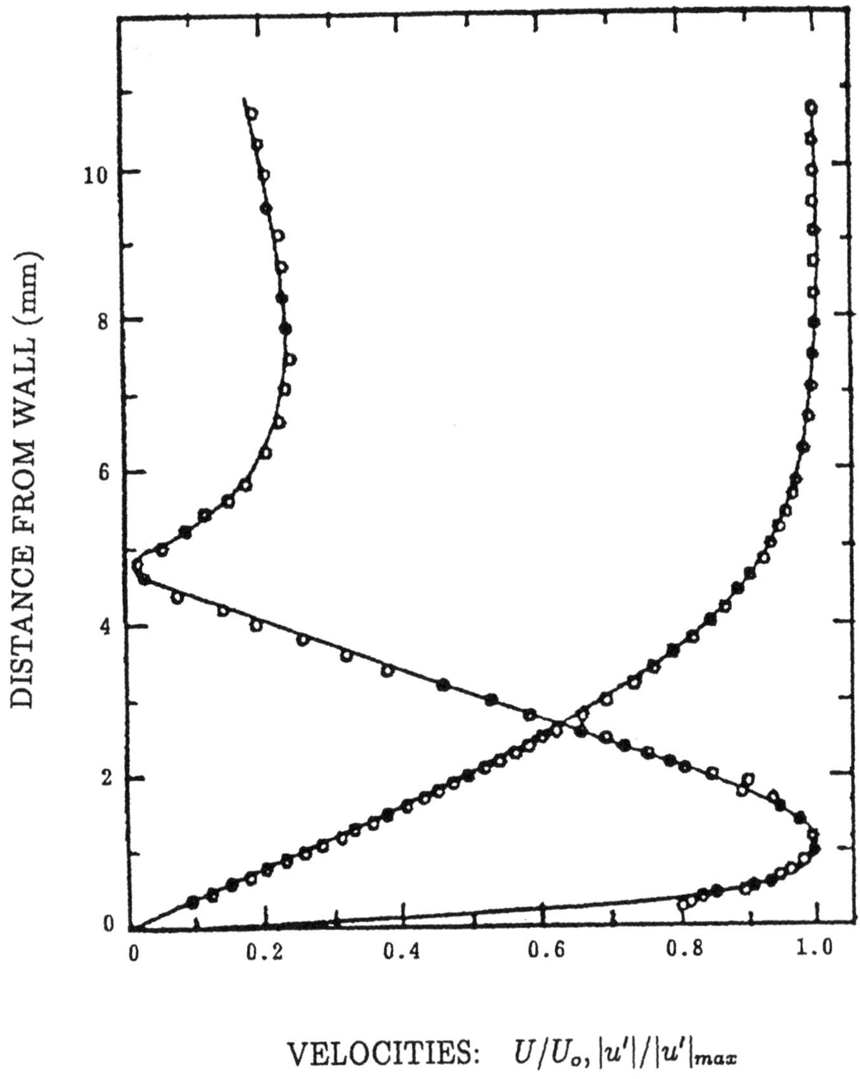

Figure 2. Comparison of theory (solid line) with experiments (circles) for mean velocity and disturbance amplitude versus y. $F = 56.4 \times 10^{-6}, R = 780$ for $f = 50hz$ and $U_o = 9.6m/s$. In this case $|u'|_{max} = 1.5\%U_o$.

COMPUTATION

PANEL SUMMARY: COMPUTATION

Steven A. Orszag

Applied and Computational Mathematics
Princeton University
Princeton, NJ 08544-1000

The past decade has seen remarkable developments in our understanding of transition to turbulence. Much of this progress has been due to a synergistic combination of novel theory with large-scale numerical computation. In this brief overview, I would like to present some of the remaining challenges to our understanding of transition and suggest some ideas for overcoming these hurdles.

As currently understood, early transition to turbulence begins with a sequence of instabilities on a succession of more and more complicated basic flows. Prototypically, the initial state is a plane, parallel shear flow. At sufficiently high Reynolds number such a flow becomes unstable to the growth of long two-dimensional waves which, if undisturbed by further perturbations, eventually equilibrate at some finite amplitude. This state can be considered as a new basic state, and it is typically unstable to short-wave three-dimensional disturbances. This secondary instability does not equilibrate at finite amplitude, and the flow appears to evolve directly into fully developed turbulence. The details of this further development are unresolved so far, and are the subjects of vigorous current investigation.

This picture is an oversimplification of the actual flow development in a real situation. Flow perturbations never arise from the truly infinitesimal perturbations assumed by linear stability theory, nor do three-dimensional perturbations wait until two-dimensional equilibrium is established before they enter the dynamics. In order to understand what is going on, however, some conceptual separation between different effects is necessary. The idea of separating transition into a succession of instabilities on more and more complicated basic flows was originally put forward by Landau, at that time with very little indication of the nature of either the succession of basic states or their instabilities.

Classical stability theory only describes the primary instability, which bears little resemblance to the final turbulent state, and the relevance of linear theory to the obviously nonlinear behavior is open to question. The solutions of the linear stability problem depend very sensitively on the nature of the basic state, and there may be more than one flow that is a sensible candidate for the basic state. In retrospect, we now know that the initial parallel flow is the wrong state to linearize about if we want to understand the origin of complex three-dimensional structure; the correct state is a nonlinear equilibrium or quasi-equilibrium arising from the primary instability.

The linear theory of this so-called secondary instability now has accurately reproduced the critical Reynolds numbers for transition in channel and boundary layer flows, together with the forms of the three-dimensional structure observed in experiments. The recently discovered instability of elliptical vortices in two-dimensional flows supplies a simple mathematical model for the secondary instability, and predicts that this kind of three-dimensional instability will be a generic feature of shear flow transition. Linear analysis of such non-parallel basic flows has led to great insight into the physics of turbulent transition.

In the remainder of this note, I will survey some of the outstanding issues that require clarification in future work.

1. Initial Conditions

The current picture of transition assumes an initial state of two-dimensional Tollmien-Schlichting waves from which the complicated transition phenomena develop. Such an initial state is very special and rarely arises spontaneously in a noninflectional flow. The typical transitional objects that arise in boundary layers and channel flows are turbulent spots, which are localized regions of disordered motion that arise seemingly independently, grow, and then dissipate while the bulk of the flow remains purely laminar most of the time. Spots appear to result from a combination of localized linear disturbances, their three-dimensional instabilities, and the turbulence that develops from the resulting small scale motion. Numerical computation is an ideal tool in which to investigate this kind of behavior and much further work should be done.

2. Numerical Accuracy and Error Estimates

It is now well-known that the accurate numerical calculation of transitional flows presents a heavy computational burden. The early stages of transition are often well resolved, but this situation changes rapidly as the flow develops into a chaotic and then turbulent regime. As turbulence develops, it is now known that the calculations may not be accurate in all detail. Nevertheless, it is assumed (perhaps I should say hoped) that because of some kind of "shadow lemma" the statistical properties of the developing turbulent flow are adequately treated by available numerical simulations. [In a problem for which there is a shadow lemma, the solution of the discretized problem is arbitrarily close to a solution of the "exact" problem for an arbitrarily long time.] The work of Zang et al. has done much to clarify the role of resolution in these transitional calculations. However, it is still necessary to determine a useful and generally applicable set of rules for designing numerical simulations of the transition process.

3. Boundary Conditions

The choice of boundary conditions in studying transition is particularly important in the study of exterior flows past bodies. Here, there are two main sources of difficulty. First, and perhaps more obvious, is the question of how to treat outflow boundary conditions downstream of the body. Usually, these outflow boundary conditions are treated using extrapolation conditions from upstream. It is now known that these extrapolation conditions are not adequate for all problems. Indeed, George Karniadakis and his collaborators at Princeton have shown that the application of downstream viscous boundary conditions give much less upstream influence than simple extrapolation. They have tested these downstream viscous conditions on unsteady flows past cylinders, flow over steps, and two- and three-dimensional jet flows with very satisfactory results.

Another type of boundary condition problem involves inflow boundary conditions. Here, the difficulty is that available experimental data is often not sufficient to accurately specify the inflow (including free-stream disturbances); this can have an enormous effect on the resulting transitional flow so significant ambiguities can remain in comparing the results of numerical computations of transitional flows to experiments. These issues require much exploration in future work.

A related boundary condition issue concerns the effect of spatially developing flows on the transition process. In particular, the question here is when one may use periodic boundary conditions to mimic flows in infinite space. Of course, to be completely realistic, one should use the true infinite space conditions but this is often out of reach on finite-sized computers. For such spatially developing flows as jets and boundary layers, the use of the appropriate inflow-outflow boundary conditions can be important. Precise determination of the effects of these boundary conditions on the transition process is an important problem.

4. Complex Geometries

Complex geometries lack the simple symmetries and invariants present in simple geometry flows so that transition in complex geometries may have quite a different structure than in simpler prototypes. In particular, it now seems that many complex geometry shear flows exhibit supercritical transitions rather than the subcritical transitions so characteristic of planar shear flows (as exemplified by plane Poiseuille flow). Preliminary results for transition in complex geometry also suggest that the characteristic transitional Reynolds numbers in these cases are often an order of magnitude or more smaller (of order 100 versus of order 1000 or more) than in simpler plane-parallel flows. These dynamical features may be directly reflected in the character of the symmetries and invariants of these more complex flows. In particular, complex geometry flows can exhibit competing instabilities of multiple, overlapping velocity profiles, an effect that requires much further exploration.

5. Chaotic Dynamics

One of the most fundamental problems of understanding transitional flows is to discover at what point the fluid motion becomes chaotic and also what are the distinguishing features of the chaotic motion. Chaotic motion appears to develop directly from the nonlinear evolution of secondary instabilities; numerical simulations of this development accurately reproduce characteristic features of real flows. The chaos observed in these developing turbulent shear flows may well turn out to be radically different from the low-order dynamical system chaos seen in constrained convecting or Taylor vortex flows. In the latter problems, the motion is strongly constrained by the boundary conditions or some force like that of an external magnetic field so that only a very small number of modes participate in the dynamics. In these highly constrained systems, chaos develops in an orderly way according to a few well-explored scenarios. A precise classification of which flows develop according to these scenarios and which are governed by "snap-through" transitions to turbulence remains an important problem.

6. Complex Physics

Most studies of transition to date have focused on the problem of transition in incompressible flow. When more complex physics is included the character of the transition process may change. Here complex physics may include stratification, variable flow properties, compressibility, external fields (like magnetic fields), and so on. Each of these complicating physical effects can introduce modifications to the fundamental linear instability modes and may also introduce entirely new modes of instability, not accounted for by incompressible dynamics. The interaction of these many degrees of freedom can present new scenarios for transition and can significantly alter the qualitative as well as the quantitative character of transition. An interesting fundamental question is whether either complex physics or complex geometry can exhibit new routes to chaos not found previously in studies of chaotic dynamics.

7. Turbulent Transport Models and Large Eddy Simulations

It would be nice to have a single computational scheme that would be capable of calculating flows all the way from their laminar state through transition to a fully developed turbulent state. Because of the high resolution requirements of direct numerical simulations of turbulent flows, especially in complex geometries, it will not soon be practicable to calculate such flows in full detail. However, we can hope that turbulent transport models, in which all effects of scales smaller than the largest scales of turbulent motion are removed from the dynamics, can be constructed which describe well the average physics of the flow through these flow regimes. Similarly, in a large-eddy simulation model, the effects of random (chaotic) motions on scales smaller than the grid size are removed from the dynamics and replaced by a transport model. The object here is to be able to calculate with a fixed resolution grid up to huge Reynolds numbers. There has been some significant progress made recently in the development of both transport models and large-eddy simulation models that can be effectively used to compute flows all the way from their laminar state through to their turbulent state. In particular, a group of us at Princeton have been attempting to extend renormalization group ideas to the treatment of these problems. The renormalization group theory of turbulence is particularly attractive for this type of transition problem because it gives a very good description of the low Reynolds number regions of turbulent flows, as exemplified by their transition regions. However, much further work remains to be done on this important class of problems. If successful, this work may lead to practical descriptions of the full transition process in complex geometries of engineering interest.

8. Relaminarization

Numerical computation should be very useful to explore the properties of relaminarization in which locally turbulent flow reorganizes itself to a laminar flow state. Fundamental progress on this problem may be connected with the differences between so-called convective instability (in which disturbance amplitudes only grow in a convective reference frame) and absolute instability (in which disturbance growth occurs in the laboratory frame). In convective instability, it is possible for a fluid element to pass from stability to instability to stability again. Interesting applications of ideas like this should be possible in the context of explaining puff and slug flows in pipes, convective instabilities in jets, and so on.

9. Generalized Elliptic Instability

The fundamental understanding of secondary instability provided by the Bayly-Pierrehumbert theory of elliptic instability suggests various generalizations. One interesting and important example concerns the character of secondary instability of base flows without closed streamlines. These problems can provide basic understanding of the role of viscosity and underlying geometric structure on the transition process and the generation of chaos.

10. Engineering Models of Transition

Current engineering practice is generally based on semi-empirical correlations or modest extensions of linear stability theory, like the so-called e^N method. With the power of available computational tools, it should be possible to quantify these methods and to provide extensions for more complex flow problems. Indeed, one of the most important uses of computation is the generation of simplified (but not oversimplified) models for complex phenomena. A recent case in point is our use of extensive direct numerical simulations of nonlinear Rayleigh-Taylor instability to develop an analytically solvable asymptotic two-fluid model for this complex nonlinear instability problem (Freed, Ofer, Shvarts, and Orszag, Proc. Comp. Turb. Symp., Springer, in press).

Transition research offers tremendous opportunities as well as challenges for numerical computation. It is becoming increasingly apparent that numerical computation is often more direct and more effective than laboratory experiments for these flows. However, accurate simulation of the complex transition physics requires attention to many details, including numerical accuracy, geometric complexity, physical modeling, and so on. With history as a guide, I am confident that the rapid progress achieved over the last decade will translate into ever more impressive development as we approach the next century.

11. Acknowledgments

This work was supported by the Air Force Office of Scientific Research under Contract F49620-87-C-0036, the Office of Naval Research under Contract N00014-82-C-0451, and DARPA under Contract N00014-86-K-0759.

NUMERICAL SIMULATION OF TRANSITION IN BOUNDARY LAYERS

H. Fasel

Aerospace and Mechanical Engineering
University of Arizona
Tucson, AZ 85721

1. Introduction

Transition in boundary layers is a spatially evolving process. In natural transition it is initiated by environmental disturbances such as pressure fluctuations, sound, vortical disturbances, etc. By virtue of certain receptivity mechanisms, which are enhanced for example by strong local wall variations, such as steps, protuberances (roughness elements) or suction slots, the disturbances generate instability waves within the boundary layers. Ideally, if no such disturbances were present, the transition process would not evolve, or if disturbances that were present initially are then suppressed, an already initiated transition process would come to a halt and the flow would relax to the laminar undisturbed state. In short, the dynamic events occurring during the transition process are initiated and derived by the continual forcing of the flow system under consideration which, in this case, is the wall boundary layer.

Because of this direct dependence of the local state of the transition development, controlled transition experiments can be and have been successfully performed where the transition process triggered by controlled forcing was studied in great detail, starting with Schubauer and Skramstad (1948). In contrast, the so-called "closed systems" behave significantly different. For example, in the Taylor-Couette flow the evolution of the disturbances is in time and not in space. Due to the *absolute* instability behavior of closed flow systems, an initial, one-time excitation may be sufficient to carry the flow all the way from the laminar to the turbulent state if the characteristic stress parameter, such as the Taylor number in Taylor-Couette flow, or the Rayleigh number in Benard convection, is large enough. "Open systems" such as the boundary layer, on the other hand are *convectively* unstable, where initial, one-time disturbances are convected downstream. Therefore, observed locally, the flow relaxes to its laminar undisturbed state after the disturbance waves have passed and if no further forcing is applied.

This fundamental physical behavior of the boundary layer has to be represented adequately in any numerical model that claims to realistically capture the many aspects of transition in this flow. In direct, numerical simulations based on the complete Navier-Stokes equations, this would require a so-called spatial model as discussed below. However, certain developments of a controlled transition process can also be captured, albeit mostly qualitatively, by the so-called temporal model, in particular the onset of the three-dimensionality (i.e., secondary instability) as shown for example by Orszag and Patera (1983), Wray and Hussaini (1986), Spalart and Yang (1987), as well as Zang and Hussaini (1988) and Laurien and Kleiser (1989).

2. Spatial Versus Temporal Model

Employing either the temporal or spatial model has different implications when used in direct numerical simulations of transition based on the complete Navier-Stokes equations. In the temporal model the flow is assumed to be spatially periodic, such that the conditions at the inflow and outflow boundary are identical. The development of the flow, for both the disturbance flow and the base flow (unless it is kept constant), is temporal, i.e., the disturbances grow or decay in time, depending upon whether the base flow is unstable or stable, respectively. The base flow is assumed to be invariant with respect to the downstream coordinate and thus nonparallel effects cannot be taken into consideration. For comparison with experiments, the phase velocity is used to correlate the temporal development of the numerical simulations with the spatial development of the experiments. Exploiting the downstream periodicity, i.e., using periodic boundary conditions, allows for an efficient use of spectral approximations in the downstream direction. Most of the transition simulations reported in literature are based on this temporal model, where for the case of controlled transition, good qualitative/quantative agreement with experiments was achieved.

However, the disadvantages of the temporal approach result from the underlying, somewhat unphysical assumptions. Therefore, in situations where the approximations are not justified, for example, when a strong spatial variation of the base flow exists (caused by strong pressure gradients, local geometric variations such as steps, humps, flow separation, etc.) or when the spatial upstream feedback is important (for examples for realistic "active" transition control simulations) the temporal model cannot be employed.

The spatial model on the other hand is physically realistic in that it allows for the base flow and disturbance flow to develop in downstream direction as in laboratory experiments or in practical applications, such as for the flow over wings, etc. In this model the nonparallel effects, and therefore effects of locally strong variations of the base flow, are included. The response (in both upstream and downstream direction) to local disturbance sources can also be studied with this model. Therefore, it is applicable for meaningful receptivity studies. On the other hand, a temporal model is not applicable for receptivity studies.

The disadvantages of the spatial model result from the fact that it is numerically considerably more difficult to deal with than the temporal model. This can to a great extent be attributed to the outflow boundary conditions which are required for this model. In addition, calculations using the spatial model are considerably more expensive in both computer time and memory than temporal calculations because they require a much larger downstream domain.

3. Spatial Calculations

In spite of these disadvantages, we have concentrated on developing numerical methods for solving the Navier-Stokes equations based on the spatial model that are applicable for realistic situations for which the temporal model would not be valid. Because of the lack of computing power, at first emphasis was on the two-dimensional transition development. In a first successful demonstration, spatially developing two-dimensional instability waves in a boundary layer were simulated using a fully-implicit finite-difference method of second order accuracy in time and space (*J. Fluid Mech.*, 1976). With increasing computer power, two-dimensional spatial simulations of increasing complexity were attempted.

Examples of such two-dimensional, spatial Navier-Stokes calculations for a flat plate boundary layer are: 1) Investigation of nonparallel effects in boundary layer stability (Konzelmann, 1982; Fasel and Konzelmann, 1989), 2) Receptivity with respect to a localized free-stream pressure fluctuation (Fasel, 1977), 3) Nonlinear disturbance development (Fasel et al., 1977; Konzelmann, 1983; see also Murdock, 1977) (for plane Poiseuille flow, see Fasel and Bestek, 1979), 4) Effect of a single roughness element (backward- facing step) on transition (Fasel et al., 1977), 5) Linear and nonlinear development of a two-dimensional wave packet (Fasel, 1983), 6) Interaction of Tollmien-Schlichting waves with a localized separation (separation bubble) (Gruber et al., 1987).

Employing more powerful supercomputers such as the CRAY-XMP or the CRAY-II even more challenging investigations were undertaken, for example, 7) Stability and transition of a boundary layer over a cavity (Currie and Fasel, 1989), 8) Three-dimensional, subharmonic and fundamental, breakdown (Fasel et al., 1987), 9) Effect of longitudinal vortices on subharmonic and fundamental breakdown (Konzelmann et al., 1988), 10) Three-dimensional breakdown in the presence of a strong adverse pressure gradient (Kloker and Fasel, 1989), 11) Later stages of fundamental breakdown (Rist, 1989; Fasel, 1990), 12) Passive and active control of two- and three-dimensional disturbances (Kral and Fasel, 1989, 1990), 13) Two- and three-dimensional transition development in compressible (subsonic and supersonic) boundary layers, (Thumm et al., 1990), 14) Nonlinear development of a two-dimensional wave packet (Konzelmann, 1990; Fasel, 1990).

In our opinion, these examples demonstrate that realistic and meaningful simulations based on the spatial model are already feasible using the current generation of supercomputers. Our latest investigations have shown that with such simulations considerable insight into the relevant physical mechanisms is possible. These simulations are highly useful in interpreting transition experiments and can support the validity of difficult transition experiments. They are a powerful tool (in addition to experiments and theory) in transition research, and therefore will play an increasingly important role in the future. With future generations of supercomputers, such simulations will become much more economical and therefore problems of even greater complexity can be attacked. Test calculations have shown that many of our numerical methods used in the examples above are well suited for parallel computing architectures, requiring relatively little overhead. Employing parallel computers would therefore also significantly reduce the computer time required for such computations.

4. Conclusion

I strongly believe that in the future numerical simulations based on the spatial approach will, to a large extent, replace the temporal simulations, in particular, since for many pressing problems (such as, for example, receptivity, etc.) only the spatial model is appropriate. With the anticipated significant increase of computing power in the near future, problems and tasks should be - and will be - addressed for which access with other tools, experimental and/or theoretical is either not feasible or very costly. Examples in this respect are supersonic/hypersonic transition, transition in chemically reacting flows, effect of roughness (single or distributed), receptivity, etc. These can be, and probably will be, areas of transition research where numerical simulations will have to push ahead, and it might be long before the results can be verified by experimental measurements or other theoretical results. This might be particularly true, for supersonic/hypersonic transition or transition in chemically reacting flows. This is in contrast to the past, where known experiments were simulated and the measurements were known before the simulation was started. Therefore, the measurements could be used for code validation. For many of the pressing problems of the future (as for the examples given) this will no longer be the case. Thus, the challenge for this type of research will be extraordinary. And because it might take a long time to verify the results of some of these very costly simulations, either by other simulations using different numerical methods or by laboratory experiments, it is important to document as much detail as possible concerning the numerical procedure and all relevant computation parameters so that possible deviations with other numerical results and/or measurements can be explained.

References

[1] Schubauer, G. B. and Skramstad, H. K. (1948), "Laminar boundary layer oscillations and stability of laminar flow," *NACA Rep. 909.*

[2] Fasel, H. (1976), "Investigation of the stability of boundary layers by a finite-difference model of the Navier-Stokes equations," *J. Fluid Mech.*, *78*, p. 355.

[3] Fasel, H. (1977), "Reaktion von zweidimensionalen, laminaren, inkompressiblen Grenzschichten auf periodische Stroerungen in der Aussenstroemung," *ZAMM 57*, p. 180.

[4] Fasel, H., Bestek, H., and Schefenacker, R. (1977). "Numerical simulation studies of transition phenomena in incompressible, two-dimensional flows," *Proc. AGARD Symp. on Laminar-Turbulent Transition*, Copenhagen.

[5] Murdock, J. W. and Taylor, T. D. (1977), "Numerical investigation of nonlinear wave interaction in a two-dimensional boundary layer," *Proc. AGARD-CP-224, Symp. on Laminar Turbulent Transition*, Copenhagen.

[6] Fasel, H. and Bestek, H. (1979), "Investigation of nonlinear, spatial disturbance amplification in plane Poiseuille flow," *Proc. IUTAM Symp. on Laminar-Turbulent Transition*, Stuttgart.

[7] Konzelmann, U. (1982), "Untersuchung der nichtparallelen Effekte bei der hydrodynamichen Stabilitaet von Grenzschichtstroemungen," Diploma project (Studienarbeit) Univ. of Stuttgart.

[8] Konzelmann, U. (1983), "Nichtlineare Wellenausbreitung in einer Grenzschichtstroemung," Diploma-Thesis, Univ. of Stuttgart.

[9] Orszag, S. A. and Patera, A. J. (1983), "Secondary instability of wall-bounded shear flows," *J. Fluid Mech.*, *128*, p. 347.

[10] Fasel, H. (1984), "Numerical simulation of nonlinear growth of wave packets in a boundary layer," *Proc. IUTAM Symp. on Turbulence and Chaotic Phenomena in Fluids*, Kyoto, Japan.

[11] Wray, A. and Hussaini, M. Y. (1984), "Numerical experiments in boundary-layer stability," *Proc. Roy. Soc.*, *A392*, p. 373.

[12] Fasel, H., Rist, U., and Konzelmann, U. (1987), "Numerical investigation of the three-dimensional development in boundary layer transition," *AIAA Paper 87-1203*, see also *AIAA J.* (1989) (to appear).

[13] Gruber, K., Bestek, H., and Fasel, H. (1987), "Interaction between a Tollmien-Schlichting wave and a laminar separation bubble," *AIAA Paper 87-1256.*

[14] Spalart, P. R. and Yang, K. S. (1987), "Numerical study of ribbon-induced transition in Blasius flow," *J. Fluid Mech.*, *178*, p. 345.

[15] Currie, J. and Fasel, H. (1989), "Numerische Simulation der instationaeren Stroemungsvorgaenge bei einer Cavityumstroemung," *ZAMM* (to appear).

[16] Fasel, H. and Konzelmann, U. (1989), "Nonparallel stability of boundary layers," (submitted for publication).

[17] Konzelmann, U., Rist, U., and Fasel, H. (1989), "Numerical investigation of the effects of longitudinal vortices on the onset of transition in a flat plate boundary layer," *Proc. AGARD Symp. on Fluid Dynamics of Three-Dimensional Turbulent Shear Flow and Transition*, Cesme, Turkey.

[18] Kral, L. and Fasel, H. (1989), "Numerical investigation of the control of the secondary instability process in boundary layers," *AIAA Paper 89-0984.*

[19] Laurien, E. and Kleiser, L. (1989), "Numerical simulation of boundary-layer transition and transition control," *J. Fluid Mech.* (to appear).

[20] Rist, U. (1989), "Numerische Untersuchung der raeumlichen, dreidimensionalen Stoerungsentwicklung beim Grenzschictumschlag," Dissertation, Univ. of Stuttgart.

[21] Fasel, H. (1990), "Numerical simulation of instability and transition in shear flows," *Proc. IUTAM Symp. on Laminar-Turbulent Transition*, Toulouse, France.

[22] Kloker, M. and Fasel, H. (1990), "Numerical simulation of two- and three-dimensional instability waves in two-dimensional boundary layers with streamwise pressure gradient," *Proc. IUTAM Symp. on Laminar-Turbulent Transition*, Toulouse, France.

[23] Konzelmann, U. (1990), "Numerische Untersuchungen zur raeumlichen Entwicklung dreidimensionaler Wellenpakete in einer Plattengrenzschichtstroemung," Dissertation, Univ. of Stuttgart.

[24] Kral, L. and Fasel, H. (1990), "Numerical investigation of the control of the three-dimensional transition process in boundary layers," *Proc. IUTAM Symp. on Laminar-Turbulent Transition*, Toulouse, France.

[25] Thumm, A., Wolz, W., and Fasel, H. (1990), "Numerical simulation of spatially growing three-dimensional disturbance waves in compressible boundary layers," *Proc. IUTAM Symp. on Laminar-Turbulent Transition*, Toulouse, France.

COMPUTATION OF INSTABILITY AND TRANSITION

P. R. Spalart

NASA Ames Research Center[1]
Moffett Field, CA

ABSTRACT

Some of the author's views on the state of transition calculations in 1989, and on desirable future trends, are presented. Aeronautical applications are kept in mind, and the emphasis is on wall-bounded flows. It is argued that some of the effort now devoted to the late stages of transition in excessively confined periodic domains should be shifted to studies of the early stages of transition (up to mildly nonlinear) and in particular to receptivity. There are many reasons besides nonlinearity to embark on a numerical study. Such work would contribute more to transition *prediction*. A worthwhile study of the early transition stages requires a larger computational domain and more difficult boundary conditions, but not a very fine grid. Intermediate strategies, more involved than the classical Orr-Sommerfeld equation but much more economical than full Navier-Stokes solutions, should not be forgotten. We must also strive to treat more realistic geometries and disturbances, and the need for compressible calculations is pressing.

1. Introduction

In 1979, Professor R. S. Shevell, who was teaching design at Stanford after being a top designer at Douglas, reminded us that "Laminar-Flow Control is so promising it has been promising for 40 years." As far as designers were concerned, the boundary layers on an airplane (especially with swept wings) were turbulent. In the last few years, we have witnessed a renewal of interest in transition prediction and control, in particular at NASA, the present workshop being one example. This renewal has two causes: the technological advances that make Laminar-Flow Control (LFC) appear practical

[1]Presently, Clark B. Millikan Visiting Professor of Aeronautics, California Institute of Technology.

at last, and the fact that at the supersonic and hypersonic speeds (and high altitudes) being contemplated for the next generation of flight vehicles much larger regions of laminar or transitional flow are expected to show up spontaneously in the wing boundary layers. LFC is particularly advantageous at these high speeds, allowing a reduction of both drag and heating.

Because the new interest in transition is to a large extent fueled by the new interest in supersonic flight, observers may get the false impression that low-Mach-number transition is a solved problem and compressibility is the only remaining barrier. The most famous "by-passes" and the most severe experimental difficulties do occur in supersonic flow, but the truth is that incompressible transition still presents a large number of major challenges including receptivity, competition between disturbances, and response to curvature and three-dimensional base flows to name a few. Engineering prediction methods are just not available. The problem was known to be difficult, but not perceived as pressing.

Because of the experimental difficulties in high-speed flow, it is anticipated that supersonic transition prediction (and much of the aircraft design) will rely heavily on numerical methods. This is a new challenge for numericists; we have to be ready to produce reliable conclusions with little or no experimental backup. The initial phase of running codes just to show agreement with previous experiments is over. Presumably the most fruitful exchanges will occur when a few generic flows are studied concurrently by experiment and calculation. However, some flows may have to be studied by calculation alone.

2. State of the Field

Given highly improved numerical algorithms and supercomputers, what can we do now that Tollmien and Schlichting could not do in 1933? How can we best use this new power? To arrive at a tractable equation Tollmien and Schlichting made the following assumptions. The mean flow was treated as spatially uniform (a) and parallel (b); the disturbances were linearized (c) and periodic in x (time developing) (d). This generated a one-dimensional eigenvalue problem, the Orr-Sommerfeld equation; single modes were considered both in the streamwise x direction (e) and the normal y direction (f). Oblique waves were treated using Squire's transformation. Relaxing assumptions is always desirable, but the order in which assumptions

(a) to (f) should be relaxed is not obvious, and relaxing them independently can be a very good strategy.

There is a wide and almost continuous range of strategies between what was possible in 1933 and a solution of the full Navier-Stokes equations. In fact the distinction between "linear *theory*" and "numerical simulations" is unfounded. The Orr-Sommerfeld equation is always solved numerically (except in the simplest cases, such as piecewise-linear velocity profiles). Thus, work at that end of the spectrum has a significant numerical content. At the other end, even the "brute-force" Navier-Stokes simulations require significant theoretical input especially for transition (as opposed to fully turbulent flows). Countless decisions must be made regarding the boundary conditions, disturbances, details of the base flow, and so on. The theoretical content of a good numerical transition study is high.

Let us review the advances made in the field during the last two or three decades. We can now obtain many of the leading y-eigenfunctions quickly and accurately (work by Jordinson, Orszag, and others), thus relaxing assumption (f). It is safer to obtain these with a matrix method instead of one (such as shooting) that relies on an initial guess especially in new situations, in order to avoid missing a new branch of unstable modes (as found by Mack in supersonic flow). The continuous spectrum complicates the situation in boundary layers (Grosch and Salwen); even though these modes are damped, they may become important to account for free-stream disturbances. Numerical methods effectively designed for the discrete spectrum (eigenmodes confined to a region a few times thicker than the boundary layer) converge quite slowly to the continuous spectrum (the phase velocity is too low). This may disqualify them for theoretical work based on eigenmodes, but should be tolerable in studies of the initial-value problem as long as it does not disrupt the numerical convergence.

Conversely, we can relax only assumption (e) by including many Fourier modes in x and z while only retaining the most unstable y-mode. In this way Gaster treated wave packets and Mack wave packets and line sources of instability (a finite-length vibrating ribbon) successfully. To relax assumption (d), we may treat spatially-developing disturbances in a parallel laminar flow, which is appropriate in a practical prediction method (assuming a slowly evolving base flow). An intermediate approach is Gaster's approximation based on the group velocity.

Assumptions (a) and (b) (which are normally tied because of the continuity condition) have been relaxed in a few studies (Bouthier, Gaster, Saric, Nayfeh, and others). This delicate subject went from fashionable to controversial; recent results by Bertolotti and Herbert (fully confirmed by the author using a very different method) indicate that at least in Blasius flow the nonparallel effects were vastly over-estimated by some authors and obscured by the arbitrariness in the definition of the growth rate. The experiments may have been tainted by nonlinear effects. In any case, too much attention should not be devoted to the very low Reynolds number region; the waves there have amplification ratios so small that eigenmode analysis is irrelevant. If nonparallel effects are small also in other flows, it is fortunate and will help keep the prediction methods relatively simple (the author's first results in an adverse-pressure-gradient Falkner-Skan flow indicate that the relative change in growth rates is indeed small). Addressing similar issues, Hall and his co-workers have discussed the subtleties of the Görtler instability.

Another assumption that may now be discarded is linearity (c) and this is where simulations have had the most impact. Herbert produced saturated 2D waves in a channel requiring a few numerical modes in x and z and steady in an appropriate reference frame. "Complete" numerical solutions of the initial-boundary-value problem are now almost routine, although still very expensive. Early examples of such studies are from Orszag's group, and Wray and Hussaini. Complete solutions require a large number of modes in all spatial directions and numerical time integration. In the early simulations periodic conditions were used in x and z, with domains just large enough to contain one wave, and such a wave or combination of three waves was introduced as the initial disturbance. The author firmly believes that this strategy, the analog of using a vibrating ribbon (if not two) in the lab, should be considered to be preliminary to a study in which the dynamics, not the user, dictates the wavelength of the flow structures. There are no ribbons on airplanes. The ribbon technique was devised to verify the results of the Orr-Sommerfeld equation, that is, the behavior of single modes. Now that we are not limited to single modes any more, why confine ourselves to the nonlinear evolution of a flow that is quite artificial? There is no point in continually repeating Wray and Hussaini's 1980 work, even with slight variations.

To allow the dynamics to select the wavelength, one needs disturbances with a noisy or random character, *and* a domain large enough to contain, say, five structures (e.g., Λ vortices) in each direction. The flow then has the choice of making four structures, or six. If using a small domain (enough for one structure) one should at least repeat the calculation with different sizes and single out the one that gives fastest growth. This would save memory, but not computer time, and is still not conclusive. For a given cost, a calculation on a large domain with a coarse grid, even if it must be stopped long before full transition because of lack of resolution, is preferable to one on a finer grid but with flow structures that are confined inside a small domain. The recent and impressive work by Gilbert and Kleiser, in which they calculated channel flow all the way through transition, does not mean that channel transition is solved. This transition is not unique. Other domain sizes and other initial disturbances would probably result in significantly different behavior.

Similarly, the theory of secondary instability (Orszag and Patera, Herbert), although very successful in explaining the sudden increase in growth rates and the role of nonlinearity, should not be expected to apply in all situations. It is tailored for vibrating-ribbon situations. In "natural" transition, the nonlinear "spark" may well occur between two TS (oblique) waves of comparable amplitude, instead of between a dominant 2D TS wave and infinitesimal oblique waves. This possibility is supported by the results of Yang, Ferziger & Spalart, for transition in a boundary layer with a large domain in both x and z, and white-noise initial disturbances.

The use of random noise has been challenged, notably by Morkovin, who would favor a more controlled approach. However, noisy conditions are an effective way to remove bias (not to mention spurious symmetries) and the best hope for identifying the dominant transition mechanism with reasonable effort. The time-consuming alternative is to run a large number of calculations, each with a carefully controlled set of disturbances. Random disturbances and large domains allow us to use all the power of full simulations. Having analyzed the results of a calculation from random initial disturbances, Yang *then* proposed a model using only two waves (one 2D, the other oblique) which captured the essential features of the flow (dislocations between patches of TS waves, non-aligned Λ vortices). On the other hand, the very simple noise distributions used so far (white in x and z, uniform or crudely shaped amplitude in y) are not the last word on the subject. There is also the option of adding random noise continually, instead of just initially; it may be more realistic and reduce the influence of the starting time. The words "natural transition" may have been used a little prematurely by Yang, Ferziger & Spalart. In the long run we will devise disturbances representative of actual disturbance environments (free-stream turbulence or sound waves, for instance); we will not be directly inputting TS waves.

In the work of Fasel's group, assumptions (a) to (f) are all relaxed since a full Blasius base flow is used and the disturbances obey the Navier-Stokes equations with inflow and outflow conditions. Recent studies with that code are three-dimensional, with a few periodic modes in the spanwise direction. Inflow-outflow codes tend to be significantly slower and use more memory than periodic codes. Outflow conditions also sometimes cause problems when the outgoing disturbances are not smooth TS waves. These difficulties must be faced if the "complete" problem is to be solved. TS waves have been introduced by Fasel and his co-workers at the inflow boundary, but also using heating strips; this was an example of receptivity. Recent work also addresses the secondary instability and the early stages of breakdown.

3. Recommendations

We may distinguish two major tasks in the transition field. The first is to "explain" transition, and/or turbulence. Linear theory could not do this; on the other hand one would like more than just the statement that "turbulence occurs because the equations are nonlinear." There is progress in that direction thanks to secondary-instability theory, and chaos theory, but a simple answer (and one that is not excessively abstract) is not likely to exist. The second task, which is equally challenging and important, is to predict with some accuracy where and if transition is going to occur in as many flows as possible. Unless there is a by-pass, this depends primarily on three mechanisms: receptivity, linear growth, and nonlinear breakdown. The amplitude at which breakdown occurs is of the order of 1%, give or take a factor 2 or 4. For prediction purposes, exploring this stage is warranted only if it covers a large region of the boundary layer (which *may* occur in hypersonic flows). Obtaining a similar bracket for the linear amplification ratio (at least for a single wave) seems possible with current techniques. In contrast, the uncertainty associated with the receptivity stage is orders of magnitude, and qualitative understanding of this process is definitely lacking. In addition to the excellent theoretical work that is now applied to this problem (e.g., triple-deck, see Smith, Goldstein, Kerschen) there is a place for careful numerical work, not because of nonlinearity, but because the spatial evolution of the laminar flow is essential, and because some of the assumptions needed by the theoreticians (e.g., large Reynolds number) may be rather sweeping. Outlining their region of validity would benefit everyone.

It is important to remember that "linearized" does not imply "single-mode," or even that eigenmodes are involved at all. In fact, a solution of the linearized disturbance equations globally (without locally-parallel-flow or similar assumptions) in a practical flow (e.g., boundary layer on a wing) is still out of reach and would be an achievement. The author has performed calculations of leading-edge contamination and cross-flow instability with the full spatially-evolving base flow, and intends to do the same for a rotating disk in the near future. This is with a flat plate, the base flow known exactly, and a spectral method. For less simple geometries and given the facts of life in numerical methods, the idea of obtaining the laminar flow with a full nonlinear method in generalized coordinates (which often requires fairly low-order time integration, approximate factorization, and explicit or hidden smoothing and artificial dissipation) and then solving the disturbance equations with a "cleaner" method (higher order in time thanks to the linearity, almost Cartesian mesh, centered schemes, much less dissipative overall) should be explored. The problem of how to introduce the disturbances would have to be addressed. Also, a finer streamwise grid may be needed to resolve TS waves than to resolve the laminar flow.

In terms of codes, the present need is for accurate and efficient compressible solvers, and codes for more complex geometries. ICASE has spectral compressible codes including periodic boundary layers. Initially, they used fully explicit time integration and were therefore rather slow. Recent advances with high-order "spectral-like" finite-difference schemes (S. Lele) makes those very competitive; in addition, spectral methods are less attractive for compressible flows since full dealiasing is not possible and also since the pressure interactions are not instantaneous. Again, combining a very elaborate spatial discretization and implicit time advance will be a challenge. For flat-plate boundary layers, periodic codes and inflow-outflow codes such as Fasel's are likely to coexist for quite some time because periodicity is so attractive numerically and because of recent advances made by the author in using periodic codes in non-periodic situations such as the attachment-line region, and even Blasius flow (work in progress). We certainly need a highly accurate code to treat airfoil flows or even just a bump on a plate; periodic spanwise conditions would be adequate. The challenge is to produce a code that gives accurate growth rates for TS waves. In any case, the perfect full-Navier-Stokes code would not solve transition by itself; it would need to be used with utmost care and constant theoretical input to begin unravelling the intricacies of the problem.

Acknowledgements

This paper was reviewed by Dr. R. Moser (NASA Ames Research Center).

NUMERICAL SIMULATION OF INSTABILITY AND TRANSITION PHYSICS

C. L. Streett

TAB / FIMD
NASA Langley Research Center
Hampton, VA 23665-5225

INTRODUCTION

The application of computation to problems in fluid dynamic stability and transition has resulted in a great deal of physical understanding of the underlying processes over the last 15 or so years. The capability to precisely adjust and monitor quantities which may be difficult or impossible to control or measure in an experiment has allowed us to explain a number of hitherto puzzling experimental observations. It now appears obvious and generally accepted that computation will play a key role in transition research in the future.

The algorithms used to carry out instability and transition simulations have in many areas reached a fairly mature state. It is now possible, for instance, to carry parallel-flow simulations of boundary-layer or channel flow transition well into the laminar breakdown stage. Parallel-flow calculations, which have yielded striking qualitative agreement with experiment, became feasible with the advent of the first true supercomputers five to ten years ago. Since the advances of that early period, however, we appear to have stalled; although much may still be learned using methods developed about a decade ago, the next surge of understanding will require a quantum leap in algorithm technology. This paper focuses on a number of issues which must be dealt with to achieve this step forward.

DISCRETIZATION METHODS

Discretization is perhaps the most fundamental algorithmic element. To an overwhelming degree, spectral methods have been the discretization method of choice for incompressible transition simulations. This is due to their proven properties of low phase error and exponential-order error decay for such problems. Collocation methods are often used for their generality and ease of application, but Galerkin methods may be employed to enforce special constraints such as in the development of divergence-free basis functions, and tau methods may be utilized in situations where fast-solvers are available. Recent advances in the application of spectral methods to more complex geometries through the use of multidomain schemes, to parallel-processor architectures, to compressible flow, and to situations requiring non-reflecting outflow boundary conditions will be discussed later.

High-order finite-difference methods, born of CFD research, are becoming more popular in transition simulations, especially for compressible flow.

These discretization scheme fall under two general categories: High-order upwind-biased schemes of the ENO type, and high-accuracy centered schemes. The former would be advantageous mainly in situations involving flowfields with shocks or slip lines. These methods also possess reasonably good phase error properties; this can be seen from their performance in standard test problems such as the traveling pulse. High-order ENO methods are currently only in the development stage for multi-dimensions; present applications are simple implementations of the 1-D operators in each direction, and it is not clear that this is the proper formulation. Additionally, the logic required to evaluate these discretizations is involved, thus making them fairly costly to implement. In general, however, these methods are fairly promising for computation of high-speed flows. As a caveat, on the other hand, we find that second-order ENO and upwind methods such as those commonly used for CFD applications are unacceptable for used in transition simulations; error levels observed in tests in the linear range indicated their phase error characteristics are such as to give rise to unacceptable departures in later stages.

High-order centered difference formulae, such as those based on compact differencing or Pade´approximates, have recently been applied to transition simulations. The advantage of these methods over spectral methods is not clear. Efficiency is the first issue; these methods require solution of an implicit system to evaluate derivatives. This produces the requirement of solving a nonlinear system in an implicit scheme, even when the PDE being solved is linear, thus fundamentally limiting the maximum time step severely. This inherently-scalar process can also render implementation expensive; vectorization can be carried out over other coordinate directions, but having to transpose entire fields to order data properly for such vectorization still adds to the cost. On the other hand, evaluation of spectral derivatives via matrix multiplies or fast transforms is efficient, straightforward, and optimized kernels exist on virtually any machine. Also, high-order centered schemes suffer from the usual difficulty of wide stencils near boundaries; providing auxiliary conditions to the interior scheme to handle the boundaries is required and impacts the overall accuracy of the discretization, as opposed to the straightforward and uniform accuracy of Chebyshev or Legendre spectral discretization. Finally, these compact difference schemes offer little to no advantage over spectral methods in the presence of unresolved high gradients such as shocks.

The use of spectral discretization offers several unique advantages in addition to its error properties. The first is the capability to monitor resolution on-the-fly. By monitoring the energy in the high coefficients of the series, one can be warned of marginal or inadequate resolution. Discretization schemes which are dissipative in the high wavenumbers will give no warning of inadequate resolution of small-scale structures until the entire solution accuracy is compromised. Another advantage lies in the inherently natural and consistently-accurate interpolation operators available for spectral discretization. Such operators are required for staggered-mesh schemes, and when grid refinement is required.

The above accuracy arguments are made with instability and transition simulations in mind; it can be argued that the advantage of spectral methods is perhaps lessened for simulations of homogeneous turbulence. In the case of transition, the high-wavenumber field is in a strongly nonequilibrium state, with a great deal of mode interaction which small phase errors can disrupt. These details are perhaps not as important in the case of fully-developed turbulence. This is particularly true for two-dimensional turbulence simulations, in which the initial energy rapidly splits into barely-interacting large- and small-scale components; the large-scale field is easily resolved, and the small scales are insensitive to the exact details of their eventual dissipation. This may explain some recent results which show fourth-order finite-difference and Fourier spectral discretizations being roughly equivalent for such problems.

One clear advantage of finite-difference methods over spectral methods lies in implementation of non-rigorous boundary conditions. Spectral methods are pathologically sensitive, requiring well-posed boundary conditions at all times. The use of second-order "soft" boundary conditions, such as sometimes used for an outflow condition in finite-difference calculations, are guaranteed to be explosively unstable for all but extremely coarse spectral discretizations. The natural dissipation of finite-difference methods make them far more tolerant of boundary condition improprieties. As will be mentioned later, some advances have been made in this area for spectral methods.

SPATIAL SIMULATIONS

We feel that the next significant advances in our understanding of transition physics and in our ability to predict transition will come with more physically realistic simulations. Although some excellent qualitative agreement with experiment has been achieved by temporal simulations, quantitative predictions of transition location, as well as inclusion of effects such as roughness and pressure gradient which fundamentally violate the parallel-flow assumption will require spatial simulations. While techniques have been employed to incorporate small amounts of meanflow non-parallelism through coordinate transformations or the use of modifying functions which render the meanflow periodic by rapidly adjusting it at the endpoints of the domain, the assumption of streamwise periodicity of the disturbances remains.

There are two algorithmic difficulties with spatial simulations: Cost, and the requirement for a general nonreflecting outflow boundary treatment. Temporal simulations have given a dependable estimate of the resolution which would be required for a spatial simulation of, say, flat plate boundary layer transition through the breakdown stage. This estimate comes from the temporal ansatz of interchanging time and downstream distance; in the simulation, the streamwise-periodic computational domain is traveling downstream at the group velocity of the disturbance. In most temporal simulations, one is only concerned then with the flowfield in a one-wavelength region; the discretization is started off coarse in the early linear stages, and refined as the simulation progresses. This sort of tailored resolution can be achieved in a spatial simulation through the use of multidomain techniques. The large streamwise extent of the simulation can be broken down into a number of regions, with increasingly-fine discretizations in the more downstream domains. The spanwise direction can be particularly coarse for a large part of the upstream region; the secondary instability can be captured with relatively few spanwise (Fourier) modes. Very fine resolutions are required in the later stages, however. Another scheme to reduce the cost of such a spatial simulation hinges on the dominantly convective instability of the early TS stage. Given an effective and accurate outflow boundary condition, the early stages of TS growth up to the secondary instability can be simulated, and data on a downstream plane stored and used as the inflow condition for simulation of the later stages. This is, of course, only viable if upstream influence on these stages is weak; this is certainly true for boundary layer flows with small pressure gradient, as shown by temporal simulations which by their nature neglect large-scale upstream influence.

The above brings to mind a conceptual difficulty with spatial simulations, truncated in streamwise extent by an outflow boundary treatment. No matter how efficacious such a condition might be, if the domain is truncated in a region where there is significant upstream transfer of energy, perhaps due to a global or absolute instability, then the physical veracity of the simulation is compromised. We have observed such a phenomenon in our outflow boundary treatment tests. The caveat here is that the nature of the instability mechanisms active in a problem must be taken into account in setting up a spatial simulation, or extensive testing with truncated domains of various lengths may be required to yield a reliable spatial simulation.

Outflow boundary treatments have been reported by a number of researchers. The difficulty with providing a condition viable for spatial simulations is that is must be capable of passing large, nonlinear disturbances with unsteady variations of all velocity components. This rules out simple steady gradient conditions, which work only when the outflow is smooth, steady, and virtually parallel, and also rules out conditions based on transparency to a simple wave of a particular wavenumber or group velocity. Any candidate condition must be subjected to rigorous tests against linear theory, in channel flow for example, with more than one eigenmode of different wavenumber and group velocity imposed at inflow. No reflection from the leading wave must be observed, and the simulation must rapidly settle into a limit cycle for a long period without buildup of energy in time. It is clear that a treatment based on continually moving the downstream boundary ahead of the leading wave will rapidly render a calculation too expensive for all but early-stage simulations. It is also recognized that low-order finite-difference discretizations are far more tolerant toward imposition of an outflow treatment than are spectral methods. We have demonstrated a treatment which meets the above criterion with spectral accuracy, and have incorporated it into an incompressible simulation code.

COMPRESSIBLE-FLOW ALGORITHMS

The state of the art in spectral simulation algorithms for compressible flow is not nearly as mature as for incompressible flow. A widely-used compromise between cost per time step and size of time step for incompressible algorithms is that of treating viscous terms implicitly to avoid their severe explicit time step restriction ($O(N^{-4})$ for Chebyshev discretization), and treating the convection terms explicitly. For compressible flow, the unsteady continuity equation introduces a new time scale, based on acoustic processes. This typically renders schemes which treat only the viscous terms implicit unstable at time steps only slightly larger than purely explicit. As of now, most compressible simulations are being carried out using fully explicit time stepping schemes. We have developed a fully implicit algorithm, based on iterative solution of the algebraic system from high-order backward time differencing the compressible Navier-Stokes equations, but our research has achieved but limited success to date. It is certainly true that developments in spectral compressible simulation algorithms are sorely needed.

CONCLUSIONS

Given the tools for incompressible spatial simulations which have been recently developed, we are prepared to attack a number of problems. Certainly an attempt to calibrate e^N-type engineering correlation methods based on inexpensive stability theory, via spatial simulations with imposed pressure gradients and non-parallel three-dimensional meanflows, would be well worthwhile and immediately applicable. Clarifying the effect of roughness could also be accomplished; the mechanism by which instability is enhanced, if that is indeed the dominant effect of roughness, could be elucidated. The process by which turbulence is self-sustained in spots or wedges could be explored. Simulation of transition in the free shear layer over a separation bubble, an event which is believed to hasten the layer's reattachment, is also within reach and would impact high-lift airfoil and massive-stall engineering.

With the development of adequate compressible algorithms, exploration of bypass mechanisms on simple bodies at high speed would be possible, in addition to uncovering the physics involved in mixing in high-speed free shear layers. Design of turbomachinery is also an area in which compressible transition simulations could make a large contribution. Finally, the use of compressible algorithms capable of resolving acoustic time scales at low speeds would be expensive, but worthwhile toward elucidating some of the physical mechanisms active in receptivity.

ACKNOWLEDGEMENT

The author would like to thank M. Y. Hussaini for his constant help and encouragement.

NUMERICAL COMPUTATION OF TRANSITION TO TURBULENCE

Thomas A. Zang
NASA Langley Research Center
Hampton, VA 23665

1 Introduction

The extreme application of numerical techniques to transition problems involves the direct solution of the full nonlinear, three-dimensional, time-dependent Navier-Stokes equations for prescribed initial and boundary conditions. All spatial and temporal scales are intended to be resolved, i.e., no modeling is employed. This position paper addresses the current and prospective uses of direct numerical simulation of transition to turbulence.

Numerical simulations of those transitional phenomena with significant to aerodynamical applications have focused on a few idealized physical problems, including

1. Taylor-Couette Flow (TCF). This is typically studied by gradually increasing the rotation rate of the inner cylinder (with a stationary outer cylinder) and finding the resulting flow: steady, periodic, doubly-periodic, or temporally chaotic. This type of problem is amenable to bifurcation analysis and the steady and periodic cases can be described well by linear and weakly nonlinear methods.

2. Free Shear Flow (FSF) The idealized mixing layer exhibits such interesting phenomena as vortex roll-up, vortex pairing, secondary instabilities, and compressibility effects.

3. Boundary-layer Flow (BLF). Transition in a boundary layer is either linearly triggered or by-pass triggered. Boundary-layer transition at a fixed Reynolds number in a low disturbance environment often passes through discernible stages of linear instability, secondary instability, tertiary instability, nonlinear breakdown, and final transition to turbulence. The first two stages have been well-described by theoretical methods, whereas the subsequent stages can only be treated by large-scale computation. Boundary-layer transition in a high-disturbance environment, on the other hand, usually occurs through so-called by-pass mechanisms; these are very poorly understood theoretically and numerical simulations of sensible model problems are sorely needed.

4. Channel Flow (CF). This is representative of transition in a confined flow. Unlike BLF, CF is truly parallel and it does admit finite-amplitude steady traveling TS waves. Otherwise, most of the physical effects are very similar to those in BLF.

The current achievements of numerical simulations have been constrained primarily by the computational resource demands arising from the limitations of existing algorithms and secondarily by inadequate theoretical guidelines.

2 Current Applications

To date, numerical simulations have been used to study mildly nonlinear effects (finite amplitude states, secondary instabilities, vortex pairing, low-order bifurcations, wave interactions); strongly nonlinear effects (vortex roll-up, eddy shocklets,temporal chaos, nonlinear breakdown); and incipient turbulence (final stages of transition, turbulent spots).

Mildly nonlinear effects can be addressed with low resolution calculations (using as few as 10^3 grid points). Finite-amplitude, steady traveling TS waves have been reproduced in CF and Taylor vortices have been simulated for TCF; secondary instabilities have been mimiced in numerical simulations of CF, BLF, and FSF. Moreover, direct simulations have recently confirmed dramatic TS-Gortler interactions in curved CF. For these problems, however, time-dependent direct simulations are relatively inefficient as a primary tool, requiring at least a substantial fraction of an hour for a single case. Parametric studies are far more efficiently conducted using simpler methods such as weakly nonlinear theory and Floquet theory. Direct simulations of these problems are now best employed in a supporting role to check the results of nonlinear theories, both to confirm that the theory (and the simulation) were done correctly for low amplitudes and to establish the limits of the theory at large amplitudes.

One mildly nonlinear application for which direct simulations need take a leading part is in establishing confidence in the dominant instability of a flow. Linear and weakly nonlinear theories generally make some assumptions about the type of waves that are present; nonlinear simulations which begin from random initial conditions permit the most physically significant instability to emerge and are, therefore, a good check on whether some unsuspected instability has been ignored in the theory: This class of computation calls for between 10^4 and 10^5 grid points.

Direct simulations necessarily are the primary tool when nonlinear effects are stronger than can reasonably be handled by existing asymptotic methods. For TCF this means the flow regimes beyond the Taylor vortex and one wavy vortex stages, and for BLF and CF it means the laminar breakdown and subsequent stages. Such dramatic calculations, however, require over 10^6 grid points and consume many tens of hours of computer time.

Transition simulations (as well as theories) can be divided into those which employ a periodic approximations, such as the parallel flow approximation in BLF and FSF and the infinite axial extent assumption in TCF, and those which are free from this constraint, such as channel flow, true growing boundary-layer flows and finite-aspect-ratio TCF. The computational savings of the parallel flow approximation are most dramatic for BLF since spatial simulations require either an order of magnitude more grid-points in the streamwise direction with a standard streamwise grid, or else a strongly non-uniform streamwise grid coupled with a more complex and expensive algorithm. The effect of boundary-layer growth can be accounted for within the parallel flow approximation, although the non-parallel effects themselves must largely be ignored. Direct numerical simulations of BLF using the parallel flow approximation have been surprisingly successful in reproducing observed phenonema: development of lambda vortices as well as detached shear layers and their subsequent roll-up in TS-induced transition, advanced stages of crossflow-induced transition, and even the evolution of turbulent spots. Indeed, these successes have seemingly pushed non-parallel effects so far into the background for some problems that one must ask "where are the non-parallel effects?" However, non-parallel simulations are likely necessary for studying Gortler-induced transition as well as most receptivity issues in BLF and they are surely quite important for supersonic flows.

3 Algorithmic State-of-the-Art

The characteristics of existing algorithms for transition simulations warrant the following conclusions:

1. Numerical simulations of transition should be at least fourth-order accurate in space. Higher-order methods, such as sixth-order compact schemes and spectral methods, do have their advocates. However, a definitive answer to the question of the most computationally efficient method is unavailable and is almost surely unattainable.

2. Computer storage limitations (2 time levels) constrain temporal accuracy to be third-order for explicit terms and second-order for implicit terms. No method has yet been found for achieving higher than third-order accuracy in time with acceptable computer storage. The resulting loss of accuracy of the

small wavelength components of the solution is an embarrassment, with unknown consequences for the temporal resolution of small-scale structures.

3. Transition simulations have been confined to methods employing tensor product and domain decomposition grids. Neither unstructured nor adaptive grids have yet been seriously employed.

4. The greatest numerical resolution is required in the last transitional stages and not in the final turbulent state.

5. The key to efficient algorithms for incompressible flow is the capability of solving implicit equations rapidly. This can be done by direct methods only in very simple geometries. Iterative methods have been incorporated in slightly more general geometries.

6. Compressible algorithms are currently explicit, which restricts their time-steps to be small and their run times long compared with incompressible algorithms.

7. Several impressive demonstrations of parallel processing on key components of transition algorithms have recently appeared.

4 Pacing Items

Algorithmic improvements which would extend the range of problems amenable to numerical simulation are:

1. General three-dimensional multigrid solvers for high-order discretizations of Poisson/Helmholtz equations. Multigrid technology has developed to point where this is an achievable goal. What is needed is a concentrated effort to incorporate this admittedly complex scheme into numerical simulation codes.

2. Efficient implicit methods for compressible flow. Both acoustic and viscous limitations are unnecessarily restrictive for explicit methods. Overcoming this appears to require the use of block implicit methods, i.e., a coupled discretization of the dependent variables.

3. High resolution shock-capturing algorithms for hypersonic flow. It is unrealistic to expect to resolve the bow shock at high Mach numbers. Modern upwind schemes need to be tuned for time-dependent transition problems rather than steady-state ones.

4. Effective data compression techniques as well as vastly increased mass storage capacity.

5. Effective asymptotic, or far-field boundary conditions for external flows.

6. Reliable outflow boundary conditions, particularly for incompressible flow.

7. Rigorous numerical analysis of splitting methods, especially for incompressible flow. This is admittedly a numerical analysis task. However, splitting methods are substantially faster than unsplit methods and better correction schemes for splitting errors should improve the performance of simulation algorithms.

5 Beckoning Problems

With existing supercomputers, the following problems are manageable, in the sense of requiring several hours per run in a parameter study, or several hundred hours for a single benchmark calculation:

1. Fully turbulent TCF (with rotating outer cylinder).

2. Spatial simulations of Gortler vortices in curved BLF.

3. Wave interactions in the presence of random noise and/or strongly nonlinearities.

4. Studies of convective versus absolute instabilities.

5. Initial evolution of obstacle-induced roughness transition.

6. Weakly nonlinear effects in compressible BLF.

6 Horizon Problems

Assuming a steady evolution of computational power and the algorithmic improvements mentioned above, the following problems should prove amenable to numerical simulations in the early 1990's.

1. Natural transition to turbulence under the parallel flow approximation in BLF and CF. The key effect here is the spontaneous emergence of turbulent spots from small rather than large disturbances.

2. Forced transition to turbulence for non-parallel BLF.

3. Transition in non-parallel three-dimensional boundary layers.
4. Roughness-induced transition to turbulence. Theoretical guidelines would be most helpful in reducing the parameter space.
5. Nonlinear breakdown in compressible transition. There is a special interest in this at high Mach number, for which there is no experimental evidence of turbulent spots.
6. Transition in separated flows.
7. Surface juncture effects, such as occur on aircraft.
8. Transitional phenomena effected by interactions with large-scale, strong shock waves.
9. Ablation effects on transition in high-speed flows.

RECEPTIVITY

PANEL SUMMARY: RECEPTIVITY

Eli Reshotko

Kent H. Smith Professor of Engineering
Case Western Reserve University
Cleveland, OH 44106

Much progress has been made over the last six years or so in the delineation and understanding of receptivity phenomena. This has come about principally through detailed analysis of a few critical experiments using the results of related asymptotic analyses. While we can now give some fairly detailed descriptions of *some* receptivity mechanisms, we are also impressed by the myriad of combinations of disturbance type and receptivity mechanism that can come into play in the transition process. All of the presentations of this panel dealt with boundary layer flows.

In Reshotko 1976, I stated: "...understanding of the mechanisms by which a forcing disturbance of a given frequency and prescribed phase velocity excites a free disturbance of the same frequency but different phase velocity is of great importance." It has now become clear that strong inhomogeneities of the mean flow lead to local regions where disturbances acquire the T-S wavelength and couple to the T-S wave. Thereafter they are mismatched and simply superpose. For boundary layers, the principal regions are the leading edge, and also at any downstream location where the boundary layer has to adjust to sudden changes in surface geometry or boundary conditions. For example, for a flat plate with a real (elliptical) leading edge, the rapid non-parallel growth of the boundary layer, the rapid profile adjustments due to varying pressure gradient, the change in surface curvature at the junction of the leading edge with the flat plate, can all result in coupling between free-stream vorticity and acoustic disturbances and T-S waves. In addition, the curvature of the leading edge can result in diffraction of the acoustic field. Thus real leading edges provide many receptivity mechanisms. What is their accumulated effect? Which one if any is dominant? The answer to these latter questions may in the end sharpen up our approaches to transition prediction.

Receptivity experiments taken alone have generally been difficult to interpret, often inconclusive and even confusing. The results of studies using analytical techniques based on triple-decking have proven very helpful in understanding mechanisms and in the intrepretation of the experiments. The Goldstein and Kerschen groups have as of now solved numerous cases for both acoustic and vortical free-stream disturbances. Since the analyses are linear, receptivity to convected turbulent fields might well be modeled by three-dimensional superposition of the convected gust disturbances treated by Kerschen. Computational Navier-Stokes approaches have not yet contributed significantly to the understanding of receptivity phenomena. Morkovin suggests that forced wave-packet techniques (experimental and theoretical) may be useful in receptivity studies since they provide the opportunity to distinguish in a spatio-temporal sense between the forcing signal and the T-S response.

Progress to date has been principally for incompressible two-dimensional flows, and then primarily to two-dimensional disturbances. The needs for extension to compressible flows, to three-dimensional flows, and to consideration of three-dimensional disturbances are apparent. The role of receptivity in the initiation of secondary instabilities must also be addressed.

Bushnell has commented at length about the sparseness of disturbance information for the flight environment. He points out that aside from acoustic and vortical disturbance fields, attention must be given to influences on the transition process of particulate matter in the flow, surface roughness and electrostatic discharges. The latter might be of some significance for supersonic and hypersonic flows.

Clearly there is much work yet to be done.

References

[1] Reshotko, E., 1976. Boundary-Layer Stability and Transition. Annual Review of Fluid Mechanics, Vol. 8, pp. 311-349.

RECEPTIVITY - AN OVERVIEW

Eli Reshotko

Kent H. Smith Professor of Engineering
Case Western Reserve University
Cleveland, OH 44106

The topic of receptivity and its importance were suggested twenty years ago by Morkovin (1969). There was relatively little early progress. The beginnings of receptivity studies are summarized in a paper published just five years ago (Reshotko, 1984). However, the content of that paper has been largely superseded by developments within the last five to six years that are highlighted in this panel presentation.

Receptivity differs from stability both physically and mathematically. Physically, it describes the signature in the boundary layer of some externally imposed disturbance and the means by which normal modes may be excited. Mathematically, the problem is non-homogeneous either in the equations or boundary conditions. Hence in contrast to the normal modes stability evaluation, receptivity is not an eigenvalue problem. The boundary layer is driven by the external forced oscillations, and its response is a "neutral" solution of the linearized disturbance equations having the same frequency as the particular forcing disturbance being considered. Each receptivity case has its own particular inhomogeneity.

Receptivity is of course studied both experimentally and analytically. The combination of the two is clearly most desirable. The experiments are directed at understanding the receptivity of a particular geometry to a particular class of environmental disturbances and many of the available analyses are attempts at direct modeling of the experiments.

In the presentations that follow, Bushnell addresses the plethora of disturbance environments of concern both in wind tunnels and in flight, Morkovin presents much of the basic phenomenology as gleaned principally from experiments, Leehey follows with more specific details of the experiments in acoustic receptivity done by his group at MIT. The session ends with presentations by Hultgren (in place of M. E. Goldstein) and Kerschen on the applications of modern asymptotic analysis in the modeling and understanding of receptivity to acoustic and vorticity disturbance fields. A presentation by Kendall, giving in conjunction with the roughness panel, on the transition on a flat plate due to disturbances from a rodded array in the tunnel settling chamber, should be considered as part of the program of this panel.

References

[1] Morkovin, M. V., 1969. Critical Evaluation of Transition from Laminar to Turbulent Shear Layers with Emphasis on Hypersonically Traveling Bodies. AFFDL-TR-68-149, Air Force Flight Dynamics Lab., Wright-Patterson AFB, Ohio.

[2] Reshotko, E., 1984. Environment and Receptivity. In Special Course on Stability and Transition of Laminar Flow, AGARD Report No. 709, pp. 4-1 to 4-11.

NOTES ON INITIAL DISTURBANCE FIELDS FOR THE TRANSITION PROBLEM

Dennis Bushnell

NASA Langley Research Center
Hampton, VA 23665-5225

1. Introduction

The transition process is a 6-stage affair which is initiated by the ambient disturbance environment. These initial disturbance fields, which can be either stream- or body-generated (or a combination thereof) are processed by the body inviscid and viscous flow fields. The resulting disturbances then constitute the initial and bounding values for the subsequent linear and nonlinear growth processes which result, ultimately, in final breakdown to turbulence, e.g., the location of transition per se. The existing state of the art in transition estimation, the e^N method (e.g., Reference 1), deals only with the amplification ratio as given by linear theory and hence can only estimate (generally to better than 0(20 percent) currently) transition location when the initial disturbance fields are the equivalent of a "dull hum," i.e., 0(.05 percent), as they appear to be in quiet, low- disturbance wind tunnels and many atmospheric flight situations. Improved prediction accuracy and/or predictions for non-canonical disturbance fields requires solution of transition as an initial-boundary value problem (via numerical simulations) and specification, for instance in flight, of the amplitude, spectra, orientation, and mode of all initial disturbance fields as a function of latitude, longitude, altitude, and time along the flight path. These notes summarize some of our present, vastly incomplete, information concerning the qualitative nature of this initial disturbance problem. Detailed quantitative estimates of these various fields are simply not available.

In most applications, more than one type of disturbance which can affect transition is present at any time, and their effects have often proven to be synergistic in the few combined disturbance cases studied. Roughness and waviness can affect boundary-layer stability in many ways, only a few of which (over particular parameter ranges) correspond to acting as a disturbance source. In general, and in the absence of intense (bypass) amplitudes, the disturbance fields which are of the most concern to transition involve frequencies which correspond to these which are amplified in the normal mode processes. Since the normal modes can include any or all of the following: (a) Tollmien-Schlichting waves, (b) cross-flow disturbance vortices, (c) Taylor-Görtler instabilities, and (d) embedded Kelvin-Helmholtz waves and receptivity processes can alter the incident wave fields, the "dangerous frequency" range can vary from zero to quite high values depending upon the mode(s) excited, which in turn is a function of the particular body geometry and flow condition under study.

Any discussion of stream and vehicle-induced disturbance fields and their internalization by the vehicle inviscid and viscous flow fields vis-à-vis the transition problem should consider the various possibilities associated with high (e.g., supersonic/hypersonic) and subsonic speeds and the various experimental apparatus employed, e.g., conventional wind tunnels of several operational types, heated wind tunnels, free-flight ranges, atmospheric free flight, water tunnels and tow tanks, and open ocean testing, as the sources, types, and resulting effects of stream and vehicle-induced disturbances (and vehicle flowfield processing thereof) can be quite different in many of these cases. Initial disturbance fields can be categorized and discussed either with respect to facility type (and the various disturbance modes which occur therein) or mode of disturbance (and the variants which occur for each test technique). The former is the more usual approach and convenient for the interpretation of transition experiments, while the latter is perhaps better suited to correlation with theory. Herein, we adopt the second approach. It should be noted that, in general, except for body vibration at low speeds (References 2, 3), attempts to determine the influence of wall vibration upon transition have yielded a null result, i.e., no discernible influence (References 4-8), except possibly for Reference 9. Another possibly dominant disturbance in Reference 9 could have been electrostatic discharge, as discussed herein.

2. Stream Turbulence

Four fluctuation modes are generally included under the category of stream turbulence: (1) vorticity fluctuations, (2) entropy disturbances or temperature spottiness, (3) noise, and (4) concentration fluctuations. The internalization of such disturbances can be quite different depending upon body geometry and speed. At subsonic speeds, the body inviscid flow effects can be approximated to first order by rapid distortion theories, whereas in the high-speed case the disturbances are processed by the body shock which can cause both amplification and intermodal transfer of disturbance energy (e.g., Reference 10). For embedded subsonic regions (e.g., blunt noses, every vehicle is blunt to some degree), the resultant acoustic modes can reflect off the body and back onto the shocks (References 11, 12) and may be re-reflected, etc., which, when one considers the relatively broad-band nature, random phasing, and amplitudes of the incident disturbance fields, may allow, in some instances, higher frequencies to form due to sums and differences of interacting and reflecting waves. Similar effects could also result from body vibration-induced acoustic fields. Incident concentration fluctuations could, at ($M \gg 1$), cause additional pressure perturbations via chemical reactions. Incident perturbations the order of the vehicle size can also cause vehicle motions which would have a first-order influence on the mean viscous flow field and therefore the normal modes.

The sources and thus the amplitude, spectral content, "organization/structure," and orientation of stream "turbulence" is a function of the flow medium/test apparatus. In the atmosphere and ocean, the usual sources are instabilities associated with "natural convection," e.g., instabilities in (thermal and/or salinity) stratified flows. Although the scales of the atmospheric and oceanic ambient turbulence are generally large, there is still some residual energy at higher frequencies (0.1m kolomogorov microscale at 0(60K ft)) (References 13, 14), and the high wavenumber content is generally increased in the vicinity of embedded localized shear layers such as the jet stream, weather fronts, ocean currents, the benthic and atmospheric boundary layers, and the proximity of the ocean surface where surface-wave perturbations, including small-scale capillary waves, can influence the ocean below the immediate surface layer and also the atmospheric flow above the surface. Clear air turbulence observations extend to 200,000-ft altitude. Intermittent atmospheric small-scale events are also possible (from internal wave breaking). References 15-22 provide typical atmospheric/oceanic ambient turbulence data.

In conventional wind tunnels, the sources of stream turbulence are multitudinous an depend upon the facility design and whether it is open or closed circuit, high speed, transonic, or low speed, and any heating processes. The stream turbulence scales are generally closer to those of the flow under study in the wind-tunnel case. The usually dominant stream- disturbance modes are vorticity and acoustic fluctuations. Sources of vorticity disturbances include ambient air turbulence (e.g., atmospheric boundary-layer motions, building air handling systems) processed by the contraction (open circuit), unsteady separated flow regions in the various diffusers and honeycomb/screen wakes (utilized to control the vorticity mode, Reference 23). The tunnel flow upstream of the test section can also be utilized to reduce the stagnation chamber vorticity fluctuation field, especially effective at high speeds (Reference 24). Unfortunately, for heated tunnels, particularly the ones generally utilized in combustion research, the upstream expansion process can *create* vorticity fluctuations through processing of the stagnation chamber temperature spottiness (Reference 25).

Heroic efforts to reduce stream fluctuating vorticity levels merely allows the next disturbance in order of importance to assume dominance of the transition process and in the low-speed case (Reference 25) as well as high speeds (Reference 27), this is usually the acoustic field. The slots and perforations utilized in transonic tunnels to absorb shock waves and other disturbances act as a particularly virulent source of acoustic disturbances (Reference 28). For the flight case, acoustic fields which can affect transition are generated by engines (compressor, jet, and shock noise) and airframe (sources include separated regions and turbulent boundary layers) (References 29, 30). Of particular interest are observations that the (usually stationary) cross-flow instability modes are not particularly influenced by sound except in the presence of roughness, a clear case of the synergism of combined disturbance fields (e.g., Reference 31 and unpublished NASA Langley Research Center data). At supersonic and hypersonic speeds, where (1) the free-stream vorticity fluctuations are small, (2) the small throat size limits the transmission of upstream noise, and (3) the flow speed mitigates against the upstream transmission of downstream (e.g., diffuser) noise the major source of stream disturbances is acoustic radiation from the turbulent nozzle wall boundary layers. The resulting disturbance levels can be tremendous (0(5 percent) in terms of $(P'_\infty/\overline{P_\infty})$ at higher Mach numbers and dominate the transition process (Reference 32). The only known cure for this particular transition "spoiler" is to establish laminar boundary layer flow on the facility nozzle walls at the acoustic origin of the test-flow region (Reference 33). Even then, care must be taken to avoid oscillating Mach waves (finite amplitude sound waves) induced by the convection of residual wall boundary-layer disturbances past "roughnesses." Such roughnesses often result from deposition of stream particulates which, as indicated in the next section, can themselves cause transition-inducing disturbances directly.

3. Particulates

Historically, particulates have been utilized or observed to damp disturbance fields, e.g., rain or water sprays are used to control acoustic fields/turbulence and particulates can provide turbulent skin-friction reduction in wall flows. For the transition problem, however, particulates can be a major source of initial disturbance energy. Sample experiences include for RV's, "Anomalous dispersion has been observed when the vehicles fly through thin, high altitude ice clouds" (Reference 34), on aircraft, "LFC performance (on the X-21) was observed to be partially degraded or erratic when penetrating light cerrius haze, even where the horizontal visibility as much as 50 miles" (Reference 35) and, in a hypersonic wind tunnel, "The only disturbances which were observed to induce transition in the stagnation region were those associated with minute dust particles in the flow" (Reference 36). Additional experiences regarding particulate- induced effects on transition are described in Reference 37 and 38.

The mechanisms by which particulate matter can affect transition are manifold and include (1) impacting/sticking to surface (roughness effect), (2) vortex/vorticity shedding while particle is immersed in or external to the boundary layer (function of several parameters including ratio of particle-to-flow density), (3) particle rotation and consequent fluid motions, induced by the boundary-layer mean shear at high speeds, (4) possible production of "reverse shocklets" within the vehicle flow field after passage through the vehicle shock, (5) rebound after impact on the vehicle surface and subsequent dynamic interaction with the vehicle bow shock itself causing formation of embedded shear layers and jets, and (6) surface "cratering" (roughness formation) due to particle impact. The available analyses of disturbance modes (2) and (3) indicate that the particle size should be an appreciable fraction of the boundary-layer thickness (References 39, 40). The physics of the high-speed case is considerably richer. The general problem of high-speed particle-body interaction has been studied in considerable detail, spurred mainly by the influence of flight through rain, dust, etc., on RV accuracy. This work (e.g., References 41-44) involves primarily rather large particles which are typically present at lower altitudes and which can impact the body, perhaps rebound and/or produce "ejecta" (erosion) from the body itself and alter the body-shock dynamics to, in many cases, first order. The case of smaller particles which, while they do not alter gross shock dynamics, can still influence transition, has not been studied to any great detail except for References 45 and 36. Reference 45 indicates, from disturbance energy considerations, that quite small particulates can provide a ready mechanism for "unit Reynolds number" effects in flight. Obviously, particulate matter is present, to a greater or lesser extent, in all test environments unless special filtering techniques are employed as in the Langley Supersonic Pilot Quiet Tunnel. A disturbance mechanism involving particulate shocklets and unique to high-speed flows containing particulates does not appear to have been noted or analyzed thus far. The mechanism is due to the greater particle inertia vis-à-vis the gas flow. Upon crossing the vehicle shock, the particle may not immediately slow down as does the gas and therefore the particle can, depending upon Knudsen number and shock strength as well as other parameters, create a "reverse particulate shocklet" while it decelerates to the local gas velocity. This dynamic shocklet provides a mechanism whereby a particle in the vehicle flow field, even outside the boundary layer, can induce disturbances within the near-wall region, i.e., in this scenario the particle does not have to be embedded within, or even be near, the boundary layer to induce transition. Estimates for the realization distance involved are available, for example, from References 46 and 47 and can extend the order of 10cm downstream of the shock at hypersonic speeds. Particulate-induced chemical reactions and their subsequent influence upon transition could also become first order depending upon particulate composition and velocity.

The occurrence of particulate matter is much better documented for the atmosphere than for experimental ground facilities. At high altitudes, (mesopheric, 0(80Km)) noctilucent clouds can occur in the upper latitudes. The clouds are composed of cosmic dust coated with ice with resultant diameter the order of 0.1 micron (e.g., References 20, 48). For lower altitudes, the particles are terrestrial in origin including the huge sizes associated with rain-inducing clouds. The non-water particulate concentration in this altitude range (Reference 49) is a function of volcanic activity (e.g., References 20 and 50-52). Of particular interest is a finding that a substantial portion of the particulate matter in certain altitude ranges having sizes of 10μ or greater (i.e., sizes of concern for transition) consists of the products of rocket exhaust. As would be expected, the atmospheric particulate data are documented best in the altitude range with maximum human commercial utilization, conventional takeoff and landing (CTOL) transport routes. Such data from the Global Atmospheric Sampling Program (GASP) effort is documented in References 53. For the wind tunnel case, the possible influence of LDV seeding upon model transition location may be of concern.

4. Electrostatics

For years, electrostatic discharges have created a variety of problems on aerospace vehicles (e.g., References 54-56). These problems range from only an annoyance to loss of life and entire vehicles. Typical effects studied include biological shock, premature ordinance ignition, in-flight fires and explosions, surface coating degradation, sensor errors, communication and navigation disruption, and digital computer errors. However, up to this point, *evidently no consideration has been given to the possibility of electrostatic discharges initiating transition in the external flow.* Various anomalous transition observations, in particular the "blunt body paradox" associated with the Mark 2 vehicles, may be connected with this electrostatic discharge phenomena.

The vehicle charging occurs due to two major sources, flight through particulates and propulsion-related processes. The static electrical field buildup on the vehicle can exceed 300kv and, on commercial aircraft, is generally more or less continuously discharged via various sharp-tipped devices to avoid "ball-lightening" discharge and resultant interference with communication/electronics. Of particular concern are dielectric surface such as coverings, radomes, antennas, and windshields, where charge can accumulate with discharge, generally via corona mechanisms, occurring to adjacent conducting surfaces. Typical discharge frequencies are 200 kHz to 30 mHz which is within the critical disturbance frequency range for many high-speed flows. At very high speeds where a plasma forms over the vehicle, the surface charge is dissipated by this external flow.

From Reference 57, "the primary interaction between the electrical discharge and the flow is through heating of the gas. The gas at the wall expands drastically and moves away from the cathode causing streamline deflection in the boundary layer. In supersonic flow, an oblique compression wave is developed at cathode leading edges through which the flow is turned." Reference 58 indicates that electrical discharge can certainly constitute a "bypass" transition trip.

5. Overall Disturbance Levels in Flight vs. Wind Tunnel

Research such as described in Reference 59 indicates the first-order influence of stream-disturbance level on transition. Since the lowest practical disturbance level attainable in conventional ground facilities is on the order of 0.05 percent, an overall measure of disturbance level in flight is of interest for purposes of crudely scoping the relationship and fidelity of such low-disturbance facilities to the actual fight environment and thereby obtaining, by inference, some expectation of the relationship between transition in flight and low-disturbance tunnels. To the best knowledge of the present author, only two flight-transition studies are available with simultaneous stream-disturbance information (References 60 and 61). The measurements of Reference 60 were obtained using a hot wire mounted on the nose of a low-speed glider (no engine noise) and yielded turbulence intensities in the range of 0.2 percent (with most of the signal at low frequencies below 200 Hz) compared to a measured most unstable vehicle boundary-layer frequency in the range of 600 Hz. The measured free-stream disturbance level increased in the presence of clouds.

In the second study (Reference 61), fluctuating pitot pressure data and cone transition data were obtained ahead of a powered fighter aircraft for both subsonic and supersonic conditions. Of particular interest in this data are the large reductions in the stream disturbances which occur when the aircraft goes supersonic, an indication that much of the disturbance field is acoustic, generated by either the engine or airframe, and radiated forward onto the test model at subsonic conditions. The supersonic flight speed stops this forward radiation and, therefore, the measurements at high speeds must be due either to disturbances convected into the probe face or probe vibration. The relatively broad-band nature of the high-speed disturbance data, with appreciable signal out to 20 kHz, is astonishing. The overall level of the stream disturbances measured in this flight program are the order of what is obtainable in the better ground facilities, both at low and high speeds and the measured transition locations correspond to data obtained in such low disturbance facilities. This finding suggests that the transition processes in flight and in low-disturbance tunnels may be of the same order, in the absence of "by-passes" or any of the myriad concatenations with which the transition problem is rife due to its tremendous parameter space.

6. Concluding Remarks

The *mean* ambient atmospheric conditions can change from day to day up to the order of 50 percent (at altitude), and therefore a zeroth-order consideration for examination of flight transition is accurate (simultaneous) mean-flow data along the lift path. Lacking this, information from the GRAM atmospheric model can be used which includes the usual (nominal) excursions for time of year and latitude/longitude (as well as altitude). The "standard atmospheric tables" are, in general, probably not accurate enough for transition estimation/data reduction.

For the flight case, our current, grossly incomplete knowledge of initial disturbance fields indicates the following:

(a) Stream fluctuation fields $(U'_\infty, P'_\infty, C'_\infty, T'_{infty})$ - microscale data indicates transition coupling may be possible up to 0(40k ft) with intermittent small scale events (internal wave breaking) extending this to higher altitudes.

(b) Particles - sources include ice clouds, volcanic and other terrestrial dust, rocket exhaust particles (from previous flights) and cosmic dust. Knudsen number and number density estimates by R. Davis of NASA Langley Research Center indicates coupling possibilities to 0(70k ft to 80k ft).

(c) Electrostatic - surface charging occurs due to particulate interactions and propulsion processes and is generally exacerbated on dielectric coatings and surface segments such as radomes and antennas with measurable high frequency discharges (220 kHz to 30 mHz) observed to 0(100k ft).

(d) Acoustic radiation from own vehicle (airframe/engine noise) - function of vehicle design, location on vehicle and speed, especially important for propulsive surfaces.

(e) Roughness - simplex guidelines as to allowable roughness sizes available via "roughness Reynolds numbers," of critical importance is direct coupling into stationary disturbances such as cross-flow and Görtler vortices. Can also exacerbate other disturbances including acting as a discharge site for electrostatic effects.

(f) Brownian motion - always present and would probably induce transition (eventually) if other initial perturbation fields were nulled.

References

[1] Bushnell, D. M.; Malik, M. R.; and Harvey, W. D.: Transition Prediction in External Flows via Linear Stability Theory. IUTAM Symposium Transsonicum III, Gottingen, West Germany, May 24-27, 1988. Springer-Verlag, 1989. (Also, NASP Technical Memorandum 1021.)

[2] Gilev, V. M. and Kozlov, V. V.: Excitation of Tollmien-Schlichting Waves in the Boundary Layer on a Vibrating Surface. *J. Applied Mech. and Tech. Phys.*, Vol. 25, No. 6, November/December 1984, pp. 874-877.

[3] Pilipenko, A. A. and Shapovalov, G. K.: Effect on Model Vibration on Transition form Laminar to Turbulent Boundary Layer Flow. *Fluid Mechanics - Soviet Research*, Vol. 17, No. 4, July-August 1988, pp. 68-76.

[4] Final Report on LFC Aircraft Design Data Laminar Flow Control Demonstration Program. NOR 67-136 (Contract AF 33(657)-13930), Northrop Corp., June 1967.

[5] Kendall, J. M.: Wind-Tunnel Experiments Relating to Supersonic and Hypersonic Boundary-Layer Transition. Presented at the 12th Aerospace Sciences Meeting, Washington, DC, January 1974, AIAA Paper No. 74-133.

[6] Potter, J. L.: Boundary Layer Transition on Supersonic Cones in an Aeroballistic Range. AIAA Paper No. 74-132.

[7] Fischer, M. C. and Wagner, R. D.: Transition and Hot-Wire Measurements in Hypersonic Helium Flow. *AIAA J.*, Vol. 10 No. 10, October 1972, pp. 1326-1332.

[8] Olson, Lawrence E.; Gregorek, Gerald M.; and Lee, John D.: The Influence of Artificially Induced Turbulence upon Boundary Layer Transition in Supersonic Flows. Aerospace Research Laboratory Report No. 71-0022, January 1971.

[9] Lemcke, Bo; Naysmith, A.; Picken, J.; and Thomann, H.: Comparison of Heat Transfer Measurements in Free Flight and in a Wind Tunnel at $M = 7$ at Similar Reynolds Numbers and Temperature Ratios. The Seventh Congress of the International Council of the Aeronautical Sciences, Rome, Italy, September 14-18, 1970. ICAS Paper No. 70-06.

[10] Anyiwo, J. C. and Bushnell, D. M.: Turbulence Amplification in Shock Wave Boundary Layer Interactions. *AIAA J.*, Vol. 20, 1982, pp. 893-899.

[11] Morkovin, Mark V.: Transition at Hypersonic Speeds. ICASE Interim Report No. 1, NASA CR-178315, May 1987.

[12] Morkovin, M. V.: Note on the Assessment of Flow Disturbances at a Blunt Body Travelling at Supersonic Speeds Owing to Flow Disturbances in Free Stream. *J. of Applied Mechanics*, June 1960, pp. 223-229.

[13] Balsley, Ben B. and Green, John L.: Coherent Radar Systems for Probing the Troposphere, Stratosphere, and Mesosphere. Fourth Symposium on Meterological Observations and Instrumentation, Denver, Colorado, April 10-14, 1978.

[14] Gage, K. S. and Balsley, B. B.: Doppler Radar Probing of the Clear Atmosphere. *American Meterological Society*, Vol. 59, No. 9, September 1978, pp. 1074-1093.

[15] Thrane, E. V.; Blix, T. A.; Hall, C.; Hancen, T. L.; von Zahu, U.; Meyer, W.; Czechowsky, P.; Schmidt, G.; Widdel, H.-U.; and Neumann, A.: Small Scale Structure and Turbulence in the Mesosphere and Lower Thermosphere at High Latitudes in Winter. *J. of Atmospheric and Terrestrial Physics*, Vol. 49, Nos. 7/8, 1987, pp. 751-762.

[16] MacCready, Paul B., Jr.: Turbulence Measurements by Sailplane. *J. of Geophysical Research*, Vol. 67, No. 3, March 1962, pp. 1041-1050.

[17] Weinstock, Jerome: Vertical Wind Shears, Turbulence and Non-Turbulence in the Troposphere and Stratosphere. Geophysical Research Letters, Vol. 7, No. 10, October 1980, pp. 749-752.

[18] Waco, David E. and Ashburn, Edward V.: Turbulence Variations during the High Altitude Clear Air Turbulence (HICAT) Program. *J. of Aircraft*, Vol. 10, No. 1, January 1973, pp. 56-58.

[19] Delay, Richard D. and Dutton, John A.: An Analysis of Conditions Associated with an Occurrence of Stratospheric CAT. *J. of the Atmospheric Sciences*, Vol. 28, October 1971, pp. 1272-1279.

[20] *Handbook of Geophysics and the Space Environment.* Scientific Editor, Adolph S. Jursa, Air Force Geophysics Laboratory, 1985.

[21] Caldwell, Douglas R.: Small-Scale Physics of the Ocean. *Reviews of Geophysics*, Vol. 25, No. 2, March 1987, pp. 183-192.

[22] Tieleman, H. W.: A Survey of the Turbulence in the Marine Surface Layer for the Operation of Low-Reynolds Number Aircraft. Department of Engineering Science and Mechanics, VPI& SU, Blacksburg, VA, March 1985, ONR Contract No. N00014-83-K-0658.

[23] Morkovin, Mark V.: On Transition Experiments at Moderate Supersonic Speeds. *J. of Aeronautical Sciences*, July 1957, pp. 480-486.

[24] Laufer, John and Marte, Jack E.: Results and a Critical Discussion of Transition Reynolds Number Measurements on Insulated Cones and Flat Plates in Supersonic Wind Tunnels. JPL Report No. 20-96, November 1955.

[25] Goldstein, M. E.: Turbulence Generated by the Interaction of Entropy Fluctuations with Non-Uniform Mean Flows. *J. Fluid Mech.*, Vol. 93, Part 2, 1976, pp. 209-224.

[26] Leehey, Patrick: Influence of Environment in Laminar Boundary Layer Control. **Viscous Flow Drag Reduction**, edited by Gary R. Hough, Vol. 72, of *Progress in Astronautics and Aeronautics*, 1980.

[27] Pate, Samuel R.: Effects of Wind-Tunnel Disturbances on Boundary-Layer Transition with Emphasis on Radiated Noise. A Review. 11th Aerodynamic Testing Conference, Colorado Springs, Colorado, March 1980, AIAA Paper No. 80-0431.

[28] Varner, M. O.: Noise Generation in Transonic Tunnels. AIAA 8th Aerodynamic Testing Conference, Bethesda, MD, July 8-10, 1974, AIAA Paper No. 74-633.

[29] Charmichael, R. F. and Pelke, D. E.: In-Flight Noise Measurements Performed on the X-21A Laminar Flow Aircraft. Northrop Corporation, NOR-64-81, April 1964.

[30] Mangiarotty, R. A.: Effects if Engine Noise on Aircraft Wing Laminar Boundary-Layer Stability. *J. Acoust. Soc. A*, 70 (1), July 1981, pp. 98-109.

[31] Rooney, T. R.; Charmichael, R. F.; and Eldred, K. M.: Investigation of Noise with Respect to the LFC, NB-66 Aircraft. Northrop Corporation, NOR-61-010, April 1961.

[32] Dougherty, N. Sam, Jr.: Influence of Wind Tunnel Noise on the Location of Boundary-Layer Transition on a Slender Cone at Mach Numbers from 0.2 to 5.5. *Volume 1 - Experimental Methods and Summary of Results.* AEDC-TR-78-44, Vol. 1, March 1980.

[33] Beckwith, Ivan E.; Creel, Theodore, R., Jr.; Chen, Fang-Jenq; and Kendall, James M.: Free-Stream Noise and Transition Measurements on a Cone in a Mach 3.5 Pilot Low-Disturbance Tunnel. NASA TP-2180, September 1983.

[34] Donaldson, Coleman duP. and Sullivan, Roger D.: A Discussion of the Effects of Surface Roughness, Tunnel Noise, and Free-Stream Vorticity on Transition and Heat Transfer on Blunt Bodies. Aeronautical Research Associates of Princeton Report No. 411, January 1980.

[35] Davis, Richard E. and Fischer, Michael C.: Cloud Particle Effects on Laminar Flow in the NASA LEFT Program: Preliminary Results. Presented at the AIAA/AHS/CAS/

DGLR/IES/ISA/ITEA/SETP/SFTE 3rd Flight Testing Conference, Las Vegas, NV, April 2-4, 1986, AIAA Paper No. 86-9811.

[36] Holden, Michael S.: Studies of Boundary Layer Transition and Surface Roughness Effects in Hypersonic Flow. AFOSR-TR-84-0251, October 1983.

[37] Holmes, Bruce J.; Obara, Clifford, J.; and Yip, Long P.: Natural Laminar Flow Experiments on Modern Airplane Surfaces. NASA TP 2256, June 1984.

[38] Lauchle, Gerald C. and Gurney, G. B.: Laminar Boundary-Layer Transition on a Heated Underwater Body. *J. Fluid Mech.*, Vol. 144, 1984, pp. 79-101.

[39] Hall, G. R.: On the Mechanics of Transition Produced by Particles Passing through an Initially Laminar Boundary Layer and the Estimated Effect on the LFC Performance of the X-21 Aircraft. Northrop Corporation Report, October 1964.

[40] Chen, C. Phillip; Goland, Yaccov; and Reshotko, Eli: Generation Rate of Turbulent Patches in the Laminar Boundary Layer of a Submersible. **Viscous Flow Drag Reduction**, Gary R. Hough, Editor, *Progress in Astronautics and Aeronautics*, Vol. 72, 1980, pp. 73-89.

[41] Hove, D. T. and Smith, A. A.: Holographic Analysis of Particle-Induced Hypersonic Bow-Shock Distortions. *AIAA J.*, Vol. 13, No. 7, July 1975, pp. 947-949.

[42] Holden, M. S.; Gustafson, G. Q.; Duryea, G. R.; and Hudack, L. T.: An Experimental Study of Particle-Induced Convective Heating Augmentation. Presented at the AIAA 9th Fluid and Plasma Dynamics Conference, San Diego, CA, July 14-16, 1976, Paper No. AIAA 76-320.

[43] Galdman, G. D. and Reinecke, W. G.: Particle Trajectories, Heating, and Breakup in Hypersonic Shock Layers. *AIAA J.*, Vol. 9, No. 6, June 1971, pp. 1040-1048.

[44] Sepri, P.; Chen, K.; and Pallone, A.: The Effect of Impact Ejecta on a Hypersonic Nosetip Environment. Presented at the AIAA

15th Aerospace Sciences Meeting, Los Angeles, CA, January 24-26, 1977, AIAA Paper No. 77-94.

[45] Pearce, Blaine E.; Sullivan, R. D.; and Donaldson, C. duP.: Cloud Particle Induced Boundary Layer Transition on Hemispherically Blunted Bodies in Supersonic Flight. A.R.A.P. Report No. 460, December 1981.

[46] Walsh, Michael J.: Influence of Particle Drag Coefficient on Particle Motion in High-Speed Flow with Typical Laser Velocimeter Applications. NASA TN D-8120, February 1976.

[47] Nicolt, W. E.; Wool, M. R.; Laub, B.; Jaffe, N. A.; and Berry, R. A.: An Investigation of Hydrometeor Shock Layer Interaction. Presented at the AIAA 9th Fluid and Plasma Dynamics Conference, San Diego, CA, July 14-16, 1976. AIAA Paper No. 76-318.

[48] Thomas, Gary E. and McKay, Christopher P.: On the Mean Particle Size and Water Content of Polar Mesospheric Clouds. Planet. *Space Sci.*, Vol. 33, No. 10, 1985, pp. 1209-1224.

[49] Swain, C. E.; Webber, W. T.; and Timmer, H. G.: Particle/Shock Layer Interaction in Hypersonic Reentry. Presented at the 13th Aerospace Sciences Meeting, Pasadena, CA, January 20-22, 1976, AIAA Paper 75-110.

[50] Gooding, James L.; Clanton, Uel S.; Gabel, Elizabeth M.; and Warren, Jack L.: El Chichon Volcanic Ash in the Stratosphere: Particle Abundances and Size Distributions after the 1982 Eruption. Geophysical Research Letters, Vol. 10, No. 11, November 1981, pp. 1033-1036.

[51] Miranda, Henry A., Jr.: Stratospheric Aerosol Sizes. Geophysical Research Letters, Vol. 1, No. 5, September 1974, pp. 201-203.

[52] Farlow, Neil H.; Ferry, Guy V.; Lem, Homer Y.; and Hayes, Dennis M.: Latitudinal Variations of Stratospheric Aerosols. *J. of Geophysical Research*, Vol. 84, No. C2, February 1979, pp. 733-743.

[53] Nastrom, Gregory D.; Holdeman, James D.; and Davis, Richard E.: Cloud-Encounter and Particle-Concentration Variabilities from GASP Data. NASA TP-1886, December 1981.

[54] Nanevicz, J. E.: Static Electricity Phenomena: Theory and Problems. Conference on Certification of Aircraft for Lightning and Atmospheric, Electricity Hazards. ONERA, Chatillon, France, (Proceedings) September 14-21, 1978.

[55] Wolfers, C. V.: Electrostatic Charge Phenomena Related to the Shuttle Orbiter Design. Space Shuttle Avionic System Engineering Support Task Order No. 00104 Design Note, December 19, 1974, McDonnell-Douglas Mechanical Services.

[56] Tanner, R. L. and Nanevicz, J. E.: Precipitation Charging and Corona-Generated Interference in Aircraft. Stanford Research Institute, AFCRL 336, TR-73, April 1961.

[57] Margalith, E. and Christiansen, Walter H.: Interactions between an E-Beam Sustained Discharge and Supersonic Flow. Proceedings of the 12th International Symposium on Shock Tubes and Waves, Jerusalem, July 1979.

[58] Klein, Enrique, J.: Excitation of Boundary-Layer Turbulence through Spark Discharges. NASA TN D-6378, May 1971.

[59] Dryden, H. L.: Transition for Laminar to Turbulent Flow. Turbulent Flows and Heat Transfer, Princeton University Press, 1959, pp. 3-74.

[60] Zanin, B. Yu.: Transition at Natural Conditions and Comparison with the Results of Wind-Tunnel Studies. Laminar-Turbulent Transition, Ed., V. V. Kozlov, IUTAM Symposium, Novosibirsk/USSR, 1984, pp. 541-546.

[61] Fisher, David F. and Dougherty, N. Sam, Jr.: In-Flight Transition Measurement on a 10° Cone at Mach Numbers from 0.5 to 2.0. NASA TP-1971, June 1982.

POSITION PAPER FOR THE BOUNDARY LAYER RECEPTIVITY PANEL

M. E. Goldstein and Lennart S. Hultgren

NASA Lewis Research Center
Cleveland, OH 44135

1. Introduction

The prediction of boundary layer transition was initially approached by asking where the laminar boundary layer first becomes unstable to what were then thought of as temporally growing instability waves (commonly referred to as Tollmien-Schlichting waves for inviscidly stable flows). However, it eventually became clear that the transition point (or better, the transition Reynolds number) is strongly affected by the specific details of the free-stream disturbance environment, and some of the research got redirected toward understanding just how it was that instability waves got generated by that environment, which was subsequently termed the receptivity phenomenon by Morkovin (1969). It was natural for experimentalists to study this phenomenon by imposing controlled, small amplitude, harmonic disturbances (i.e., disturbances of a single frequency, say ω) on their flows, which then generated a spatially growing shear wave of the type referred to simply as an amplifying wave by plasma physicists who, incidentally, recognized very early on that these spatially growing waves are intimately connected to temporally growing instabilities. Even the Schubauer and Skramstad experiment, which first demonstrated the existence of the Tollmien-Schlichting waves and (to our knowledge) was conducted solely for that purpose, was in fact a kind of receptivity experiment that generated spatially amplifying waves of this type.

More recent experiments directed toward understanding the receptivity phenomenon can roughly be grouped into two categories: (a) those in which the disturbances are introduced locally within the boundary layer and thus include the Schubauer-Skramstad experiment, or perhaps just outside it as in the Nishioka and Morkovin experiment; and (b) those in which the disturbances are introduced at some distance from the boundary layer and therefore must generate intermediate free-stream disturbances (usually of the acoustic or vortical type), which in turn generate the boundary-layer instability waves. The latter experiments are generally much more difficult to control than the former, and this situation has often led to conflicting interpretations of the results.

The subject of boundary layer receptivity deals with the early stages of transition where the unsteady motion is of small amplitude and can therefore be treated as a small perturbation of an appropriate mean flow. The unsteady flow will then exhibit the same harmonic time dependence as the externally imposed forcing (say $e^{-\omega t_*}$, where t_* denotes time). The role of this (receptivity) stage in the overall transition process was discussed in some detail by Heinrich, Choudhari, and Kerschen at the 1st National Fluid Dynamics Conference in Cincinnati.

2. Rational Asymptotic Analysis of Receptivity

The transition process can roughly be divided into three stages: a generally localized receptivity stage followed by a slow (viscous) linear downstream growth of the resulting disturbance (the disturbance may in some situations actually decay for a considerable distance before amplification occurs), and finally, if a sufficient amplitude is achieved, nonlinear breakdown of the disturbance flow initiated by rapid (inviscid) secondary instability.

We concentrate here on the receptivity stage, and then only on the second group of receptivity experiments alluded to above where the source is located at some distance from the boundary layer. Except for the early pioneering work at supersonic speed by Les Mack and his colleagues at the Jet Propulsion Laboratory, most such receptivity experiments were carried out at very low Mach numbers where the free-stream disturbance wavelengths are typically very large compared with the Tollmien-Schlichting wavelength. In fact, acoustic disturbances, which are very efficient Tollmien-Schlichting wave generators (primarily because of their large spanwise coherence), have effectively infinite wavelengths in the zero-Mach-number limit, on which (with a few brief exceptions) we now concentrate.

Since the boundary layer on which the instability waves ride is only defined in the infinite-Reynolds-number limit, it is appropriate to approach this problem analytically by seeking asymptotic expansions in inverse powers of some suitable Reynolds number. Much of the relevant theory is based on this approach.

Figure 1 is a stability diagram for viscous or Tollmien-Schlichting-type instability on a Blasius boundary layer. Here we are plotting the frequency parameter $F = \omega\nu/U_\infty^2$ vs. the length Reynolds number or normalized distance from the leading edge. The solid line is the neutral stability curve, with the flow being unstable inside the loop and stable outside. A spatially amplifying wave of constant frequency will move downstream along a horizontal line (such as the dashed line shown in the figure), first decaying until it encounters the lower branch of the neutral curve, and then growing until it encounters the upper branch (if nonlinear effects do not first intervene) – at which point it again begins to decay.

An appropriate expansion Reynolds number for the asymptotic solution might be the critical (length) Reynolds number Re_c at which the flow first become unstable at the imposed frequency ω. However, the frequency parameter F is an equally appropriate expansion parameter, since the classical high-Reynolds-number, long-wavelength asymptotic solutions of Heisenburg, Lin, Tollmien, and others clearly show that $F \propto \mathrm{Re}_c^{\frac{3}{4}}$ as $\mathrm{Re}_c \to \infty$ (see Reid 1965). But, for reasons that will become clear subsequently, it is often convenient to take the actual expansion parameter (say ϵ) to be $F^{\frac{1}{6}}$ or, alternatively, $\mathrm{Re}_x^{-\frac{1}{8}}$, since an $O(1)$ constant is irrelevant in defining an expansion parameter. The latter ($\epsilon = \mathrm{Re}_x^{-\frac{1}{8}}$) is appropriate when the coupling takes place near or almost anywhere upstream of the lower branch, but not, as we shall see, in the main unstable region. The former ($\epsilon = F^{\frac{1}{6}}$) is appropriate almost everywhere, but other powers of F or Re_x may be more appropriate when strong adverse pressure gradients play an important role in the coupling process. It is worth noting that Re_x becomes infinite as $\epsilon \to 0$ at all streamwise distances of the order of a convective wavelength U_∞/ω or larger, since Re_x is equal to $x^*\omega/U_\infty$ divided by ϵ^6.

The asymptotic approach has the advantage of providing a succinct description of the relevant physical phenomena and sometimes leads to useful scaling laws for the unsteady motion. It also provides a "rational" method for lumping the receptivity effects into a so-called coupling coefficient that determines the amplitude of the Tollmien-Schlichting wave. This tends to make the results more universally applicable, since the downstream (linear) evolution of the Tollmien-Schlichting wave usually depends on such things as externally imposed pressure gradients that vary on a scale that is large compared with the Tollmien-Schlichting wavelength and do not have (with at least one important exception, which we discuss below) a significant influence on the receptivity phenomenon. The coupling coefficient, on the other hand, usually depends on local conditions, on the scale of the Tollmien-Schlichting wavelength, which do not significantly influence the subsequent growth of the Tollmien-Schlichting wave.

3. Receptivity Mechanisms

It is reasonable to suppose that the free-stream disturbances are basically inviscid and of small amplitude. Then it is appropriate to require that their streamwise spatial scale (or wavelength) remain of the order of the convective wavelength U_∞/ω (or even larger) in the limit as the small viscous parameter ϵ goes to zero, since disturbances of smaller scale decay exponentially fast according to linear inviscid theory. Then the ratio of the Tollmien-Schlichting wavelength to the free-stream disturbance wavelength will be $0(\epsilon)$. This means that the free-stream wavelength will be very long compared with the Tollmien-Schlichting wavelength in this limit. Reshotko pointed out in his now classical annual review article on boundary layer transition that the explanation of the wavelength-adjustment mechanism was the main unresolved issue in the receptivity problem. It has now become clear that the adjustment can be attributed to (i) rapid streamwise variations in the mean boundary layer flow (which invalidate the parallel-flow assumption of the Orr-Sommerfeld equation) and (ii) sudden changes in surface boundary conditions (which produce diffracted disturbances that directly couple to the Tollmien-Schlichting waves). Sufficiently rapid variations in the mean boundary layer flow can occur (a) at a leading edge where the boundary layer is thin and rapidly growing, or (b) further downstream when the boundary layer is forced to make a sudden adjustment by small but abrupt changes in surface geometry and/or boundary conditions. The important common feature is that there is some region of the flow where the correct asymptotic approximation to the Navier-Stokes equations is the unsteady boundary layer equation instead of the Orr-Sommerfeld equation. In case (a), this region is the leading-edge region itself, and in case (b), it is the lower deck of the triple-deck structure that provides the correct high-Reynolds-number approximation to the full Navier-Stokes equations. It is worth noting that appropriate sudden boundary layer adjustment can also be forced by shock boundary layer interaction or by the rapid static pressure variations that occur when the boundary layer is on the verge of separation.

4. Concluding Remarks

Finally, it is worth noting that receptivity theory is complementary to stability theory, which studies the evolution of disturbances already present in the boundary layer. Transition prediction methods combining receptivity with linear stability theory would account not only for the influence of free-stream disturbances but also for the boundary layer characteristics upstream of the neutral-stability point.

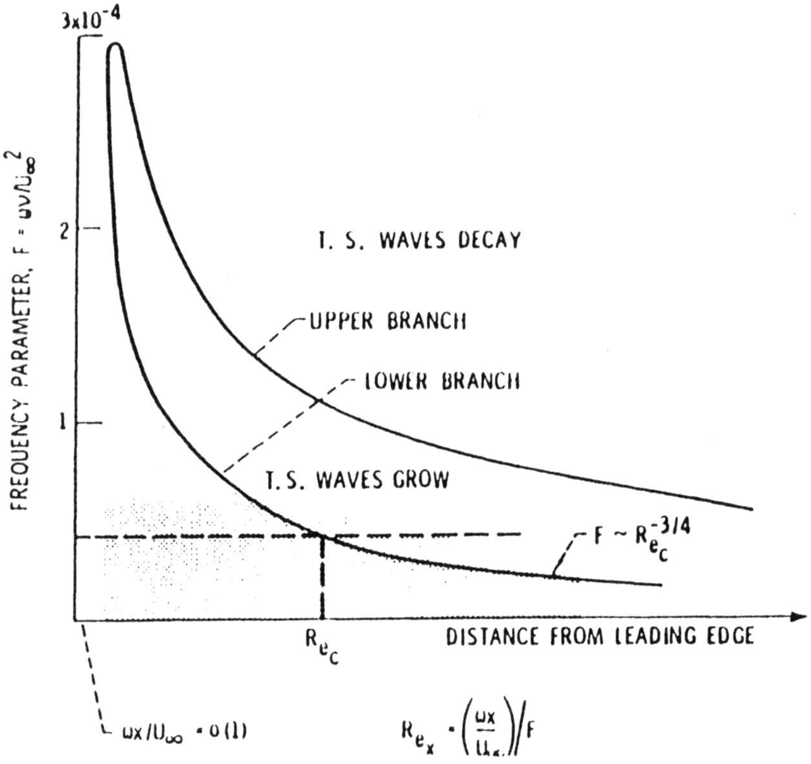

Figure 1 Stability diagram for Tollmien-Schlichting wave

RECEPTIVITY OF BOUNDARY LAYERS TO ACOUSTIC AND VORTICAL FREE-STREAM DISTURBANCES

E. J. Kerschen

University of Arizona
Tucson, Arizona

ABSTRACT

The receptivity mechanisms by which free-stream disturbances generate instability waves in laminar boundary layers are discussed. Free-stream disturbances have wavelengths which are generally much longer than those of instability waves. Hence, the transfer of energy from the free-stream disturbance to the instability wave requires a wavelength conversion mechanism. Recent analyses using asymptotic methods have shown that the wavelength conversion takes place in regions of the boundary layer where the mean flow adjusts on a short streamwise length scale. This paper reviews recent progress in the theoretical understanding of these phenomena and suggests directions for future research in this area.

1. Introduction

The accurate prediction of boundary layer transition and the need to control this phenomenon are becoming increasingly important in the development of advanced aeronautical vehicles. These requirements have focused renewed attention on the physics of the transition process. It was known for many years that boundary layer transition is significantly influenced by the free-stream disturbance environment. However, the physical mechanisms by which energy is transferred from the long wavelength free-stream disturbances to the much shorter wavelength instability waves were not understood. This came to be known as the receptivity problem.

Recent work by Goldstein [1,2,3] and extensions by the author [4,5] have shown that the receptivity process occurs in regions of the boundary layer where the mean flow exhibits rapid changes in the streamwise direction. This occurs (a) near the body leading edge and (b) in any region farther downstream where some local feature forces the boundary layer to adjust on a short streamwise length scale. The rapid streamwise adjustment requires that non-parallel mean flow effects be included at leading order, in contrast to the parallel flow assumption of classical Orr-Sommerfeld stability theory. We consider receptivity produced by small-amplitude acoustic and vortical free-stream disturbances interacting with the leading edge of a flat plate and with a localized mean flow adjustment downstream of the leading edge.

2. Leading Edge Receptivity

In this section, we examine receptivity in the leading edge region of the Blasius boundary layer. Results are presented for both free-stream acoustic waves and convected gusts. The analysis considers low Mach number, two-dimensional flow. The flow far from the plate consists of a small amplitude, harmonic disturbance superposed on a uniform mean flow U_0. The Reynolds number is assumed large, and hence the outer problem for the unsteady flow corresponds to the inviscid interaction of the small amplitude free-stream disturbance with the semi-infinite flat plate. This outer solution provides the distributions of pressure and slip velocity which drive the unsteady motion in the boundary layer on the plate. Since the free-stream disturbances are assumed to be of small amplitude, linear superposition is valid. Hence, the velocity of the incident disturbance can be separated into components parallel and perpendicular to the plate surface. The parallel component contributes directly to the slip velocity on the plate surface, while the perpendicular component contributes to the slip velocity via its scattering by the plate surface. Near the leading edge, this scattered component has a square root singularity corresponding to inviscid flow around the sharp edge, while far downstream the scattered component takes a simple form appropriate to a doubly infinite plate.

Goldstein's [1] analysis of the unsteady flow in the boundary layer shows that the asymptotic solution for $\epsilon \ll 1$ ($\epsilon^{-6} = U_0^2/\omega\nu$) contains two distinct streamwise regions. The receptivity occurs in the first region near the leading edge, where $X = x'\omega/U_0 = O(1)$ and the motion satisfies the linearized unsteady boundary layer equation (LUBLE). Farther downstream where $x = \epsilon^2 X = O(1)$, the unsteady flow satisfies the classical large Reynolds number, small wavenumber approximation to the Orr-Sommerfeld equation (OSE). Goldstein examines the LUBLE for $X \gg 1$ and the OSE for $x \ll 1$, and shows that the first asymptotic eigenfunction of the LUBLE, with amplitude C_1, matches onto the T-S wave which eventually exhibits growth farther downstream. Hence the amplitude of the T-S wave is linearly proportional to C_1, which we call the "receptivity coefficient." To find C_1, a numerical solution of the LUBLE is required. Unfortunately, for $X \gg 1$ the first eigenfunction makes only an exponentially small contribution to the solution. In order to extract C_1 from the numerical calculation, the equation is analytically continued into the complex plane and solved along a ray where the first eigenfunction dominates the solution.

Figure 1 is a plot of the receptivity coefficient for a convected gust as a function of disturbance orientation. The velocity and frequency of the free-stream disturbance are held constant as the orientation is varied. Free-stream convected gusts can be interpreted as a two-dimensional model of free-stream turbulence. The convected gust is vortical, has no pressure fluctuations, and the velocity fluctuations are perpendicular to the wavevector. The maximum and minimum levels of receptivity are found for free-stream velocity perturbations parallel ($\theta = 90°$) and perpendicular to the plate surface, respectively. The weak receptivity levels produced by the perpendicular component of free-stream velocity are rather surprising, since this component produces both a singular flow field near the leading edge and a significant slip velocity in the doubly infinite plate limit far downstream. Essentially, O(1) receptivity levels are produced by the individual contributions (such as the singular flow near the leading edge) to the slip velocity, but for the convected gust the amplitudes and phases of these various contributions lead to a high degree of cancellation. Hence, the receptivity for convected gusts is produced mainly by the parallel component of the free-stream disturbance velocity.

The receptivity produced by free-stream acoustic waves is illustrated in Fig. 2, for two values of the Mach number, $M = 0.01$ and 0.1. Here, $\theta = 0°$ corresponds to incident acoustic waves propagating downstream parallel to the plate surface, and was the case considered by Goldstein. The receptivity for this case is found to be only about 1/4 of that produced by the convected gust slip velocity component parallel to the plate surface (Fig. 1, $\theta = 90^o$). The explanation for this may be the slower phase speed of the convected gust compared to the acoustic wave. However, the receptivity to acoustic waves rises rapidly as the incidence angle increases, reaching a maximum value at $\theta = 180^o$, which corresponds to an acoustic wave propagating from downstream infinity toward the leading edge. The high receptivity for oblique acoustic waves is caused by the perpendicular component of the free-stream disturbance velocity. This component produces a flow around the leading edge whose strength increases as $M^{-1/2}$ as the Mach number decreases and the acoustic wavelength becomes large compared to the hydrodynamic length scale of the problem.

The above result applies only for cases where the acoustic wavelength is short compared to both the plate length and the distance to any surrounding surfaces. To address the effect of nearby surfaces, we have analyzed the receptivity to upstream traveling waves for the case of a semi-infinite flat plate centered within a channel. The slip velocity was calculated using a Wiener-Hopf analysis and confirmed for low frequencies by a matched asymptotic expansion analysis (MAE). Figure 3 contains results as a function of channel width H for a free-stream Mach number of 0.1. The receptivity coefficient grows rapidly with increasing H and reaches a maximum $C_1 = 35$ at $\omega H/c = \pi$. For higher values of $\omega H/c$, C_1 exhibits a regular pattern of weakly damped oscillations, gradually approaching the isolated plate result $C_1 = 20$ (see Fig. 2). The explanation for this behavior can be found in the alternate cut-on of upstream and downstream traveling acoustic modes.

3. Localized Receptivity Mechanisms

In this section, we examine receptivity mechanisms which occur when some local feature downstream of the leading edge causes the boundary layer to adjust on a short streamwise length scale. Examples of local features are short scale wall humps or suction strips or shock - boundary layer interaction. The asymptotic descriptions of the mean and unsteady flow components in the local region involve triple-deck structures. For sufficiently weak mean flow disturbances (small hump height, for example), the mean flow triple-deck equations can be linearized and solved in closed form. The unsteady flow can then be solved via Fourier transforms, and the instability wave is given by a residue in the complex wavenumber plane.

For localized receptivity mechanisms, the wavelength conversion process is produced by the unsteady flow conforming to the local short-scale geometry. Essentially, the long wavelength, unsteady, free-stream disturbance temporally modulates the local short scale mean flow, generating frequency-wavelength combinations that match the instability wave. The detailed physics of the modulation processes are somewhat different for acoustic and convected free-stream disturbances, and will be described separately.

First consider the case of acoustic free-stream disturbances [2]. The free-stream pressure fluctuations penetrate the boundary layer and produce velocity fluctuations, which are brought to rest at the wall by a thin Stokes wave of the same thickness as the lower deck. The receptivity to acoustic waves arises in the adjustment of the Stokes wave to the local geometry, and hence the source of the instability wave is near the wall, deep inside the mean boundary layer. Receptivity can also be produced by an acoustic wave scattering off a short scale variation in wall impedance [4]. This mechanism is operative even in the absence of a mean flow adjustment.

Now consider the localized receptivity process for convected gusts [5]. In contrast to an acoustic wave, a convected gust contains no pressure fluctuations. Hence, away from the local region, the free-stream velocity fluctuations are transmitted to the boundary layer only by viscous diffusion. At high Reynolds numbers, this process is very weak and the unsteady motion is mainly concentrated at the outer edge of the boundary layer. In fact, the amplitude of the unsteady disturbance in the lower deck near the wall is exponentially small. Hence, the near wall interaction is not a significant source of receptivity in this case. However, the hump produces short scale mean flow gradients in the upper deck just outside the mean boundary layer. The inviscid interaction of the convected gust with these mean flow gradients produces short scale pressure fluctuations, which penetrate the boundary layer and generate instability waves. Although the physical mechanisms for the convected gust and the acoustic wave are quite different, the receptivity coefficients are found to be of the same order of magnitude.

For localized, two-dimensional receptivity mechanisms, we have

$$u_{TS} = u_{slip}\, F(\alpha_0)\, \Lambda(\alpha_0)\, \phi(y)\, e^{i(\alpha_0 x - \omega t)} \tag{1}$$

where u_{TS} is the Tollmien-Schlichting wave streamwise velocity fluctuation with mode shape $\phi(y)$ and wavenumber α_0, u_{slip} is the amplitude of the slip velocity produced by the free-stream disturbance, $F(\alpha_0)$ is the spatial spectrum of the local inhomogeneity (wall hump, etc.) evaluated at the wavenumber of the instability wave, and $\Lambda(\alpha_0)$ is the characteristic function for the specific type of receptivity mechanism under consideration. For an acoustic wave, the slip velocity is $2u_{ac}\cos\theta$, where $\theta = 0°$ corresponds to propagation parallel to the wall and $\theta = 90°$ corresponds to an acoustic wave normally incident on the wall. For a convected gust, it can be shown that the slip velocity equals the velocity of the upstream gust for all orientations of the gust wavevector.

The Λ functions for free-stream acoustic waves and convected gusts interacting with a localized wall hump are plotted as a function of the normalized frequency in Fig. 4. The functions are quite similar for low frequencies. At moderate frequencies the Λ associated with acoustic waves is somewhat larger than that associated with convected gusts. The acoustic Λ approaches a constant at high frequencies, since the amplitude of the Stokes wave is independent of frequency, while the convected gust Λ decreases at high frequencies, since only weak pressure fluctuations are produced by the inviscid distortion of a very short wavelength gust.

Equation (1) also shows that the geometry of the wall hump or other local disturbance significantly influences the receptivity coefficient. For example, a Mylar strip produces the highest receptivity coefficient when its length is a half integer multiple of the instability wavelength. The impact of receptivity on laminar flow control via suction is discussed in Heinrich et. al. [4].

Nishioka and Morkovin [6] have emphasized the role of streamwise gradients of unsteady free-stream pressure as a source of receptivity. This concept is perfectly consistent with the above discussion. In fact, the localized receptivity to convected gusts is a direct example of instability wave generation by the action of streamwise gradients of unsteady pressure. In this case the unsteady pressure fluctuations are caused by the interaction of a naturally occurring free-stream disturbance with a local mean flow perturbation. The excitation of instability waves by an external source, such as a loudspeaker located just outside the boundary layer, can also be analyzed in a similar manner [5,7].

4. Future Directions

Over the past few years, asymptotic analyses have dramatically advanced our understanding of boundary layer receptivity. These studies have identified important receptivity mechanisms for two-dimensional, low speed flows. The influence of leading edge thickness is probably the most important open question for two-dimensional, low speed flows. The leading edge radius enters through the parameter $\omega r_0/U$, which is O(1) in most low speed aerodynamic applications. Even in the case of laboratory experiments, this parameter may be large enough to cause substantial deviations from the zero-thickness, flat plate results discussed above.

Three-dimensionality and compressibility are also important issues. For the case of localized acoustic wave receptivity mechanisms, the author and Choudhari [8] have examined subsonic compressibility effects for two-dimensional interactions, and three-dimensional effects for incompressible flow. We have also initiated extensions to vortical free-stream disturbances and supersonic flow [9]. At high speeds, receptivity to entropic free-stream disturbances is likely to be important as well. Extensions of the leading edge receptivity analysis to three-dimensional free-stream disturbances and supersonic flow should also be developed. Eventually, the role of receptivity in the initiation of secondary instabilities must be addressed.

There is also a great need for computational and experimental investigations of boundary layer receptivity. The asymptotic approach is useful for identifying dominant physical mechanisms, but has limitations with respect to finite Reynolds number effects. In addition, we have utilized linear representations in a number of places. A variety of methods, some involving a combination of numerics and asymptotics such as in [10], and others fully numerical, are needed to more fully delineate the phenomena. Finally, the accuracy and limitations of the recently developed theories must be tested in carefully designed experimental investigations [11]. Such experiments are also likely to identify areas where further study is needed.

Acknowledgements

The author wishes to acknowledge the contributions of his students, Roland Heinrich and Meelan Choudhari, to the research described in this paper. Financial support was provided by the National Science Foundation under grant MEA-8351929 and by the McDonnell Douglas Research Laboratory.

References

1. Goldstein, M.E., "The Evolution of Tollmien-Schlichting Waves near a Leading Edge," *J. Fluid Mech.*, Vol. 127, 1983, pp. 59-81.

2. Goldstein, M.E., "Scattering of Acoustic Waves into Tollmien-Schlichting Waves by Small Streamwise Variations in Surface Geometry," *J. Fluid Mech.*, Vol. 154, 1985, pp. 509-529.

3. Goldstein, M.E. and Hultgren, L.S., "Boundary-layer Receptivity to Long-Wave Free-Stream Disturbances," *Ann. Rev. Fluid Mech.*, Vol. 21, 1989, pp. 137-166.

4. Heinrich, R.A., Choudhari, M. and Kerschen, E.J., "A Comparison of Boundary Layer Receptivity Mechanisms," AIAA Paper 88-3758, 1988.

5. Kerschen, E.J., "Boundary Layer Receptivity," AIAA Paper 89-1109, 1989.

6. Nishioka, M. and Morkovin, M.V., "Boundary Layer Receptivity to Unsteady Pressure Gradients: Experiments and Overview," *J. Fluid Mech.*, Vol. 171, 1986, pp. 219-261.

7. Heinrich, R.A.E., Gatski, T.B. and Kerschen, E.J., "Boundary Layer Receptivity to Unsteady Free-Stream Pressure Gradients," Proceedings of ICASE / NASA Langley Instability and Transition Workshop, Springer-Verlag, 1989.

8. Choudhari, Meelan, "Boundary Layer Receptivity in Laminar Flow Control Applications," Ph.D. Dissertation, University of Arizona, 1989.

9. Choudhari, Meelan and Kerschen, E.J., "Receptivity to Three-Dimensional Convected Gusts," Proceedings of ICASE / NASA Langley Instability and Transition Workshop, Springer-Verlag, 1989.

10. Bodonyi, R.J., Welch, W.J.C., Duck, P.W., and Tadjfar, M., "A Numerical Study of the Interaction Between Unsteady Freestream Disturbances and Localized Variations in Surface Geometry," to appear in *J. Fluid Mech.*, 1989.

11. Wlezien, R.W., "Measurement of Boundary Layer Receptivity at Suction Surfaces," AIAA Paper 89-1006, 1989.

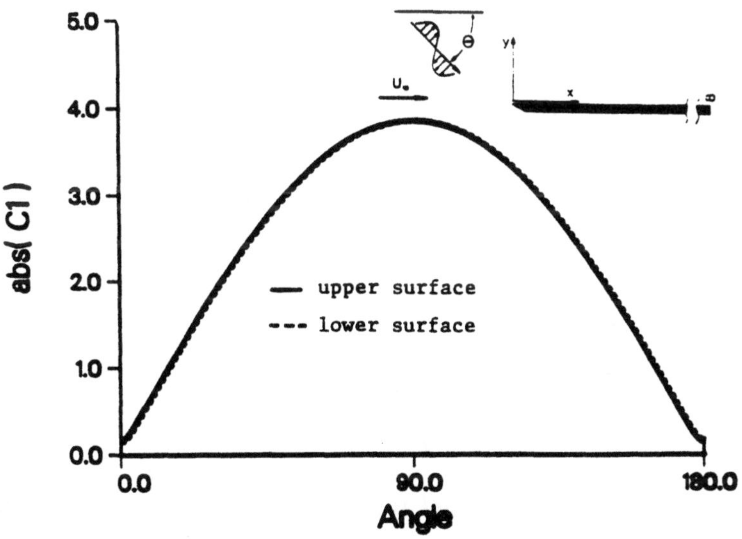

Fig. 1. Leading edge receptivity coefficient for a convected gust. Upper surface, ——— ; lower surface, - - - - .

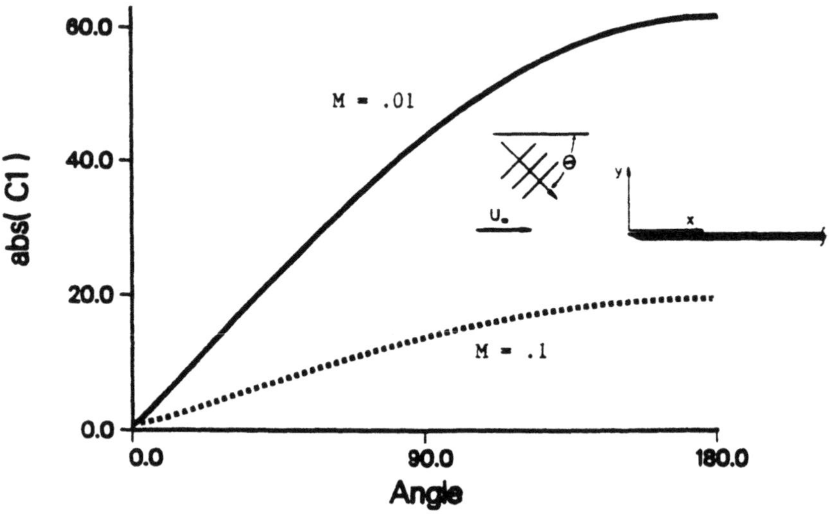

Fig. 2. Leading edge receptivity coefficient for a plane acoustic wave.

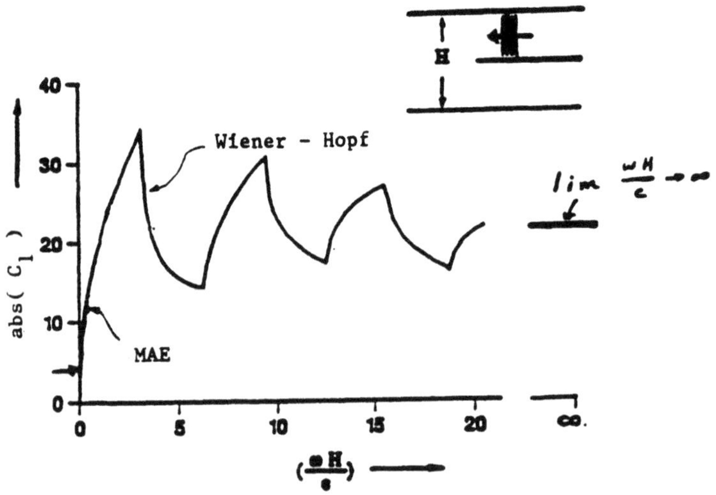

Fig. 3. Effects of finite channel width on the leading edge receptivity coefficient for an upstream traveling acoustic wave.

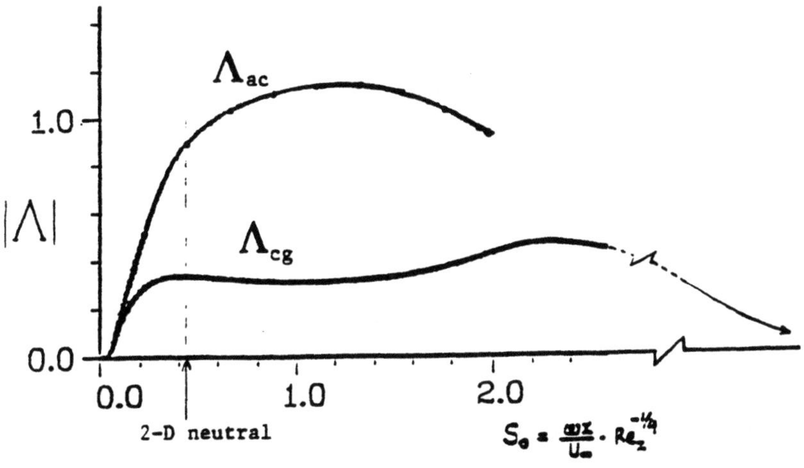

Fig. 4. The efficiency functions Λ_{ac} and Λ_{cg} for an acoustic wave and a convected gust interacting with a wall hump ($S_0 = Re_L^{-1/4} \omega L/U$).

RECEPTIVITY INVESTIGATIONS

Patrick Leehey

Massachusetts Institute of Technology
Cambridge, MA 02139

1. Early Experiments with Tollmien-Schlichting (T-S) Waves

For the most part, our investigations have been experimental. They have involved the use of a low-noise, low-turbulence, subsonic wind tunnel with a test cross-section 15″ by 15″. Laminar boundary layers are grown on a polished aluminum flat plate with internal damping. The plates are all 1/2″ thick and are fitted with a 6:1 semi-elliptical leading edge. The first experiments, reported by Shapiro [1977], involved excitation by a nearly plain, pure tone, acoustic wave travelling downstream over the test plate. A now classic standing wave pattern was found in the boundary layer (Figures 1 and 2). The wavelength of the standing wave was that of a Tollmien-Schlichting (T-S) wave at the frequency of the acoustic wave.

Relative to the scale of the T-S wave, a downstream (or upstream) travelling acoustic wave can be represented as an incompressible Stokes wave. It is then an elementary demonstration to show that the presence of the standing wave pattern indicates that the amplitudes of the Stokes wave and the T-S wave are essentially the same. This denotes what we call a 1:1 receptivity of the acoustic wave by the T-S wave. Shapiro also measured T-S growth rate and found as well that there was a significant amplitude range for linear response to acoustic excitation.

2. Later Experiments: Cancellation of T-S Waves

This work was later extended, principally by Gedney (see Leehey, Gedney, and Her [1985]). The Stokes mode was extracted from the data experimentally. This exposed the T-S eigenmode essentially correct in both amplitude and phase (Figure 3). Gedney [1983] also succeeded in cancelling the sound excited T-S wave with plate vibration (Figure 4). He pointed out that the success of the cancellation depended on the linearity and two-dimensionality of the process; no increase in the natural transition Reynolds number was achieved.

Because of the elliptical leading edge, an adverse pressure gradient exists in the near neighborhood of the critical Reynolds number. Since the pressure gradient scales inviscidly to lowest order, Gedney demonstrated experimentally how the receptivity could be significantly altered simply by changing the tunnel speed. This is a unit Reynolds number effect. Gedney also demonstrated that there is significant oscillatory cross-flow at the leading edge which may play a role in the receptivity problem.

3. Receptivity Computations

Our next effort was computational, Lee [1985]. Lee obtained a finite difference solution to the Navier-Stokes equations in a rectangular domain in which the effect of pressure gradient was introduced but not the corresponding effect of change in body geometry. A uniform oscillatory excitation was superimposed on an experimental mean profile at the upstream computational boundary. The uniform oscillation evolved gradually into a Stokes wave (Figure 5), but did not further develop within the computational domain to a T-S eigenmode form. Thus, gradient alone seems insufficient to produce the wavenumber conversion essential to the receptivity process.

4. Acoustic Radiation from T-S Waves

Our most recent work was carried out by LaFontaine and Leehey [1988]. The test plate was truncated at the rear to form a airfoil with a chord of 22 cm. The experimental arrangement is shown in Figure 6. The top side of the airfoil is in laminar flow. The bottom side was tripped to form a turbulent boundary layer. Tests were conducted in a quasi open jet in order to permit far field acoustic measurements. A typical radiated noise spectrum is shown in Figure 7. The spectrum has a characteristic "rooster-tail" form previously noted by Arbey and Bataille [1983]. This part of the radiation field is dominated by the laminar flow near the trailing edge. The broad hump corresponds to the frequency domain of unstable T-S waves. The rooster-tail spikes are equally spaced in frequency. Here, two physical mechanisms are involved. One is the conversion of T-S wave energy to sound at the trailing edge. The other is the receptivity of this same sound somewhere near the leading edge of the airfoil. Figure 8 shows a typical directivity pattern of the sound field.

Note, however, that the entire airfoil is within the sound near field at this frequency. Clearly, a feedback mechanism between radiation and receptivity is involved. The need for phase reinforcement at the receptivity point then suggests that an integer number of T-S wavelengths exist at any one time between the receptivity point and the trailing edge. Since the instability wave speed is nearly constant for the most amplified waves, this implies that the frequencies subject to feedback enhancement are spaced equally apart, thus producing the rooster-tail pattern. From measured instability wave speeds, the position from the receptivity point was inferred. This point appeared to move aft with increasing flow speed but was located at low speed just behind the point where the semi-elliptical nose of the airfoil meets the flat portion.

A narrow piezo-polymer strip was stretched spanwise across the airfoil. Its vibratory excitation was controlled by the signal from a nearby hotwire in the laminar boundary layer. As shown in Figure 9, the rooster-tail radiated noise spectrum could be dramatically decreased by a proper phase adjustment of the hotwire signal even though the strip was not optimized to vibrate at a half wavelength of any of the T-S waves involved.

References

[1] Arbey, H. and Bataille, J. (1983), "Noise generated by airfoil profiles placed in a uniform laminar flow." *J. Fluid Mech.*, **134**, pp. 33-47.

[2] Gedney, Charles J. (1983), "The cancellation of a sound-excited Tollmien-Schlichting wave with plate vibration." *Phys. Fluids*, **26**, pp. 1158-1159.

[3] LaFontaine, M. R. and Leehey, P., "An experimental investigation of trailing-edge noise from a laminar airfoil." MIT Acoustics and Vibration Laboratory Report, in preparation.

[4] Lee, S. H. (1985), "The influence of pressure gradient upon the receptivity of a laminar boundary layer." Ph.D. Thesis, Massachusetts Institute of Technology, Department of Ocean Engineering.

[5] Leehey, P., Gedney, C. J., and Her, J. Y. (1985), "The receptivity of a laminar boundary layer to external disturbances." In Laminar-Turbulent Transition IUTAM Symposium Novosibirsk 1984 (Kozlov, V. V., ed.), pp. 283-294, Springer-Verlag, Berlin, Heidelberg.

[6] Shapiro, P. J. (1977), "The influence of sound upon laminar boundary layer instability." MIT Acoustics and Vibration Laboratory Report No. 83458-83560-1.

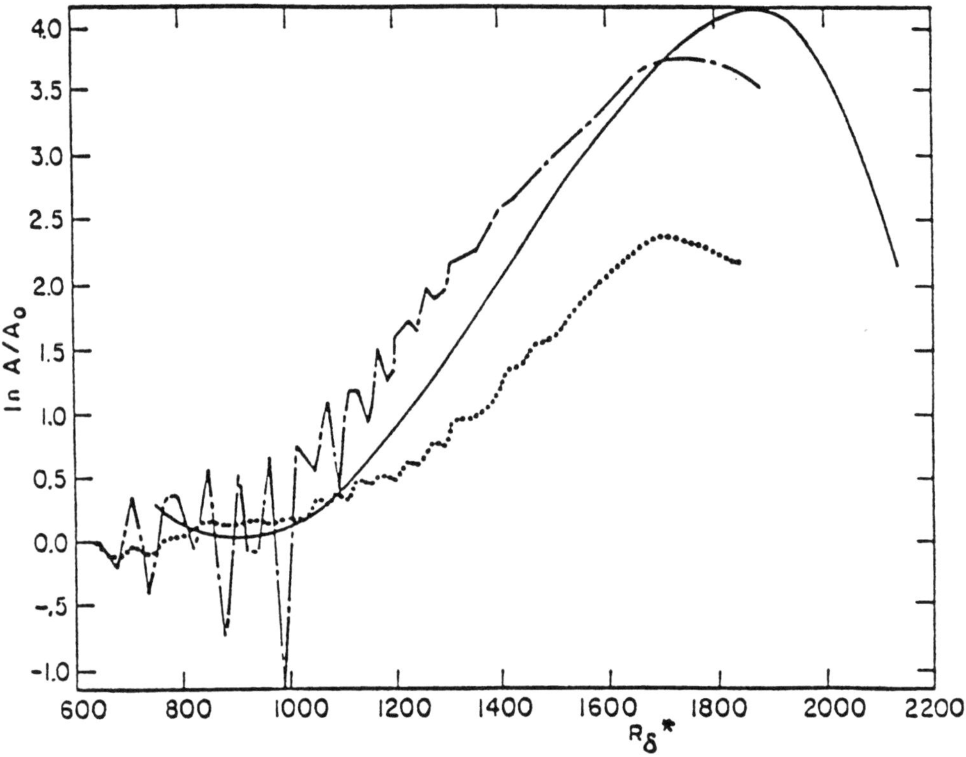
4.0
3.5
3.0
2.5
2.0
1.5
1.0
0.5
0.0
-.5
-1.0
ln A/A₀
600
800
1000
1200
1400
1600
1800
2000
2200
Rδ*

Figure 1

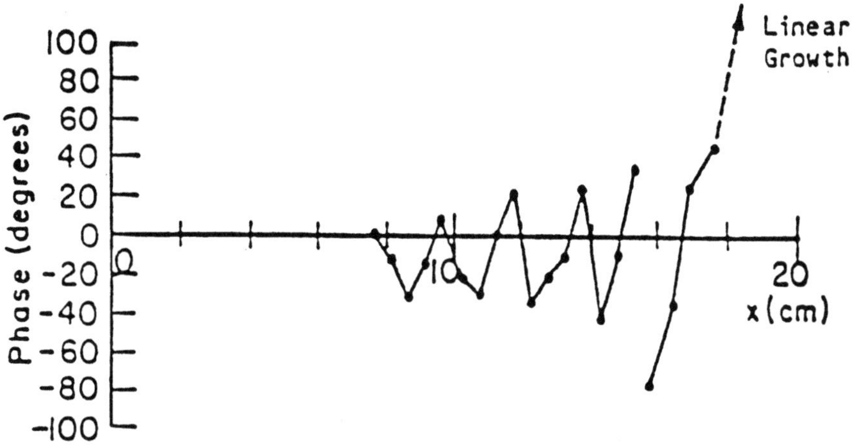

Linear
Growth
100
80
60
40
20
0
-20
-40
-60
-80
-100
Phase (degrees)
0
10
20
x(cm)

Figure 2

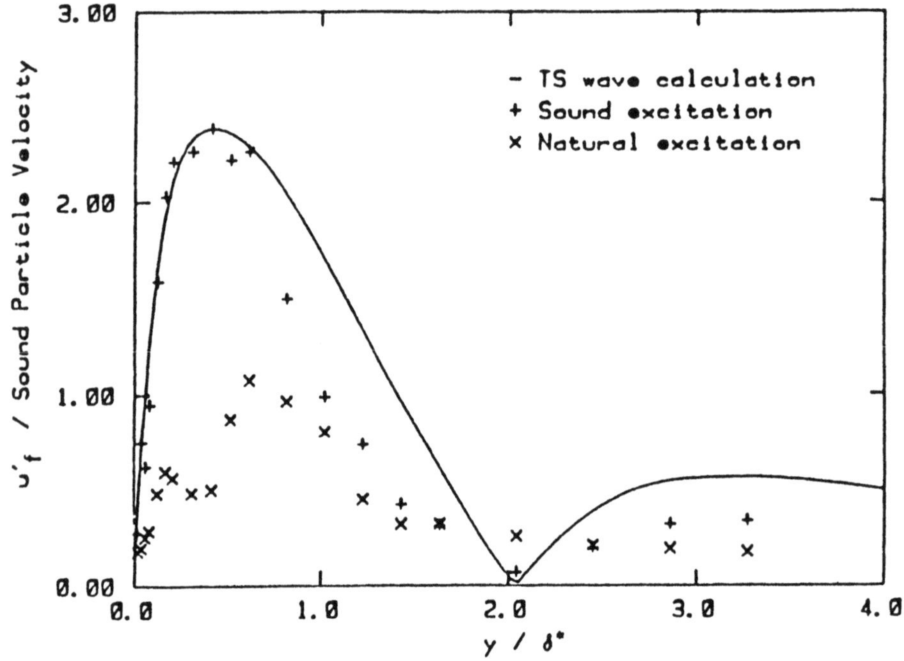

Figure 3. Velocity fluctuations (at 200 Hz) as functions of y for sound excitation. The sound excited results do not include the Stokes shear-wave (x = 63.5 cm, U_∞ = 18.5 m/s).

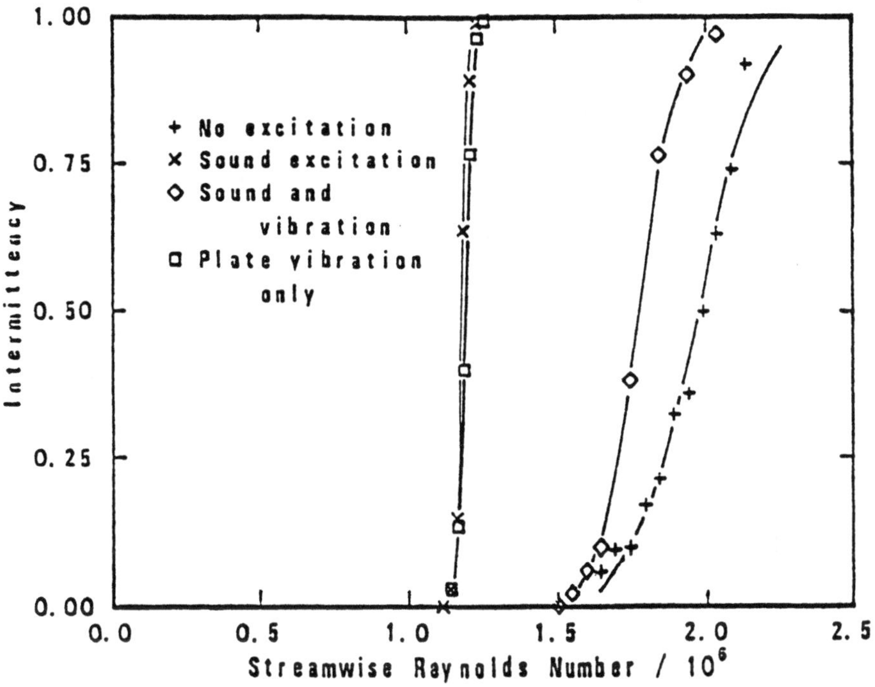

FIGURE 4. Intermittency as a function of streamwise Reynolds number (R_x).

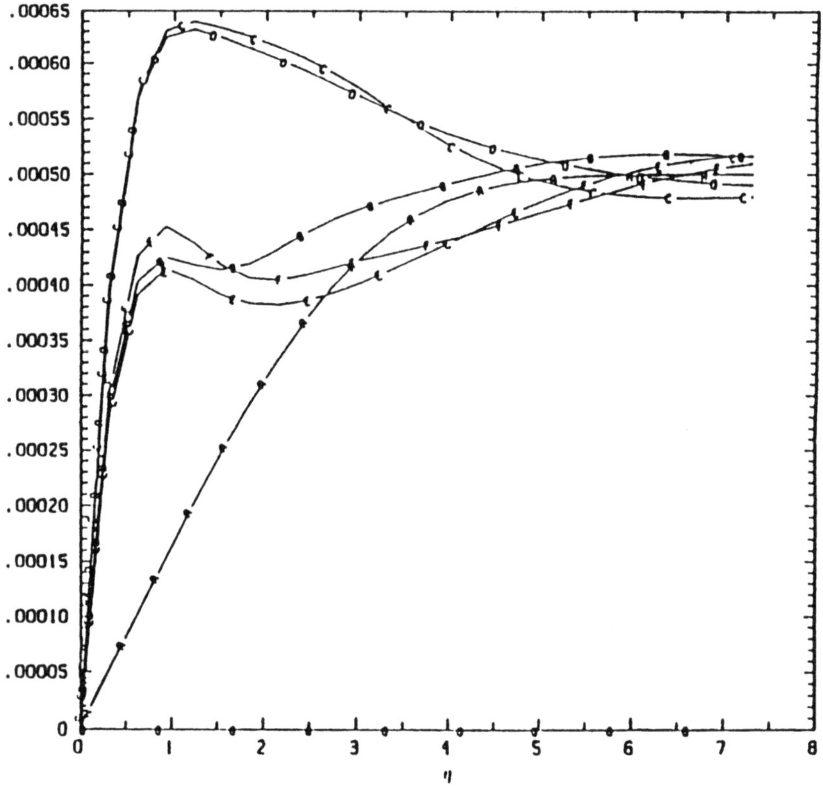

u

Figure 5

Fourier amplitudes along constant ξ
u; Lines A~F correspond to i=1(19)116.
(Receptivity problem for R = 600~860, β=131.6×10⁻⁶).

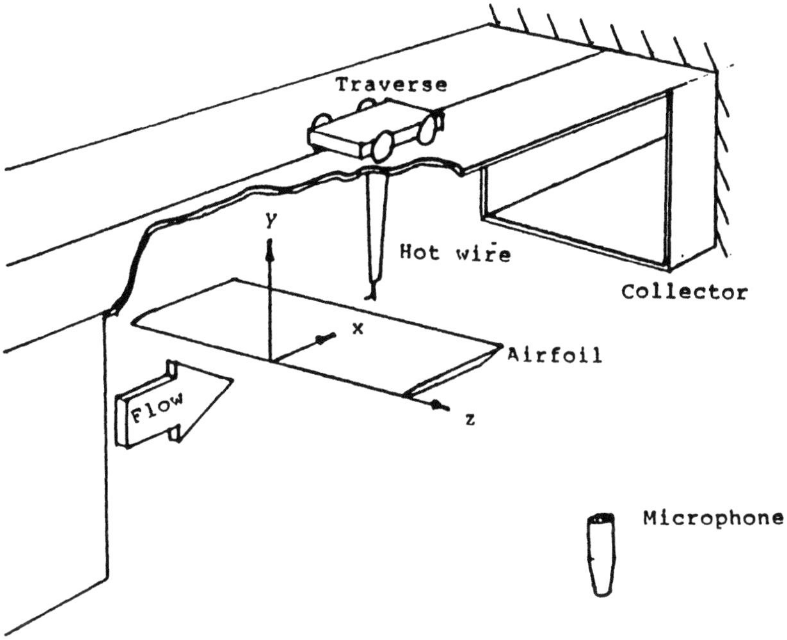

Figure 6. Wind-tunnel test set-up in open duct configuration.

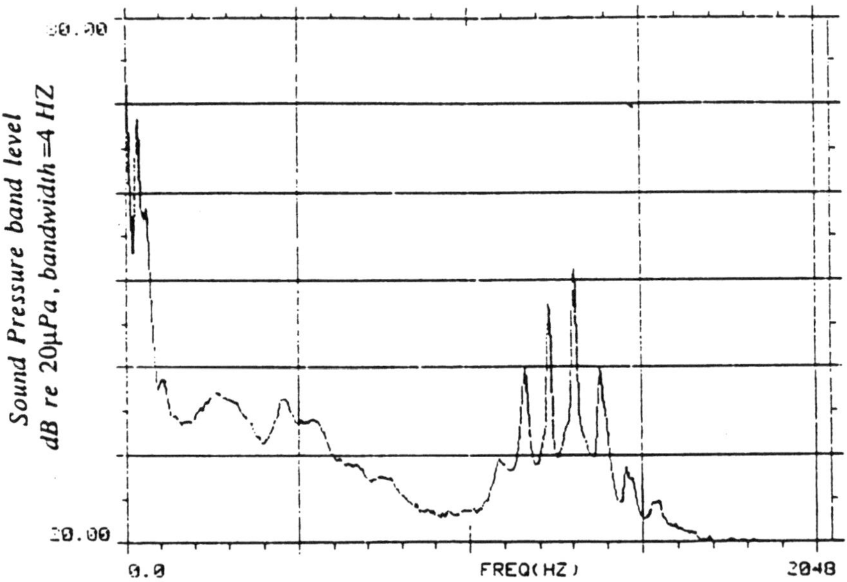

Figure 7. Pressure band levels of trailing-edge noise of U = 34.4 m/s.

Figure 8. Directivity of trailing-edge noise for U =26 m/s, f=600 HZ.

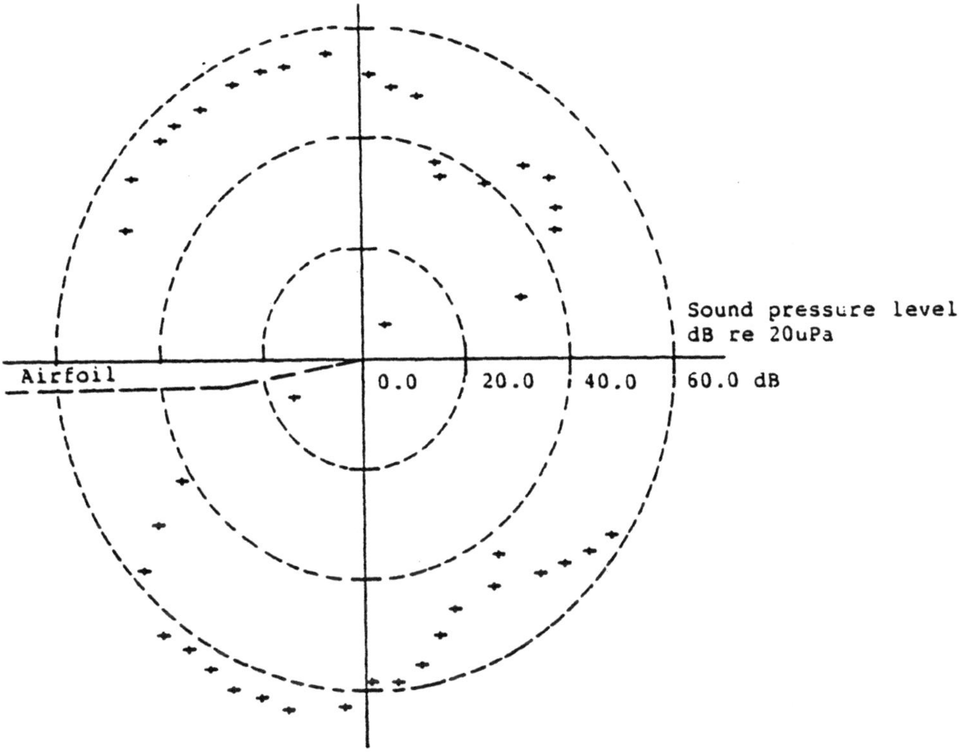

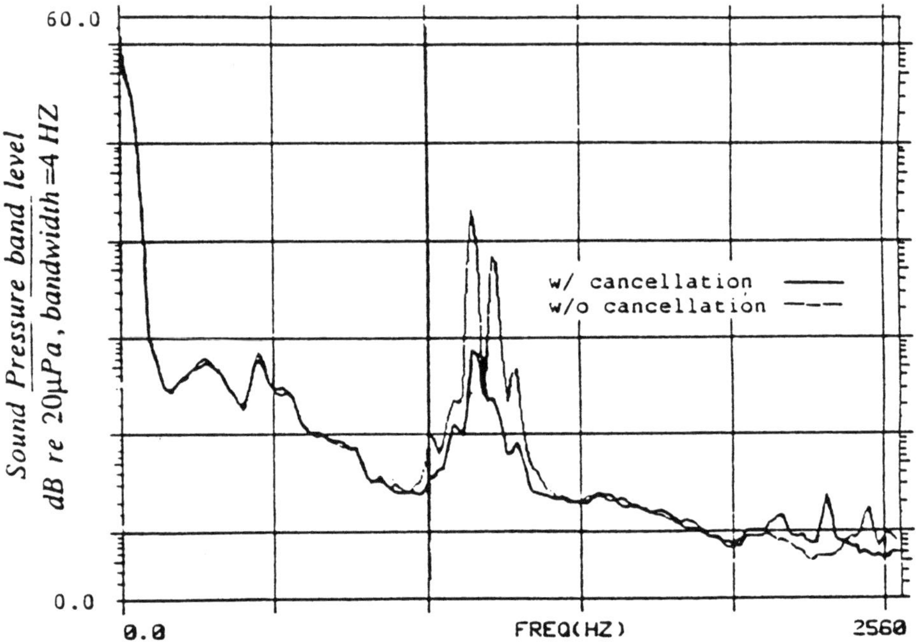

Figure 9. Broadband trailing-edge noise cancellation, U =32.2 m/s, x_{ribbon}=5 3/8 inches.

ROUGHNESS

PANEL SUMMARY: ROUGHNESS

Mark V. Morkovin

Professor Emeritus
Illinois Institute of Technology
Chicago, IL 60616

1. General Observations

This segment reflects discussions during and after the Roughness Panel presentations. Since our assigned task was more tutorial in nature than that of other panels, we discuss here at some length aspects of the phenomena which we feel are important for fundamental research into probable mechanisms of roughness-induced transition. This difficult, sadly neglected research area is primarily experimental, although computational experiments may help to clarify some features of the mechanisms; see Section 3. There is need for improved spatial (and temporal) resolution of field sensors, which should include at least two velocity components. To identify important flow structures we need at least two-sensor measurements of space-time correlations and coherence and simultaneous (say u, w) time traces over arrays of sensors, thus extending the approaches of Kendall (1981, 1985, 1987). Given the existence of long "incubation distances" (Section 5 of Morkovin's panel contribution) which make judgment of cause and effect very difficult, such approaches are essential for deeper understanding. The recent explosion of capabilities in instrumentation and in data acquisition, processing and analysis places new insights within our reach for the first time.

Detail field explorations require instrumentation strategies (including visualization) to capture salient features of the phenomena, including local and sporadic events. Guidance for such strategies must come from larger concepts and we therefore advance here some speculative scenaria concerning likely contributing mechanisms. The common element of streamwise vorticity structures in transition caused by roughness, by free-stream turbulence, by Landahl's 1980 algebraically-growing bypass mode, and by near-wall ejections in developed turbulent wall layers may provide useful similarities in mechanisms–or equally useful contrast if different evolutions take place in different regions of the parameter space.

Since vorticity patterns generated by small roughness for $Re_k < 300$ are steady, the unsteadiness must come from environmental disturbances–the *co-responsibility for transition* which had been dismissed until recently. In particular, intermittency of free-stream turbulence could well supply the initiating "lift-up" in Landahl's scenario, in presence of streamwise vortex structures near the wall, whether brought about by receptivity processes to free-stream turbulence or by roughness.

We believe that not only does the unsteadiness of free-stream disturbances combine with $x - z$ nonhomogeneity of the rough surfaces to open bypass roads to turbulence but that the regions of lower velocity (relative to smooth boundary layers) make these disturbances more effective–i.e., that they increase the receptivity to the bypass modes. We recall that a similar increase in the receptivity due to small roughness to the tuned non-bypass TS mode was strongly indicated by the experiments of Corke et al. (1986).

2. Distributed Roughness and Receptivity to Free-Stream Turbulence

In his presentation, Professor Reshotko included a progress report on the application of the velocity-measuring technique of index matching described in Tadjfar et al. (1985). The fluid used has an index of refraction that matches the index of glass spheres (attached to a glass plate in a regular pattern), thereby making it possible to measure velocity components and fluctuations within the array, using laser anemometry. This is the only current procedure which can give us information on the flow below the roughness heights and in between elements. While very promising, it is also very demanding.

Reshotko's comments on experiments with distributed roughness emphasized the importance of the leading edge of the roughness distribution and the increasing energy content in a non-TS broad, low-frequency band of u' fluctuations as x increased. Details of the many experiments and his overall views are found in Reshotko (1984).

Kendall's presentation of his experiments on receptivity to mild free-stream turbulence in a laminar flat-plate boundary layer was shifted to the Roughness Session because of the aforementioned similarity of the observed streamwise structures. For details and interpretation see Kendall (1985, 1987) and discussion in Morkovin and Reshotko (1990, Sections 2.4-2.8). These experiments verify and document the behavior of (unpublished) "Klebanoff disturbance modes" in a Blasius boundary layer. They are surely dominated by low-frequency streamwise vorticity structures and hence much closer to the motions associated with distributed roughness than to the linearized receptivity modes of Kerschen (described by him in the Receptivity Panel). For as yet unclear reasons, the narrow Klebanoff structures *grow stronger* proportionately to $x^{\frac{1}{2}}$! Without "free-stream" forcing longitudinal vortices represent the *slowest-decaying* non TS mode solutions of three-dimensional Orr-Sommerfeld equations in the limit of infinite x - wave length (Herbert, 1989). Is the observed increase in $u'(x)$ due to non-parallel boundary-layer growth, cumulative effects of the forcing, or some other cause?

When the finer scales near the wall approach a nonlinear regime, they may undergo the low-level, unsteady, straining bypass transition of Landahl, Breuer, and others, described on last page of Morkovin's contribution to this panel. Alternately, the onset of turbulence intrinsic to the layer could come from secondary and higher instabilities of the strongest, nearly steady, streamwise vortices (rather akin to breakdown processes of finite-amplitude Görtler vortices, e.g., Swearingen and Blackwelder, 1987).

Still another mechanism may be present in the evolution to turbulence (and in the sustenance of turbulence in developed layers)– amalgamation of nearly streamwise vortices into larger ones as their scale increases downstream and away from the wall. This seems almost necessary for turbulence initiated by distributed roughness at smaller Re_k where the scales have to grow to a dynamically significant range. This hard-to-document process has been identified in stagnation regions of cylindrical bodies (e.g., Morkovin, 1979), in flat-plate boundary layers (e.g., Anders and Blackwelder, 1980), and in multiple-sensor conditional temperature measurements in slightly heated turbulent boundary layers. We surmise that inherent variation (non-homogeneity) of roughness size and density (or incoming free-stream turbulence) facilitate such mergings whereby stronger (larger) vortices (or vortex pairs) sporadically capture smaller (weaker) ones. This irregular process seems to be reflected as well in the streamwise decreasing number of spanwise minima and maxima of the time-averaged velocity U in fixed y traverses (Anders and Blackwelder, 1980).

3. Possible Numerical Experimentation

The amalgamation and other scenaria can be usefully tested through Navier-Stokes simulations à la Spalart-Yang (1987) and Singer et al. (1987) in *absence of TS* forcing. To *explore the dynamics* of distributions of streamwise vorticity near the wall we do not need to satisfy the boundary conditions at the actual corrugated surface, but impose them at the wall plane $y = 0$. In the randomly conditioned temporal simulation of Spalart and Young, the most physical boundary conditions would probably be the no-slip conditions (which also imply $\partial v/\partial y = 0$) with a constant-in-time random plus-minus distribution of $\partial u/\partial y$ and $\partial w/\partial y$ at a wall grid with a mesh on the order of $0.2k$ to k. Starting at $t = 0$ there should be a slow buildup of the least damped v and ω_x fields. At some disturbance-amplitude threshold, these damped modes (with u disturbances of 0(Re) larger than v and w disturbances, Herbert, 1989) are likely to follow one of the preceding scenaria. A more constraining boundary condition would correspond to a replacement of the surface corrugations by a random fixed pattern of injection-suction with zero mean flux–i.e., by random plus-minus white-noise values of v over the wall grid, with u and w zero. This scheme would probably converge more rapidly and save computing time.

Data processing and analysis presents here problems similar to those in physical experiments and suggestions on processing in Section 1 should be helpful. The addition of a passive scalar-tracer (say by mild wall heating $T_w \neq T_\infty$) can make the geometries of amalgamation or secondary instabilities more discernible. Except for the initiation of the fields at $t = 0$ and slow diffusion thereafter, there is no direct time dependence. Can the numerics trigger sufficient inhomogeneity in time, to precipitate such three-dimensional instabilities? As mentioned in Section 1, it may be necessary to invoke the "cooperative" intermittency aspect of free-stream disturbances to initiate the growth locally in x, z, t, especially with respect to Landahl's lift-up scenario. The intermittency could be simulated by sending through suitable wave packets with Gaussian envelopes $\exp\{-a(x - U_c t)^2 - by^2 - cz^2\}$, where the first approximations to the convection velocity U_c would range from $0.5U_e$ to U_e. This corresponds to the experimental technique of repetitive-impulse wave packet excitation (pioneered by Gaster and Kovasznay) which allows multiple-probe, conditional sampling for extraction of cause and effect.

4. Isolated Roughness Elements; Supersonic Speeds

The main issues associated with the effects of single 2D and 3D roughness elements at low speeds are presented in a compact manner in Sections 3 and 4 of Morkovin's contribution to this panel. See also the summary in Figures 1 and 2. Except for the conditions and details of the final breakdown to turbulence, the flow evolution can be considered conceptually understood. A 116-page report by Klebanoff et al. (1987) on new experiments (including slow modulation of the stream velocity) adds additional insights on single 3D roughness effects.

Almost without exception the bluff single 3D roughness elements investigated in detail in transition to turbulence were symmetric with respect to the spanwise z direction. As seen in Figure 2, they generate symmetric counter-rotating legs of the horseshoe vortices, of the spiral vortices, and of the loop (hairpin) vortices, at the cost of a substantial drag increase. In contrast, the slender, fence-like vortex generators nearly aligned with oncoming velocity (used for invigorating the near-wall region of turbulent boundary layers against approaching separation) initiate asymmetric streamwise vortices rotating almost exclusively in one direction.

Such vortex generators placed in laminar boundary layers are very efficient in bringing about turbulent wedges (Norman, 1972) by an as yet undocumented process. It seems highly desirable to investigate the details of this bypass transition, especially since it may be less sensitive to Mach number. A nearly streamwise rectangular fence normal to the wall acts as a miniature half wing at small angles of attack: the vortex it generates represents the tip vortex associated with an asymmetric force in z direction on the protuberance. When the height of the trailing edge is shorter than the leading edge of the fence, the vortex is weakened, but it is directed toward the wall, where the bypass activity should take place.

Two-dimensional single roughnesses lose their tripping effectiveness as M increases past unity because separated mixing layers become progressively more stable (see physical reasons in Morkovin, 1987). In contrast to low speeds, traveling skew waves become the most unstable modes in separated as well as attached supersonic layers because of the alleviating sweepback effect. Processes associated with nearly streamwise vortices (i.e., with nearly 90° of sweepback) such as the above vortex-generator bypasses and Görtler instability may therefore be expected to be least affected by increasing Mach number. Morrisette et al. (1969) reported how ineffective symmetric roughness shapes are in tripping turbulence in hypersonic boundary layers despite their high drag. The vortex-generator bypass may retain its tripping effectiveness precisely because it does not depend on eigenfunction formation of primary and higher instabilities which is inhibited at supersonic speeds by limited upstream and cross-stream influence between fluid elements.

References

[1] Anders, J. B. and Blackwelder, R. F., 1980. Longitudinal vortices in transitional boundary layer, 110-119 of *Laminar-Turbulent Transition*, Proc. IUTAM First Symp., Stuttgart, Germany, Eppler, R. and Fasel, H., editors, Springer Publ.

[2] Herbert, Th., 1989. Exploring transition by computer, to appear in J. of Appl. Num. Math.

[3] Kendall, J. M., 1985. Experimental study of disturbances produced in a pre-transitional laminar boundary layer by weak freestream turbulence, AIAA Paper 85-1695.

[4] Kendall, J. M., 1987. Experiments on TS wave induction by free-stream turbulence, Bull. Am. Physics Soc., Vol. 32, No. 11, Paper GE6, p. 2079.

[5] Klebanoff, P. S., Cleveland, W. G., Jr., and Tidstrom, K. D., 1987. Interaction of a three-dimensional roughness element with a laminar boundary layer, Arnold Engin. Dev. Center, Rept. AEDC-TR-87-7, 116 pp.

[6] Morkovin, M. V., 1979. On the question of instabilities upstream of cylindrical bodes, 106 pages, NASA Contractor Rept. 3231.

[7] Morkovin, M. V., 1987. Transition at hypersonic speeds, ICASE Interim Report No. 1, 24 pages.

[8] Morkovin, M. V. and Reshotko, E., 1990. Dialogue on progress and issues in stability and transition research, Introductory Paper in *Laminar-Turbulent Transition*, Proc. IUTAM 3rd Symp. Toulouse, France, D. Arnal and R. Michel, editors, Springer Publ.

[9] Morrisette, E. L., Stone, D. R., and Whitehead, A. H., Jr., 1969. Boundary layer tripping with emphasis on hypersonic flows, in *Symp. on Viscous Drag Reduction*, Wells, C. S., editor, Plenum Press.

[10] Swearingen, J. D. and Blackwelder, R. F., 1987. The growth and breakdown of streamwise vortices in the presence of a wall, J. Fluid Mech., Vol. 182, pp. 255-290.

[11] Tadjfar, M., Reshotko, E., Dybbs, A., and Edwards, R. V., 1985. Velocity measurements within boundary-layer roughness using index matching, *Inter. Symp. on Laser Anemometry*, ASME Fed. Publ., Vol. 33, pp. 59-73.

ON RECEPTIVITY TO ENVIRONMENTAL DISTURBANCES

Mark V. Morkovin

Professor Emeritus
Illinois Institute of Technology
Chicago, IL 60616

1. Instability Dehomogenization of Shear-Layer Vorticity

The key characteristic of turbulence is 3D vorticity with a range of scales. The initially laminar shear layers generally vary slowly in the streamwise direction x and the spanwise direction z. They tend to be quasi-two-dimensional and quasi-homogeneous. During the transition process they become *dehomogenized* in x and z as well as unsteady in t through the *forcing* action of *environmental disturbances* in conjunction with various *more rapid spatial variations* in properties of the boundaries: slope, curvature, temperature, porosity, etc. The no-slip boundary conditions automatically *convert* any *steady or unsteady pressure gradients* $\partial p/\partial x$ and $\partial p/\partial z$ into density-weighted *vorticity outflows* (per unit wall area) of ω_z and $-\omega_x$ respectively. This is of course the principal (viscous) mechanism by which *irrotational* flow fields, *including sound*, are converted into steady *shear layers* and unsteady *vorticity waves*. It plays a role in most receptivity mechanisms whether for the *primary* or *secondary* and *higher instabilities.*

2. Conditions for Stimulation of Unstable Normal Modes

Through the *primary* linear *instability*, the original *vorticity* distribution $\vec{\Omega}$ is *restructured* with increasing "*lumpization*", e.g., of ω_x in the z-inhomogeneities of Görtler vortices, and of ω_z in the cats-eyes x-inhomogeneities of TS and inflectional vorticity waves, etc. In the latter space-time periodic cases, there is also an average *transfer* of mean-square vorticity *from* the *mean* Ω_z *to* the *unsteady vorticity wave* of magnitude $-2\overline{v\omega_z}\cdot\partial\Omega_z/\partial y$, Nishioka and Morkovin (1986), hereafter referred to as NM. (Commonly in primary instabilities the total vorticity is conserved.) For the *forcing* disturbance field, u_f, v_f, p_f, to *nurture* an unstable TS wave there must be a predominance of positive transport contributions $\overline{v_f\omega_{zTS}}$ over the shear layer, together with a positive flux $2/\rho \cdot \overline{\omega_{zTS}\partial p_f/\partial x}$ at the wall. In spatially unstable flows this indicates that the periodic forcing disturbance field *must have or acquire characteristic lengths on the order of the TS wavelength*, λ_{TS}, at least locally over the region of "seeding" of the ultimately exponentially growing TS waves.

3. Minute Wall Change as Example of TS-Wave Forcing by Sound

The growing u' fluctuations in *Figure 1* from NM *illustrate local* flat-plate *receptivity seeding* of Aizin-Polyakov by an extremely thin strip of Mylar tape ($\lambda_{TS}/4$ in width, at the wall) when irradiated by an upstream-traveling acoustic wave of intensity u'_{ac}. In the absence of the strip the above correlations between the forcing field and TS waves of the same frequency were negligible since $\lambda_{ac} \gg \lambda_{TS}$. The short-scale non-homogeneous variation of Ω_z due to the strip is imprinted on the forced fixed-frequency inner-boundary-layer dynamics (describable by the triple-deck technique, Goldstein and Hultgren (1989)) and brings about the unsteady TS x-dehomogenization. A classic TS wave, *proportional to the acoustic amplitude*, emerges over a short active seeding distance on the order of λ_{TS}. Farther downstream the mismatch of characteristic lengths (and wave speeds) is reestablished and the forcing wave and the surviving part of the response merely superpose.

4. Rapid and Moderate x-Variations in Mean and Disturbance Fields → Competing Receptivity Paths

Evidently when such mismatches occur, *special x-dependences* within the *whole system* of the shear layer, the body that generates it and the disturbance field, can provide *one* or *several receptivity* paths. As signaled in (1), sharper variations in wall slope, curvature, temperature, porosity, etc. are expected to act similarly to the Aizin-Polyakov strip, with different proportionality coefficients yielding a measure of their "efficiency," Goldstein (1985), Heinrich, *et al.* (1988), Goldstein and Hultgren (1989). The sharper the variation of a function of x, the broader its Fourier spectrum and the more likely it is to contain a contribution (though perhaps smaller) at the TS wave number, k_{TS}.

An airfoil-like body with *continuous curvature diffracts sound-waves* so that the amplitude A of the *acoustic* (or near-field) *pressure gradient* varies with x. Thus $A(x)$ acquires *additional length* characteristics besides λ_{ac}. A match with λ_{TS} is possible if the Fourier transform of $A(x)$ contains significant contributions at k_{TS}. Two types of (albeit 3D) TS geneses over a flat plate were *demonstrated* in substantial detail *experimentally* in NM. There the $A(x)$ dependence was due to the spreading of the field from a 3D source at a height of 6δ above the plate, rather than due to continuous wall curvature. Enhancing and interference patters were observed.

5. Role of the Leading Edge and the Issue of its Idealization

Idealization of the *flat plate* as one of *zero thickness* leads to the sharpest, in fact a singular variation in x at the leading edge, LE. When the upstream traveling acoustic wave of *Figure 1* reaches the edge it diffracts around it, creating large oscillatory v_f. The large unsteady vorticity fields, so generated, decay rapidly downstream at these low Re numbers yet according to the theory of Heinrich, *et al.* (1988), their remnants should be very effective in stimulating TS waves at x locations past $\mathrm{Re}_{cr}(f)$. Yet in *Figure 1*, where the strip is in the TS amplifying region, there is no evidence of any LE-generated amplifying wave upstream of the strip. This may be due to the differences between the actual sound fields in Polyakov's wind tunnel and the theoretical model (e.g., the fields in the other half channel formed by the plate) or due to overidealization.

All real-world plates have finite thickness, finite radius of curvature at LE, and a finite stagnation-line boundary-layer thickness there, e.g., Kachanov, *et al.* (1978) (also discussed in NM). The favorable pressure gradient at the nose is followed by an adverse one on the flat region downstream of the nose. There the inflected BL profile thickens and approaches rather rapidly the neutral stability conditions. *Is the sequence of receptivities and decays* near the stagnation line, on the nose where the straining field promotes $A(x)$ dependence of most disturbances and in the nearly separating BL on the flat (where there is discontinuous curvature *and* an especially interactive BL, Goldstein (1987)) *equivalent* to the sequence of processes *in the limit of zero thickness of the theory*? Would the net receptivity coefficients at Re_{cr} be comparable?

6. Experimental Difficulties; Detailed Validation of Receptivity Paths

Ultimately some *convincing agreement between* careful, difficult *experiments* and *matching* careful, difficult *theory*, with computations, has to be achieved to enable us to progress to *reliable comparisons* of *competing receptivity paths* (needed for applications). For such comparisons it is imperative that the theoretical and experimental coefficients be unequivocally defined and documented. *Thus far*, with minor exceptions receptivity *experiments* have been *inconclusive, even confusing. They suffer* from (a) lack of quantitative characterization of the forcing fields along the outer edge of the BL; (b) inadequate local information on the fluctuations inside the BL in the region where the unstable response starts growing; (c) lack of documentation of the near-singular diffraction fields near the LE and/or of the presence and effects of vibrations; (d) excessive forcing disturbance levels. For instance, attention to (a) and (b) should *help to validate directly* the characteristics of the locally disturbed velocity and pressure fields provided by the triple-deck theory in a *recheck of the experiment* in *Figure 1.* The inconclusiveness from tenuous inferences from quite inconsistent information concerning the *effects of leading edges* and their *vibrations is detailed in NM.*

7. Receptivity to Free-Stream Turbulence

The *experimental* conditions are especially challenging in studying receptivity paths to mean-flow *convected weak turbulence fields* which are initially intermittent and inhomogeneous in x, y, and z. It is these aperiodic *instantaneous local distributions*, if sufficiently intense, which may bring about higher instabilities and even local onsets of vigorous turbulence on the local scales $< \delta$ in a bypass mode of transition. In contrast, weak turbulent fields may be *invigorated* by the *mean-flow strains* due to the body shape and due to $\partial U/\partial y$ in the shear layer itself. Some *may* thus be amplified (algebraically?) past a nonlinear threshold and exhibit a *bypass* pattern not yet documented theoretically or experimentally. (Such strained decaying turbulent wave packets are suspected of causing local onsets which commonly seed turbulent patches in 2D Poiseuille ducts at Re $\sim$ 1500 to 2500). *Non-decaying* turbulent fields convected at a *flat plate* which do not undergo a bypass may well be approximated by a

3D superposition of "*convected gusts*" (i.e., periodic incompressible shear waves of various orientations and wave numbers) of Kerschen (1989) and Heinrich, *et al.* (1988). The peculiar nature of receptivity to turbulence and evidence of possible bypass will be conveyed at this session by *J. Kendall.*

Gaster (1975) and Gaster and Grant (1975) pioneered an experimental and theoretical *technique* of *forced repetitive "impulsive" wave packets* which allows extraction of commonly recurring unsteady 3D patterns out of noisy environments. The technique should be useful for studying spatio-temporal development if isolated turbulence wave packets (e.g., turbulent vortex rings) were made to enter or pass just outside the boundary layer of a body or a flat plate. Multiple-probe sensing and Gaster's signal purification by "phase averaging" over repetitive trials would be de rigueur.

8. Nonperiodic Excitation

Nishioka and Morkovin (1986) described the *difficulty* of *decomposing* the *periodic experimental signals* into the forcing field, the particular solution response, and the proportionate TS response to a continuous single-frequency acoustic disturbance. It is noteworthy that the Gaster-type repetitive *"impulsive" sound wave-packet excitation* could provide much extra information, because it is likely to *reveal* the distinct *spatio-temporal development* of the two responses and thus identify the components.

9. What are "Natural" Disturbances?

Historically, engineers used the designation of "*natural disturbances*" and "*natural transition*" for conditions indigenous to a *given facility or flight* which were *not intentionally altered by* introduction of *flow manipulators* for the specific test – e.g., roughness, acoustic irradiation, etc. When an experimenter permanently added a screen in the settling chamber or cleaned existing ones, the disturbance complex changed, *Schubauer and Skramstad*, SS (1943), Klebanoff, *et al.* (1962), but the conditions remained "*natural*". (One suspects that the usage of the subjective term "natural" included a hidden wish that the results of the particular experiment would be more widely applicable as if they stemmed from a natural law.) SS intentionally *perturbed* their boundary layer with a vibrating ribbon, loudspeaker and 3D pulsations through a hole in the wall to improve experimental characterization of the Tollmien-Schubauer-Schlichting TS waves and to *identify for the first time* the *receptivity* to unsteady pressure fields. (Prandtl referred to the waves as Tollmien's, whilst Theodorsen referred to them as Schubauer's.)

The fact is, and will remain, that the forcing *"natural" disturbances* in ground facilities and flight *comprise unknown different mixtures* of the disturbance classes (with their structures and intensities) in *Table 1.* There is some evidence that all of them can influence the type and length of the road to turbulence. The small initial *preamplified characteristics* of these disturbances (such as the distribution and orientation of spectral intensities in the different classes analyzed for receptivity by Heinrich, *et al.* (1988) are *unlikely to become quantifiable* even by judicious usage of multiple sensors in basic experiments. Our *assured ignorance* of the relevant *input* into instability amplifications in "natural" transition places *limitations* on the usage of the results of receptivity theory and experimentation and on the degree of predictability of transition in general. This involves the broader philosophy of transition research and the philosophy of transition-sensitive design.

As I harken back 9.5 years to the state of receptivity ignorance in which I heard in Novosibirsk the early, rather confused account of the experiment in *Figure 1*, I have no doubts that the intervening revelations outlined here have been profound and that we need much, much more basic understanding before we can wisely apply it. Beyond this summary, receptivity mechanisms to many disturbed fields in *Table 1* remain a challenge; from that to ω_x, to entropy fields, to particulates and aerosols, and to all the changes in receptivity with Mach number.

References

[1] GASTER, M., 1975. A theoretical model of a wave packet in the boundary layer on a flat plate. Proc. Roy. Soc., Vol. A 347, 271-289.

[2] GASTER, M. and GRANT, I., 1975. An experimental investigation of the formation and development on a wave packet in a laminar boundary layer. Proc. Roy. Soc., Vol. A 347, 253-269.

[3] GOLDSTEIN, M. E., 1985. Scattering of acoustic waves into Tollmien-Schlichting waves by small streamwise variations in surface geometry. J. Fluid Mech., Vol. 184, 509-530.

[4] GOLDSTEIN, M. E., 1987. Generation of Tollmien-Schlichting waves on interactive marginally separated flows. J. Fluid Mech., Vol. 181, 485-518.

[5] GOLDSTEIN, M. E. and HULTGREN, L. S., 1989. Boundary-layer receptivity to long-wave freestream disturbances. Ann. Rev. Fluid Mech., Vol. 21, 137-166.

[6] HEINRICH, R. A., CHOUDHARI, M., and KERSCHEN, E. J., 1988. A comparison of boundary-layer receptivity mechanisms. AIAA Paper 88-3758.

[7] KACHANOV, YU.S., KOZLOV, V. V., and LEVCHENKO, V. YA., 1978. Origin of Tollmien-Schlichting waves in boundary layers under the influence of external disturbances. Izv. Akad. Nauk SSSR, Mekh, Zhid. i Gaza, Vol. 5, 85-94 (in Russian). Transl. Fluid Dyn., 1979, 704-711.

[8] KERSCHEN, E. J., 1989. Boundary-layer receptivity. AIAA Paper 89-1109.

[9] KLEBANOFF, P. S., TIDSTROM, K. D., and SARGENT, L. M., 1962. The three-dimensional nature of boundary-layer stability. J. Fluid Mech., Vol. 12, 1-34.

[10] NISHIOKA, M. and MORKOVIN, M. V., 1986. Boundary-layer receptivity to unsteady pressure gradients: experiments and overview. J. Fluid Mech., Vol. 171, 219-261.

[11] SCHUBAUER, G. B. and SKRAMSTAD, H. K., 1948. Laminary boundary-layer oscillations and transition on a flat plate. NACA Tech. Report 909.

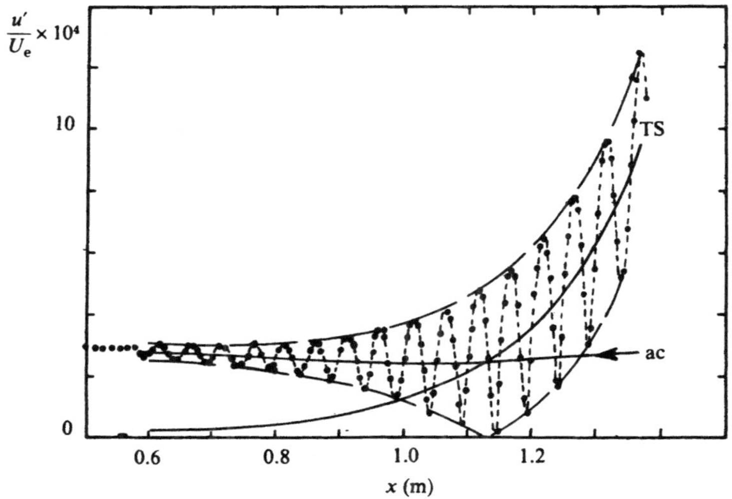

FIGURE 1. Hot-wire measurements of total signal u' yielding an interference pattern between a forcing acoustic wave of 138 Hz at nearly constant 104 dB and the growing TS response on a flat plate downstream of a strip of Mylar tape 12 mm wide and 34.5 ± 2.5 μm high. As indicated, the tape is 0.565 m from the leading edge at a local $Re_{\delta*}$ of 1550 and $\delta^* \approx 1$ mm. Aizin & Polyakov (1979).

DISTURBED ENVIRONMENT in OPEN-FLOW SYSTEMS

PROBLEM of DIAGNOSTICS of ONCOMING FIELDS and IMPURITIES

SPECTRA! minimum of 2 INSTRUMENTS to separate free-stream Tu and p' fields

STRAINING of DISTURBANCE FIELDS (as they approach the body and shear layer) is part of the problem, especially near stagnation regions, sharp edges, upstream contraction; also by the shear in the boundary layer itself.

● **IRROTATIONAL UNSTEADY PRESSURE FIELDS** $p'(x,y,z,t)$	(a) isentropic p', ρ', T' : Acoustic **HYPERBOLIC** behavior especially at higher f : propagating wind-tunnel modes, resonances. (b) constant-ρ (near-field sound~pseudosound) : **ELLIPTIC** at lower f. Wind Tunnels : SEPARATION in DIFFUSORS, Helmholtz instabilities.
○ **WALL VIBRATIONS** : receptivity similar to p' fields.	
● **Vector VORTICITY FIELDS,** ω', **True TURBULENCE**	(a) **QUASI-STEADY STREAMWISE** ω_x' : essentially **PARABOLIC**, due to upstream bends, guide vanes, screens etc. Important sources : LEADING-EDGE IMPERFECTIONS, SURFACE ROUGHNESS. (b) **UNSTEADY** $\omega_x', \omega_y', \omega_z'$ with associated V' field, **PARABOLIC**; Larger-scale free-stream Tu structures → intermittency in x, t; the sharp x, t dependence for each structure may induce TS wave packets across streamlines or when penetrating shear layer? Tu amplitude decay in x important, but much slower. ω modes are forceable by vibrating ribbon, pressure holes, inhomogeneous time-dependent wall suction and heating.
● **ENTROPY FIELDS** σ' (Temperature-density fluctuations at constant p)	Arise in any heat addition; (if explosive → transient p' field is added); propagate essentially **PARABOLICALLY** along main stream lines. Do induce unstable vorticity modes especially at higher Mach numbers.
● **PARTICULATES and AEROSOLS**	Are known to expedite transition to turbulence, but how? Presumably by generating local, traveling vorticity fields near walls.

○ **EACH DISTURBANCE FIELD** may have distinct **MULTIPLE** linear and nonlinear **RECEPTIVITY PATHS → COMPETITION** within and among the DISTURBANCE FIELDS

which is most dangerous for a given configuration?

TABLE 1

ON ROUGHNESS - INDUCED TRANSITION: FACTS, VIEWS, AND SPECULATIONS

Mark V. Morkovin

Professor Emeritus
Illinois Institute of Technology
Chicago, IL 60616

1. The Roughness Issue: Our Task

The *mechanisms* whereby small *distributed roughness* promote transition to turbulence are *unknown* and represent bypasses to known instability scenarios, including the e^N design method. Mechanisms associated with *single-roughness elements* are partially understood (at least at low speeds) but probably *not quantifiable*. Yet, when favorable-pressure gradient laminar-flow airfoils were designed, roughness was the primary agent which rendered them nonlaminar in practice. Nearly blind correlation from a limited bases of experimental data still governs roughness sensitive designs. The topic was put on the agenda of the workshop first because of its obvious practical importance. Secondly, it is hoped that our vastly improved instrumentation and computer capabilities in the 1990's may lead to *research into the mechanisms themselves* and thereby establish a more rational basis for design correlations.

The task of the *Panel on Roughness* is primarily to provide information and "feel" on the strange type of flow associated with distributed and isolated roughness, i.e., provide a target for our imagination and intuition, rather than to present competing views and techniques. I will outline what we understand in general, Morkovin (1985), and discuss briefly an experiment where distributed roughness surprisingly promoted growth of TS waves, Corke, et al. (1986) and speculate. Jim Kendall will describe low Re flow fields he observed on arrays of spheres, Kendall (1981). His wakes of streamwise vorticity possessed no intrinsic unsteadiness and were therefore not clearly connected to any transition farther downstream. Yet, at some higher Reynolds numbers, such flow fields evolve to transition so that a glimpse of their more intimate anatomy may help us speculate on possible processes which could bring about unsteadiness compatible with turbulent motion on the boundary-layer scale δ. Eli Reshotko will speak about novel experiments on just such higher - Re sphere arrays, with unsteadiness.

2. Passivity of Roughness; Direct and Co-Promotor of Linear Instabilities

Recalling the discussion of x, z, and t dehomogenization processes in my comments on receptivity, CR, Kendall's experience makes it clear that *roughness* is a *steady passive device.* It brings about x (and z) dehomogenization as did the Mylar strip of CR Figure 1; in both cases any *unsteadiness* has to be *forced* either by *sound or* freestream *turbulence.* The Aizin-Polyakov experiment and receptivity theory generally put to rest an old belief that the effect of roughness cannot be enhanced by freestream unsteadiness. Roughness-induced separation is especially sensitive to unsteady environmental forcing.

At lower Re, the specific surface geometry alters the *steady continuous distribution of vorticity* in the boundary layer. In *Görtler* and *cross-flow instabilities* such steady disturbances can be selectively filtered and amplified. They thus provide an important direct input into *primary-instability growth* of streamwise vorticity ω_x or nearly streamwise vorticity with cross-stream wavelengths on the order of $2\delta - 4\delta$. Such vorticity formations, when large enough, are then subject to various *unsteady secondary* instabilities and lead in short order to turbulent bursts. An analysis of this *primary receptivity role* would relate the surface shape to the wave-number spectra of the amplifiable vorticity disturbances.

3. Single-Roughness 2D Wall Wakes, Inflectional Amplifiers

The key feature of *single-roughness flows* centers on the *separated wall-wakes* and their special characteristics: the vortex sheets which become thinner with increasing Re, the lengths of the recirculation zones, internal secondary flows, etc. *Figure 1a* defines the prototypical geometry and the usual terminology. Effects of roughness of height k should depend on its relative (blocking) height, k/δ^* (δ^* - boundary-layer displacement thickness) or its dimensionless Reynolds disturbance to the flow near the wall, Re_k, or both. $\mathrm{Re}_k = kU_k/\nu$, where U_k is the *undisturbed velocity* at height k and ν the kinematic viscosity; for Blasius boundary layers and small k values $\mathrm{Re}_k = 0.57\ \mathrm{Re}_{\delta^*}(k/\delta^*)^2$. If u^* is Prandtl's friction velocity at the wall, $(\tau_w/\rho)^{\frac{1}{2}}$, the *alternate wall variables* u^+ and y^+ are u/u^* and $yu^*/_\nu$, respectively. Again, for small k and Blasius profiles, U_k equals $ku^{*2}/_\nu$ and $\mathrm{Re}_k = (k^+)^2$, the square of the governing viscous parameter for rough turbulent wall flows. Many engineers consider a wall hydraulically smooth when Re_k remains below 25.

Two-dimensional isolated *roughness*, such as a spanwise trip wire, causes local upstream and downstream separations, the latter extending 40k-50k before reattachment to the wall. The separated inflectional velocity profiles amplify TS waves very much more rapidly than the undisturbed boundary layer, Klebanoff and Tidstrom (1972). This *powerful overamplification* makes trip wires most efficient in causing early transition – through a well understood and hence partially controllable mechanism. We note that the *receptivity* of such devices to sound and turbulence is *doubly efficient*, because of its compactness, Morkovin (1989), and the separation line itself. Furthermore, *Figure 1c* reminds us that slight spanwise nonuniformities in the upcoming BL and/or the height k invariably cause some *local spanwise vents* in the separated "cavities." The ensuing tongues of ω_x accelerate the onset of secondary instabilities.

Flows associated with 2D roughness have *relevance* to the problem of *distributed roughness* because in most such experiments, strips or sheets of abrasive paper fastened to the wall provide the roughened surface. In so far as the substrate and granules form quasi-two-dimensional protuberances at their leading and trailing edges, some of the powerful 2D-isolated-roughness mechanism is operative. These leading and trailing-edge effects of roughness patches have complicated the correlations of data from past experiments. Thus $(k/\delta^*)_0$ and Re_{k0} values at *the leading edge*, x_{k0}, have *special risk significance* beyond the connotations of k/δ^* and Re_k values for an isolated 3D roughness element or for an element in the middle of a roughness patch, Reshotko and Leventhal (1981), Leventhal (1981).

4. 3D Single Roughness: Short Leaky Separated Pockets

As can be seen from the successive rows of sketches in *Figure 2*, the nature of locally separated regions at a *3D isolated roughness* (with nearly equal height, span, and length) *differs greatly* from that for 2D isolated roughness, at all Reynolds numbers. The rear pocket leaks steadily and is 8 to 10 times shorter, *Figure 1a*! The front separation generates a horseshoe vortex which wraps around the roughness, forming streamwise vortices which induce downwash along the centerlines of the wake. For Re_k below about 300, the separated rear shear layer and the head and legs of the horseshoe vortex are stable. The three projections in the first row of *Figure 2* hint at the fascinating almost singular complexity of these flows especially near the rear stagnation point at x_r, as revealed by Mochizuki (1961a, 1961b) and Gregory and Walker (1956).

As seen in the second row of *Figure 2* even the short separated rear shear layer becomes unstable. It starts shedding downstream-moving vortex loops looking like hairpins with the head pointing downstream past a critical Re_k of about 300. The vortex rotation is opposite to that of the horseshoe vortices which remain at the wall, somewhat to the side of the stretching hairpin arms. The rotation of the hairpin head generates an upward force which progressively moves the head of the loop to the edge of the boundary layer where it decays as it floats downstream. The *lift-up* tends to *decrease* the nonlinear vortical *interactions* near the wall. At elevations near 2k, rms fluctuation levels of $0.04U_e$ were observed without the boundary layer becoming turbulent upstream of $3x_k$, Norman (1972). Acarlar and Smith (1987) described in detail and interpreted the kinematics of the unsteady interlacing everstretching loops. The frequency of the shedding scales on the thickness of the separating vortex sheet and is therefore rather high. Hence no TS instability is induced in the neighboring boundary layer even past its own critical Reynolds number Re_{crTS}.

There is a significant *near-universality* in the *topology of the evolution of the vortex formations* sketched in the first two rows of *Figure 2*; the chimney flow within the pocket, escaping it "singularly" in the form of the spiral filament pair (which induced an uplift at the centerline at larger x); the main horseshoe vortex and the hairpin eddies with their own uplift, Morkovin (1972). R. Norman and the author observed visually for days the same evolutionary sequences in symmetric boundary-layer flows around spheres, hemispheres, cones, short cylinders (upright and on the side), parallelopipeds and "thinner fences," as the wind speed U_e increased. The Reynolds numbers of the events changed with shape but the dominant topology appeared invariant. (Asymmetric fences, alias one-sided vortex generators, cause very early local transition, evidently without the amplifier action of *Figure 1a*. The upstream head of the horseshoe vortex and the spiraling chimney flows are present at isolated buildings in atmospheric boundary layers.)

5. Sudden Breakdown of Hairpin Eddies: Mechanisms and Parameter Correlations

As Re_k increases, the horseshoe vortices and the stretching interlacing hairpin vortices grow stronger. At Re_k values from 450 to 600 (depending on x_k, protuberance shape and the pressure gradient), the interaction of the full multiple vortex system with the boundary layer causes a sudden local onset of turbulence at first rather far downstream: x_{tr} occurs at $2x_k$ to $5x_k$ or even farther! In other words the *incubation distance* from the offending roughness to the roughness-conditioned transition is *initially very long.* Furthermore this induction of turbulence bypasses the quasi-two-dimensional TS mechanism. According to Klebanoff (1983), the ***substitute instability*** which spawns the hairpin vortices, combines the effects of *primary TS* and *secondary K-type* instabilities on the road to turbulence on smooth walls. As Re_k grows still higher, x_{tr} moves very rapidly upstream; for an increment of 10 in Re_k, the forward movement of x_{tr} may cover more than half the distance to x_k.

To organize the transition information with the many parameters ($k, x_k, \partial p/_{\partial x}$, etc.), the sharply dropping values of Re_{tr}, plotted against Re_k, are usually extrapolated linearly to the limiting "critical" loci, $\mathrm{Re}_{tr} = Ux_{k/\nu}$, where transition would occur at the position x_k of the roughness as U_e is varied. Curves of such "*effective*" Re_{tr} and Ue_{kcr} condense much *useful engineering information*, first presented for "incompressible" flows by Tani in 1961 (revised and extended in 1970). Since separated layers become rather stabilized at supersonic speeds the elusive conditions under which high-speed boundary layers get tripped by 3D roughness is even less understood. However, the phenomenon of the sharp motion of transition in upstream direction seems to persist but at different parameter values.

Such *correlations* work at best for single-roughness elements and do not tell us *what* the *threshold mechanisms* associated with the rapid upstream movement of transition might be. Bits of information at low speeds suggest that the ***sudden turbulent "bursts"*** may occur at a relatively low elevation of about k, *off centerline*, close to where the legs of the hairpin vortices interlace and stretch but *not* out of interaction range with the horseshoe vortex. The last row of *Figure 2* also suggests that *turbulent spots* spread by lateral contamination from these foci, when Re of the neighboring laminar layer exceeds $Re_{t\,min}$. (Otherwise turbulence is confined to the parabolically widening wake of the element itself.) The periodic spots crowd into a turbulent near-wedge with half-angle of $9° - 11°$ at low speeds and $4° - 6°$ at supersonic speeds (again useful engineering information). However, in the absence of our understanding the underlying mechanisms, the prospects of extending the correlation schemes to distributed roughness hold little hope, see Section 0.04.10 a and b of Morkovin (1985).

To sum up: transition induced by a single 3D roughness element occurs first at large incubation distances downstream, as an interaction involving a streamwise vortex system, and is then very sensitive to even small increases in Re_k, moving rapidly to the near-wake of the element. The process is therefore vastly different from induction of transition by a single 2D roughness. While the TS instability is bypassed, the overall process still involves an instability albeit rather inefficient because of the shortness and narrowness of the separated pocket. The significant features of *relevance to distributed 3D roughness* are (a) the existence of complex, predominantly streamwise vorticity patterns and (b) the existence of long induction distances from the process-initiating roughness to the onset of turbulence. Effect (b) makes the judgement of cause and effect very difficult and dims further the prospects for successful schemes of parameter correlation.

6. On Flows over Distributed Roughness

What happens to the vortex systems of feature (a) discussed above when the roughness element has neighbors as in distributed roughness, and what k and Re_k can we use to characterize the collection of the disparate roughness elements? Since the elements are more or less randomly distributed, there will certainly be some infringement of the streamwise vortices from an upstream element upon the horseshoe vortex head wrapped around a downstream element, a potentially destabilizing occurrence. However, if the elements are packed rather closely there will be little upstream fetch to generate a strong horseshoe vortex; with increasing grain density, there is likely to be partially dead fluid in the free space between elements. In any case, much of the the *fluid* in the crooked valleys is likely to have *very low velocity (inertia)* so that it may be *influenced rather easily by free-stream velocity and pressure disturbances.* Nearer to the peaks of the elements the fluid will have considerable streamwise vorticity with spanwise scales on the order of k, i.e., rather small relative to δ. Commonly occurring statistical nonuniformities in the roughness-element distribution may create larger spanwise scales, which could promote *amalgamation* of the *smaller streamwise vortex structures.* Such amalgamation is occasionally seen in smooth boundary layers with streamwise vorticity.

Nominal designations of the k for a large conglomeration of grains spread on a surface (abrasive paper, snad paper, etc.) obviously *miss* the relevant *effect of grain density.* For smaller densities, the road to turbulence could be dominated by a small number of high local maxima, $k_{\max}$, and breakdowns of their hairpin eddies. The *most damaging* should be those *near the leading edge* of the body (or the leading edge of the roughness sheet) because Re_k is largest and so is k relative to δ^*. If $k_{\max}$ were three to five times k_{nom}, roughness with reported Re_k of say 150 could be harboring turbulence-seeding roughness peaks. (For rough (anisotropic?) surfaces made by rubbing with abrasives, $k_{\max}$ were reported about ten times measured k_{rms}!) To spot sporadic local breakdowns within high frequency wave packets one would need *spatial resolution* of the *sensor finer than k* because of the narrowness of the roughness wake. In the one case where smoke-wire visualization was utilized, Corke, et al. (1986), largest nominal Re_k did not exceed 150 and no spotty breakdowns were seen. Whatever the merits of the scenario involving dominant peaks, its discussion concretizes for us some of the troubling issues, in particular the *difficulty with standardization* and *comparison between experiments.* In this respect Kendall's 1981 selection of regular arrays of spheres as a target of study carries great merit, however, unrealistic it might be for most applications.

7. What do the Experiments Tell Us? Speculative Interpretation

Microscopic studies of actual transition over distributed roughness are few, mostly prompted by Eli Reshotko and myself. All had to grapple with the aforementioned problem of the leading edge of the roughness. The Case-Western group examined mostly TS-subcritical cases, to be discussed shortly. Corke, et al. (1986) compared TS-postcritical flow fields over a smooth surface and over the same roughness sheet (roughness protruding or recessed relative to the oncoming smooth plate) – in the same disturbance environment, see top and bottom of Figure 3. The IIT group observed no bypasses, presumably because their nominal Re_k values did not exceed 150. Instead, they *documented TS growth rates* over both the protruding and the recessed roughness ***substantially in excess*** of the theoretical and the observed *rates for the smooth Blasius profile.* Since λ_{TS} was 50 to 60 times k_{nom}, the small-scale randomization by the roughness did not hamper the tuned TS process. Evidently mean and unsteady behavior below the roughness peaks enhanced the TS growth. (Hopefully, this region will be illuminated by new Case-Western techniques.) This region is apparently *very susceptible* (receptive) to *external disturbances* and may possibly contribute to their amplification as TS waves, albeit irregular near the surface. Alternately, the physical x range of receptivity is wide and it is not impossible that "constructive interference" between seeded TS waves is present. There was also *increased three dimensionality* due to the roughness and consequently an *earlier* onset of *secondary* Herbert-type *instability* was observed.

Despite many important implications of the Corke and BSM findings, they throw no light on the bypass mechanisms associated with distributed roughness. In the Case-Western group, neither Shin, et al. (1982) nor Reshotko and Leventhal identified any significant TS growth, thus confirming the bypass nature of their transitions. Their hot wires and hot films recorded *continuous growth* in a *broad band of frequencies far below* the TS range. This growth led to turbulent episodes without discernible breakdown patterns. In his critical overview of this research Reshotko (1984) wrote of an indication that the departure from laminar flow was "explosive." Since there is a mathematical connotation to the term, it would be desirable to concretize the explosive evolution of the observed characteristics and compare it to the suddenness of onset of turbulence in other transition fields.

The important clue provided by the Case-Western experiments is the broadband growth of low frequencies. The broad-band feature invariably indicates randomization. The geometry and the low-frequency content point to three-dimensional motions. We know that Fourier components of motion with phase fronts at small angles to the x direction are registered at low frequencies f by the *one-dimensional hot wire.* The sensor, unable to distinguish between skew orientations of velocity fields, superposes their signal contributions onto those of lower-frequency x-Fourier components yielding spectra which vary inversely with f. (All hot-wire spectra suffer from this mapping of higher-dimensional information onto the one-dimensional axis, which cannot be inverted unless we know that the field is isotropic.) In view of the lessons of *Figure 2* and the observed low-frequency content it seems reasonable, to *search* for possible *modes of growth* of fine-scale, *nearly streamwise vorticity*, generated by the distributed roughness at Re_k values encountered near its leading edge, x_{k0}, *in presence of unsteady environmental disturbances.*

Another possibly related *bypass*, also involving generation of *narrow streamwise structures*, was analyzed by *Landahl* (1980) and studied experimentally, theoretically and numerically (Navier-Stokes simulation) by *K. Breuer* (1988) for *smooth walls.* Its premise is an unsteady initial lift-up, i.e., a localized up-down movement near the wall. The inviscid linear theory indicates an algebraic growth due to the straining in the boundary layer: it could possibly cross a nonlinear threshold. Experiments involving up-down motions of a membrane in the wall indeed conform to the linear theory when the disturbances are weak and lead to sudden transition when a disturbance threshold is exceeded. The *question* remains *what* could *prompt* such *time-dependent uplifting* motions in, say the 2D Poiseuille flow, where bypasses reign supreme? In that case, diverse clues point to ingested freestream turbulence, with ω_x components due to stretching in the contraction and unsteadiness (intermittency?). "Natural" sporadic uplifting in presence of distributed roughness may possibly be easier to generate: unsteady pressure fields or turbulence are likely to induce more v on corrugated boundaries than on a flat wall.

Eli Reshotko commented that *our panel is in complete agreement* namely that we are truly ignorant about bypass transition over surfaces with distributed roughness. My nose is pointed toward finer streamwise boundary-layer structures under the influence of BL strain and unsteady environmental disturbances.

References

[1] ACARLAR, M. S. and SMITH, C. R., 1987. A study of hairpin vortices in a laminar boundary layer. Part 1. Hairpin vortices generated by a hemisphere protuberance. J. Fluid Mech., Vol. 175, 1-41.

[2] BREUER, K. S., 1988. *The development of a localized disturbance in a boundary layer.* Mass. Inst. Tech. FDRL Report 88-1, also Ph.D. Thesis, Dept. Aero. and Astro. 183 pages.

[3] CORKE, T. C., BAR SEVER, A., and MORKOVIN, M. V., 1986. Experiments on transition enhancement by distributed roughness. Phys. Fluids, Vol. 29, 3199-3213.

[4] GREGORY, N. and WALKER, W. S., 1956. The effect on transition of isolated surface excrescences in the boundary layer. Aero. Res. Council R and M, No. 2779.

[5] KENDALL, J. M., 1981. Laminar boundary layer velocity distortion by surface roughness: effect upon stability. AIAA Paper 81-0195.

[6] KLEBANOFF, P. S., 1983. Private Communication.

[7] KLEBANOFF, P. S. and TIDSTROM, K. D., 1972. Mechanism by which a two-dimensional roughness element induces boundary layer transition. Phys. Fluids, Vol. 15, 1173-1188.

[8] LANDAHL, M. T., 1980. A note on algebraic instability of inviscid parallel shear flows. J. Fluid Mech., Vol. 98, 243-251.

[9] LEVENTHAL, L. and RESHOTKO, E., 1981. Preliminary experimental study of disturbances in a laminar boundary layer due to distributed surface roughness. Case Western Reserve University Report FAS/TR-81-155. Also, Leventhal, L., M.S. Thesis, Case Western Reserve University.

[10] MOCHIZUKI, M., 1961. Smoke observation on boundary-layer transition caused by a spherical element, J. Phys. Soc. Japan, Vol. 16, 995-1008.

[11] MOCHIZUKI, M., 1961. Hot-wire investigations of smoke patterns caused by spherical roughness element, Nat. Sci. Reports, Ochanomizu University Tokyo, Japan, Vol. 12, No. 2, 1961.

[12] MORKOVIN, M. V., 1972. An approach to flow engineering via functional flow modules. DFVLR, Porz-Wahn FB 72-27, 32 pages.

[13] MORKOVIN, M. V., 1985. *Guide to experiments on instability and laminar-turbulent transition.* Text for AIAA Short Course, with L. Mack, June 1985, Cincinnati, OH. (Also in hands of participants of short courses April 1987 at University of Texas, Austin, TX. Prof. R. Panton, organizer, and March 1988, NASA Ames, Dr. M. Rubesin, organizer). 35 pages on roughness effects in Sections 0.03.07, 0.04.08, c, d, and g, 0.04.09 a to d, and 0.04.10, a to d, plus Figures 0.27, 0.32-0.37.

[14] MORKOVIN, M. V., 1989. On receptivity to environmental disturbances. Proc. Instability and Transition Workshop, ICASE/NASA Langley Research Center, Springer-Verlag.

[15] NORMAN, R. S., 1972. On obstacle-generated secondary flows in laminar boundary layers and transition to turbulence. Ph.D. Thesis MMAE Dept., Ill. Inst. Tech., Chicago, IL 60616.

[16] RESHOTKO, E., 1984. Disturbances in a laminar boundary layer due to distributed surface roughness, in *Turbulence and Chaotic Phenomena in Fluids*, T. Tatsumi, Ed., Elsevier Sci. Publ.

[17] RESHOTKO, E. and LEVENTHAL, L., 1981. Preliminary experimental study of disturbances in a laminar boundary layer due to distributed surface roughness. AIAA Paper 81-1224.

[18] SHIN, H-W., PRAHL, J. M., and RESHOTKO, E., 1982. Experimental study of the effects of surface roughness on laminar boundary layer stability in water. Case Western Reserve University Report FTAS/TR-82-157. Also, Shin, H-W., Ph.D. Thesis, Case Western University.

[19] TANI, I. Effect of two-dimensional and isolated roughness on laminar flow, *Boundary Layer and Flow Control*, Vol. 2, 637-656, Lachmann, G. V., ed., Pergamon Press, 1961.

[20] TANI, I., 1970. Remarks on laminar instability leading to transition in boundary layer flows. Spec. Rept. Grad. School of Aerospace Engineering, Cornell University; expanded in German: ZAMM 1973, Vol. 53, 25-32.

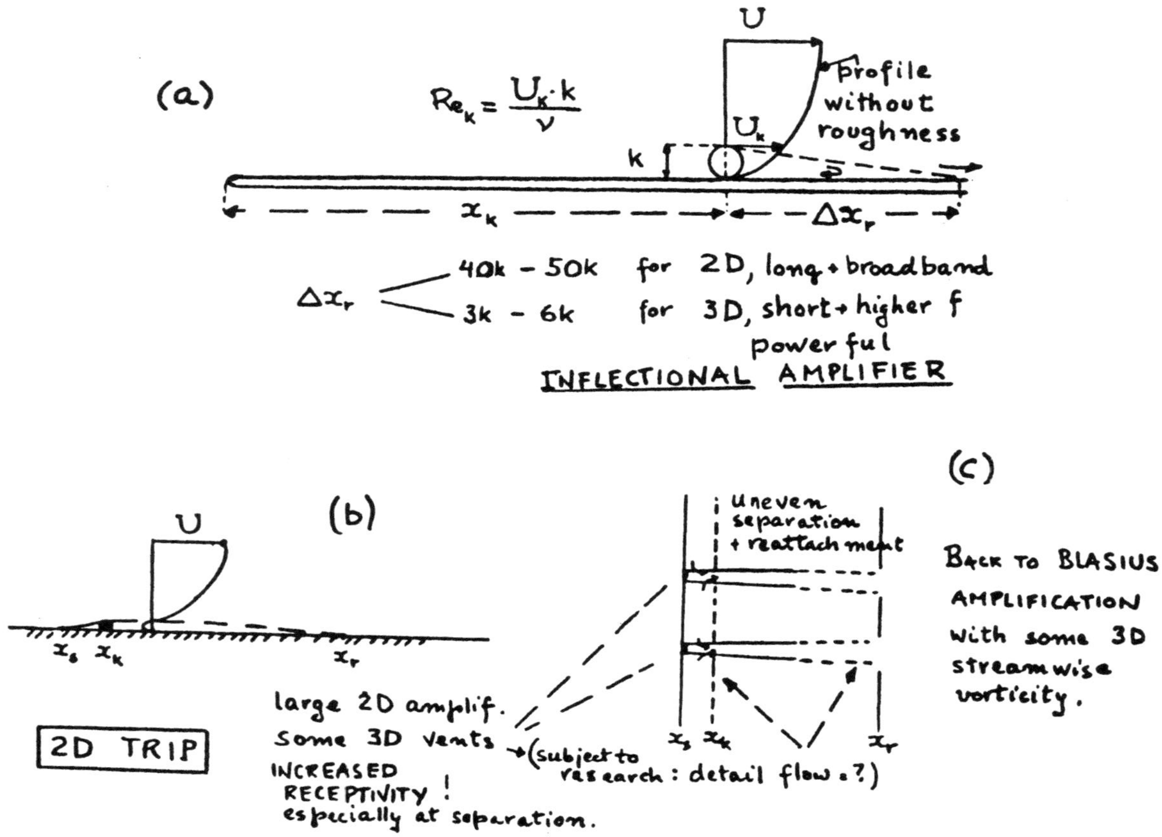

FIG 1 SINGLE-ROUGHNESS FLOWS AND DEFINITIONS

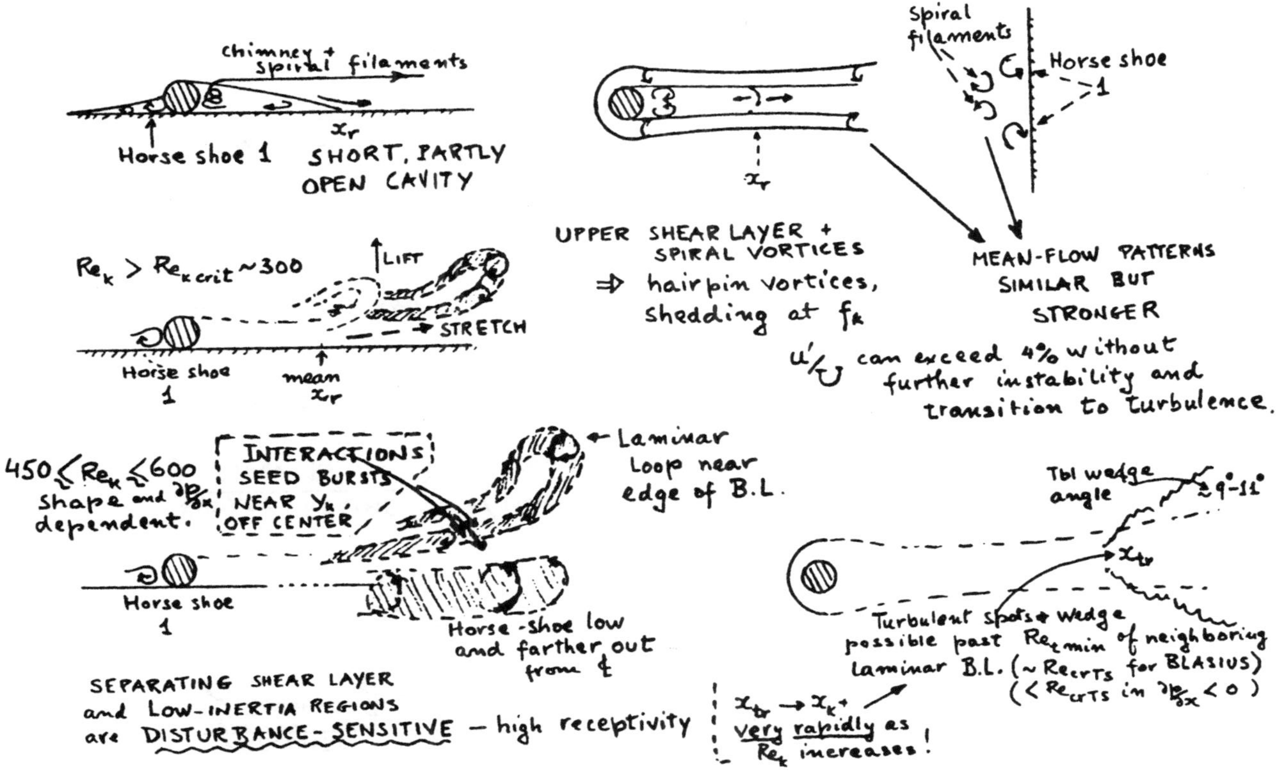

FIG 2 FLOWS AROUND 3D SINGLE-ROUGHNESS ELEMENTS

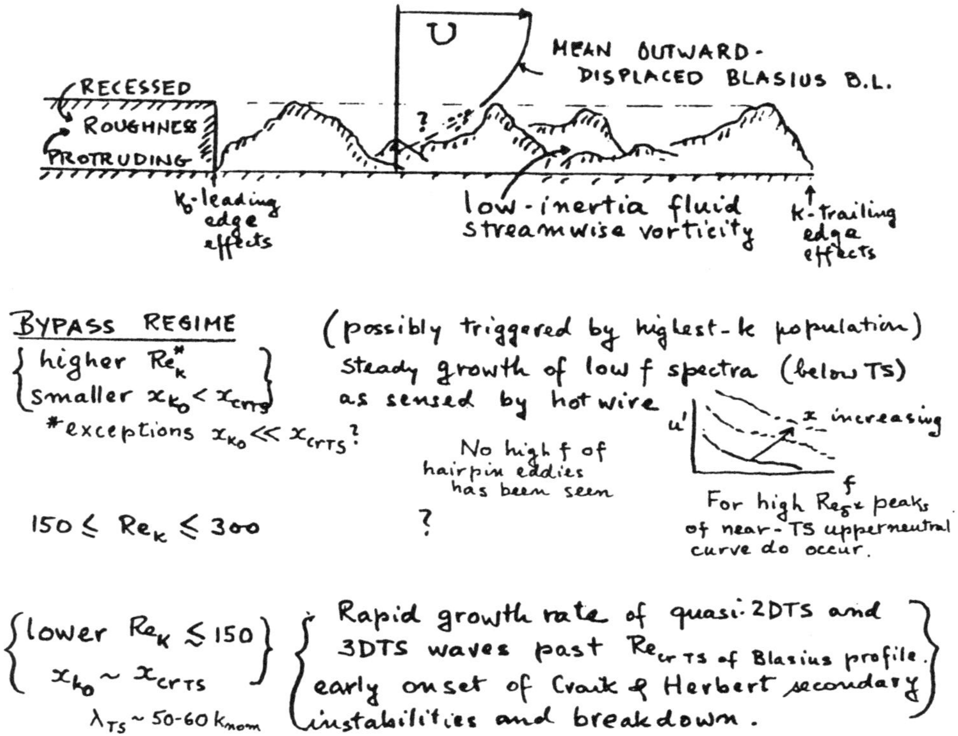

FIG 3 EFFECTS OF DISTRIBUTED ROUGHNESS

u_{TS} in complex notation: wave no = $k_{TSr} + ik_{TSi}$; angular frequency ω_r

$$u_{TS}(x) = u_{TS}(x_0)\exp i\{k_{TSr}(x-x_0) - \omega_r t\}\underbrace{\exp\{-k_{TSi}(x-x_0)\}}_{} +$$

TABULATED TS GROWTH, homogeneous soln.

$$+\int_{x_0}^{x} \underbrace{\Delta u_{TS}(x')}_{\text{RECEPTIVITY SEEDING}} \underbrace{\exp i\{k_{TSr}(x-x') - \omega_r t\}}_{} \exp\{-k_{TSi}(x-x')\}dx'$$

RECEPTIVITY SEEDING — NON-HOMOGENEOUS with account of phase interaction

For unsteady pressure-gradient disturbances non-homo contribution far in excess of homo rate has been measured, Nishioka + Morkovin 1986.
This probably true for roughness-enhanced receptivity.
Also k_{TSi} of BL could possibly exceed k_{TSi} BLASIUS.

FIG 3 (CONTINUED) EFFECTS OF DISTRIBUTED ROUGHNESS

THE EFFECT OF SMALL-SCALE ROUGHNESS ON THE MEAN FLOW PROFILE OF A LAMINAR BOUNDARY LAYER

James M. Kendall

Jet Propulsion Laboratory
Pasadena, CA 91109

1. Introduction

The processes by which distributed surface roughness accelerate boundary layer transition remain unclear. Experimental studies aimed at clarification were carried out during the early 1980s by three groups who maintained intercommunication. In each case, Blasius-like boundary layers were selected for study for reasons of simplicity and standardization. The experiments by Leventhal [1] and by Corke, Bar-Sever, and Morkovin [2] concerned measurement of the growth of naturally-occurring flow fluctuations within laminar layers during passage over sandpaper roughness. In contrast, this author [3] chose to measure only the mean velocity alteration due to roughness, with the expectation that the distorted profile could be analyzed for incremental growth using stability theory.

The first-mentioned experiments demonstrated that, as expected, fluctuation growth rates exceeded Tollmien-Schlichting-instability values, and in particular that low frequency components expected to be stable were not. Neither investigation turned up evidence of modification of the mean profile which would imply increased T-S amplification for stations ahead of the start of transition. Corke, et al. speculated on other causes of increased growth, citing: (a) streamwise vorticity due to the individual elements, (b) the "malleability" of the low-inertia fluid near the wall, and (c) energy injection by forcing.

The experiments conducted by this author concerned smaller-scale roughness than that of Corke, et al., which in turn was smaller than that of Leventhal. Here, examination was made of the profile distortion due to a single roughness element, to two such elements, and to many elements. Because the velocity alteration to be measured was small, signal-averaging of the velocity difference due to the roughness being alternately present or absent was employed in order to obtain good accuracy. This was accomplished for single elements by retracting the element flush with the surface, and for an extended field of elements by attaching these to a turntable set flush with the surface, a half-revolution of which carried the roughness out of the region of influence of the sensor. The largest roughness field which could be accommodated was small in comparison with the plate dimensions and small in comparison with the fields studied in the other two experiments.

The primary objective was to examine the distributed-roughness profiles for evidence of instability-enhancing form and a secondary one was to determine whether a superposition of the wakes of many single elements might describe the distortion due to distributed-roughness. Neither objective was met, but it may be worthwhile to reconsider some of the results.

2. Single-Element and Superposition-Related Results

The single element roughness consisted of a cylinder with axis normal to the plate, the height being 0.45 Blasius units, and the diameter-to-height ratio being two. It was installed at the station $Re_x = 0.5\times 10^6$, and the Reynolds number based on height and velocity at that height was 48. The wake was surveyed at various stations as far downstream as 480 heights, equivalent to 125 displacement thicknesses. Near-wake and far-wake regions were distinguished, the former possessing a velocity defect, and the other a velocity excess (with respect to the unperturbed value) at the centerline and a defect on either side of this. Iso-perturbation contours in the cross plane $x/k = 80$ are shown in Figure 1. This distribution may be the consequence of a longitudinal vortex pair originating at the element. For stations beyond 200 heights, the perturbation, whether defect or excess, declined very slowly, or perhaps not at all, and the growth of vertical and lateral scales also was very slow. The single-element wake is therefore extremely persistent.

The interaction of the wake of one element with that of another or with the wakes of a multitude of others were observed. The presence of a second element ahead of, and laterally aligned with, the first was found to produce little change in the wake of the second, provided that the streamwise separation equalled two diameters or more. Further, the wake of a single element amid a field of approximately 10^4 others of nearly equal height remained compact and distinct. Thus, the concept of describing distributed roughness distortion by superposition may have validity but was not tested directly.

3. Distributed Roughness Effects

Distributed roughness profile results are shown in Figures 2 and 3, the first consisting of the profile above the center of a 5.0-cm wide by 16-cm long array of 0.3-mm beads placed on a 0.6-mm square grid pattern, and also that above the smooth wall. The second gives the differential velocity and includes the computed perturbation which would result from a vertical translation of a Blasius profile over a distance chosen for best fit. The most evident results of roughness is that over the outer two-thirds of the layer, the flow is merely displaced outward by a distance which turns out to be significantly greater than the volumetric average thickness of the beads. An inflection point is apparent near the wall, and the innermost few points shown an upturn in velocity. The latter is not understood and perhaps raises the possibility of experimental error, even though various validation tests disclosed no cause. While the upturn seems questionable, the occurrence of an inflection point within that general region is not, inasmuch as the profile must deflect in order to attain zero velocity at the wall. By the same reasoning, inflection points must have been present in the pretransitional profiles of the other two experiments as well.

The reason for not meeting the objective of estimating T-S amplification from the rough-wall profiles is now clear: much of the distortion occurred close to the wall and below the element tops, where the present technique was inapplicable. The amplification cannot be computed without complete specification of the profile, and thus the objective was ill-conceived. However, a profile obtained downstream of a roughness area did prove adequate for analysis, and this showed a very large increase in amplification rate. This result may be significant in the case of roughness which is variable along the streamwise dimension.

One result not reported earlier now seems significant. In studying the streamwise development of profiles above a 5-cm wide by 7.7-cm long area of # 80 grit applied to the turntable, it was found that the velocity deficit built up slowly, reached a peak aft of midlength, and then declined. The peak values of the profiles, which resembled that of Figure 2 in a general way, are given in Figure 4. The protracted build-up is not surprising on the basis of the accumulation of defect, but the upstream influence of the aft boundary is. A pressure gradient due to the outward displacement of the layer was believed to be responsible, and this might constitute a general feature of roughness fields of finite extent or spatial variability.

4. Hypotheses and Suggestions

Distributed roughness of small height perhaps acts in the following ways to accelerate transition:

1. Any given element generates a persistent wake composed of a counter-rotating vortex pair. The lateral and vertical scales grow slowly, and upon reaching values commensurate with either the layer thickness or the T-S wavelength may affect amplification on account of the profile distortion and wavefront warping.

2. Spatial variability of the "density", i.e., size and concentration of elements, of distributed roughness may affect stability. A spanwise variability will result in streamwise vorticity on account of the displacement effect. A patch of roughness can be likened to a surface protrusion, the height of which is related to the density and spatial extent of the roughness.

3. Two-dimensional variability of roughness density also appears to be capable of increasing T-S amplification rates.

4. Fluid sheltered beneath the crests, the layer referred to as "malleable" by Corke, et al., represents a significant modification of the boundary condition to be applied at the wall in stability analysis. The inflection point which resides at the outer edge of this represents an unknown. It may not be important inasmuch as it is close to the wall.

5. Roughness of any type can be expected to enhance the receptivity to external perturbations, thereby contributing to fluctuation growth.

It would be appropriate to conduct the following investigations for the case of zero-pressure-gradient shear flows:

1. Determine analytically or numerically the asymptotic structure of the single-element wake for an infinitely-thick, constant-shear flow. A related problem is that of the stability of a layer to purely longitudinal waves.
2. Measure or compute the mean-flow perturbation downstream of distributed roughness, the density of which is variable in both the lateral and streamwise dimensions.
3. Perform a "classical" stability experiment for a rough wall, i.e., study the growth of controlled disturbances for a surface with a large area of uniformly distributed roughness, and also for roughness which is spatially variable with specified pattern in both directions.
4. Study the receptivity to sound and to freestream turbulence for a surface with a single three-dimensional element and with distributed roughness.

References

[1] Leventhal, L., "Preliminary Experimental Study of Disturbances in a Laminar Boundary Layer Due to Distributed Surface Roughness," M.S. Thesis, Case Western Reserve University, Cleveland, OH, December 1980.

[2] Corke, T. C., Bar-Sever, A., and Morkovin, M. V., "Experiments on Transition Enhancement by Distributed Roughness," Phys. Fluids, Vol. 29, No. 10, pp. 3199-3213, 1986.

[3] Kendall, J. M., "Laminar Boundary Layer Velocity Distortion by Surface Roughness: Effect upon Stability," AIAA Paper 81-0195, St. Louis, January, 1981.

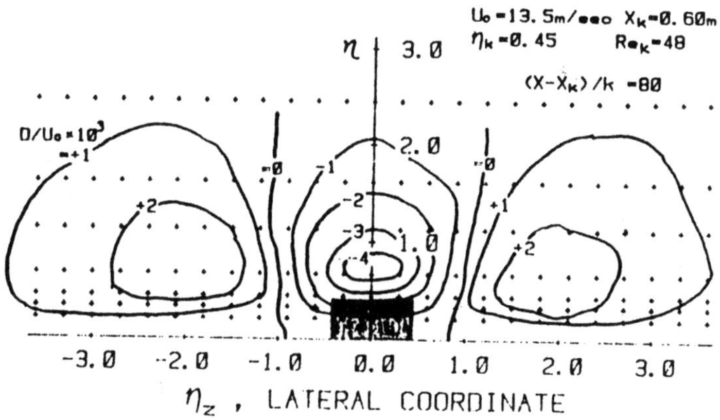

Fig. 1. Iso-Defect Contours in a Blasius Layer Behind a Small Protuberance.

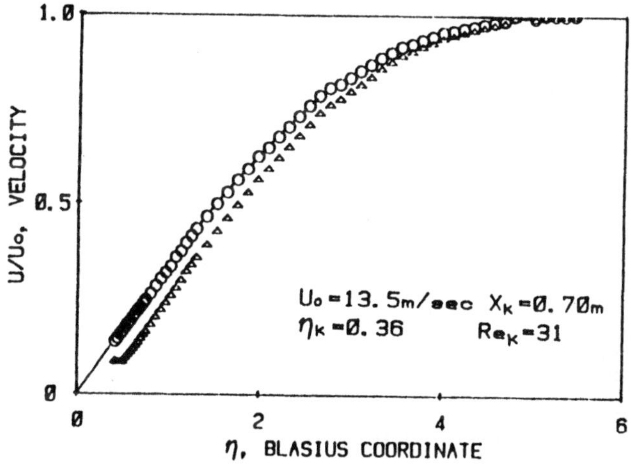

Fig. 2. Mean Velocity Profile above a Smooth and a Rough Flat Plate.

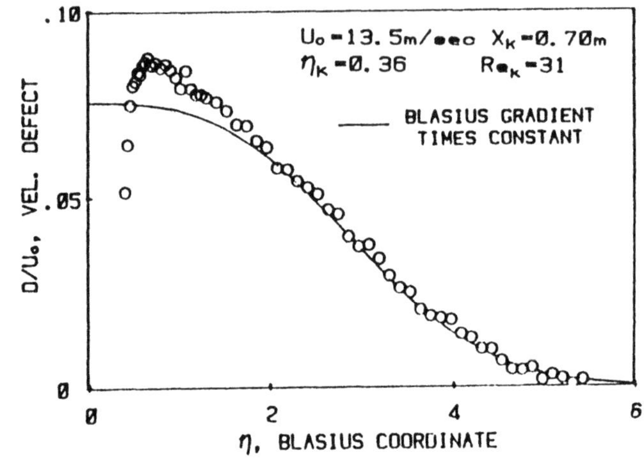

Fig. 3. Velocity Difference Between the Two Curves of Fig. 2.

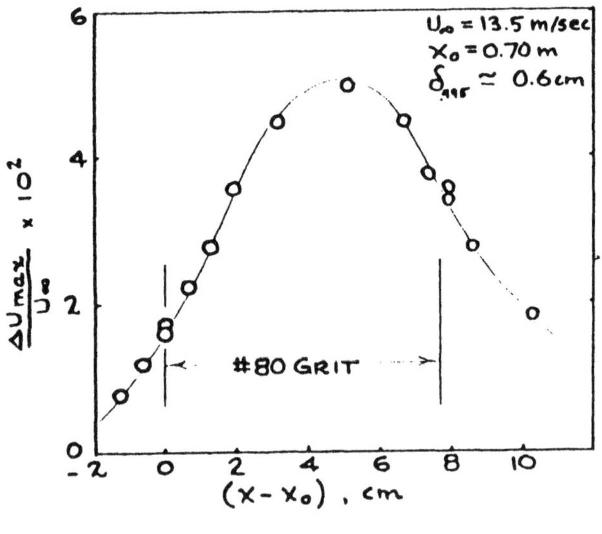

Fig. 4. Peak Velocity Defect along Length of Roughness.

EXPERIMENTS

GROUP SUMMARY: EXPERIMENTS

Stephen P. Wilkinson

NASA Langley Research Center
Hampton, VA 23665

Various brief experimental efforts were initiated during the course of the workshop with the dual aim of both advancing transition research as well as fostering mechanisms of continuing interaction among the quests. This effort was largely successful, allowing presentation at this time of four papers representing diverse individual and collaborative efforts in the area of low speed transition experiments along with a discussion of the problem of transition control.

The first paper by Bandyopadhyay and Walton presents stability data obtained for large amplitude disturbances in the inlet region of a transitional pipe flow. Employing various boundary layer trip devices, they examine the mean and disturbance flow fields with both pitot tubes and a hot wire anemometer. One of their key findings based on hot wire spectral analysis is the nearly universal amplification of dominant frequencies in the input disturbance spectrum. The following paper by Wilkinson et al. describes application of the smoke-wire flow visualization technique to the fully three-dimensional, cross-flow dominated boundary layer of a disk rotating in a quiescent ambient. They present photographic flow visualization data on the development of a wave packet due to impulsive suction through a small surface orifice along with data for a stationary crossflow vortex anchored to the same orifice. The third paper by Mousseux et al. presents various data relating to the behavior of a laminar separation bubble on a low Reynolds number airfoil. Using surface mounted thin-film arrays, surface pressure taps, oil-flow visualization and airfoil design/analysis code, they are able to document the beginning and end of the bubble with good agreement among the measurement techniques. They also present cross-correlation and coherence plots for hot-film array data within the bubble. In the final paper, Gad-el-Hak develops a general approach to the problem of transition control. By identifying four basic stability modifiers (wall motion, suction, body shaping, and heat transfer), the equations can be analyzed to show how linear amplification of disturbances can be inhibited. The problems of wave cancellation for transition control and transition augmentation are also discussed.

FLOW VISUALIZATION OF A WAVE PACKET ON A ROTATING DISK

S. P. Wilkinson

NASA Langley Research Center
Hampton, VA 23665

A. E. Blanchard, G. Selby

Old Dominion University
Norfolk, VA 23529

M. Gaster

University of Cambridge
Cambridge, ENGLAND

T. Tritz

Montana State University
Bozeman, MT 59717

M. Gad-el-Hak

University of Notre Dame
Notre Dame, IN 46556

ABSTRACT

A wave packet on a disk rotating in still air has been investigated using the smoke wire flow visualization technique. The packet was initiated by a short duration suction pulse through a small surface orifice at a radius corresponding to a sub-critical Reynolds number equal to 256. Photographs show that the initial disturbance evolves into a wave packet consisting of three or four waves which spread and grow along a path dictated by the stationary crossflow vortex emanating from the orifice. The data suggests that the packet may have a higher growth rate than the stationary vortex and further experimental and theoretical study is indicated.

1. Introduction

The propagation and growth of waves in three-dimensional (3D) flows is an important topic in laminar flow transition, playing a major role in applications ranging from swept wings on conventional aircraft to aerospace plane (NASP) configurations. As opposed to two-dimensional (2D) mean flows where TS waves and their breakdown are the primary physical phenomena of interest, 3D flows are typically dominated by crossflow instabilities. While the major impetus for the study of 3D flow stability is airplane performance and economics, the case of a disk rotating in still air, which exemplifies the crossflow instability, represents complementary opportunities for study. It is a simple, canonical 3D flow with good experimental control and is also an ideal candidate for code validation studies. Examples of this include the Malik, Wilkinson, and Orszag (1981) study of stationary disturbances in rotating disk flow in which the measured disturbance growth rates substantiated theoretical predictions. Wilkinson and Malik (1985) followed with more detailed hot-wire experiments showing that the source of the classical stationary vortices were most likely minute, unavoidable roughness sites on the surface. Mack (1985) theoretically studied the disturbance pattern emanating from a stationary point source and found good agreement with the experimental results of Wilkinson and Malik (1985) for a stationary roughness element. Although the nose of the neutral stability loop clearly shows the importance of modes with zero phase velocity (Malik (1986)) there is no special significance of the stationary modes. The stationary pattern of lines appearing on the rotating disk arises because roughness excites these unstable modes, but there may be other modes with greater amplification that are more important in the transition process. By exciting the flow by a pulsed point source, modes of all frequencies are created and the most significant ones will then naturally evolve as the resulting wave packet propagates away from the source. Our experiments, as will be shown, suggest that at large radii travelling waves begin to dominate.

In the present study, the work Wilkinson and Malik (1985) is extended to the case of dynamic disturbance sources, in particular, a short duration suction pulse (delta function) applied at the surface in a manner similar to the injection pulse employed by Gaster and Grant (1975). As in that work, the initial experimental challenge was to characterize the structure and evolution of the generated wave packet. Gaster and Grant (1975) produced a perspective view of the packet from numerous hot-wire signal traces obtained for various locations on the flat plate and delay times after the initiating pulse. The present study embarked on a similar path, obtaining clear indications of a highly repeatable travelling disturbance in the flow. It is difficult, however, to visualize the structure of a disturbance in a fully 3D mean flow from single-probe hot wire data. Flow visualization was therefore selected as a more efficient initial path to follow which would as well provide verification of the phenomenon by an independent technique. The aim of the current paper, therefore, is to present preliminary data on the structure and development of the wave packet through smoke wire flow visualization. This data can be used in a follow on study to assist in the interpretation of hot-wire flowfield data.

2. Experimental Apparatus

The rotating-disk apparatus for this study was a modified version of equipment described earlier in Wilkinson and Malik (1985) and Waitz and Wilkinson (1988). It is consisted of a 13-inch diameter, 1.25 inch thick glass disk mounted on a rotary air bearing driven by an integral brushless DC motor (Figure 1) operating at a constant speed of 1170 RPM. This arrangement provided a system with exceptionally low amplitude vibration modes (Wilkinson and Malik (1983)). Preparation of the disk for the current experiments was as follows. Two diametrically opposed 0.25-inch diameter holes were drilled through the disk at a radius of 3.54 inches as shown in Figure 1. The Reynolds number of the holes based on the disk surface velocity and the local radius was $R = r\sqrt{\frac{\omega}{\nu}} = 256$ where ω is the rotational speed, ν the kinematic viscosity and r and the local radius. The minimum critical Reynolds number for stationary disturbances in the disk flow is 285 (Malik (1986)). Each hole was covered by masking tape and the disk was spray-painted with multiple coats of flat black lacquer to enhance photographic contrast for the smoke flow visualization. The lacquered surface was hand-sanded with 600-grit sandpaper attached to a hand-sized wooden block with a local circular motion. There were visible, swirling striations in the surface due to the sanding; however, these were undetectable by touch alone. As will be shown, the smoke flow visualization indicated no discrete flow perturbations due to irregularities in the painted surface. Brass inserts with 0.01 inch diameter orifices (0.14δ, where δ is the boundary layer thickness) were inserted into each hole to a depth of 1/16 inch below the surface. The recessed regions were then filled with molten wax (hard type) with a small rod in one orifice to keep it clear. The other orifice was filled and inactive for the experiments. For dynamic balance considerations, both holes needed to be prepared similarly. After hardening, the rod was removed, the wax was polished flush with the painted surface, and the orifice cleaned with compressed air. The edges of the wax were feathered into the surrounding painted region to provide a smooth transition between the adjacent materials. Flow visualization showed no discrete influence of the wax edges.

On the lower side of the disk, 0.090-inch inner diameter tubes were used to connect the hole in the disk to an axial hole drilled through the body of the airbearing turntable (Figure 1). The disk holes were joined to the connecting tubes through small sealed epoxy cavities. The other end of the connecting tube was attached to a metal adapter at the center of the turntable with internal passages leading to the axial hole.

An electronic pulse generator, amplifier and "horn-driver" speaker were used to produce the acoustic suction impulse. As shown in Figure 1, it was positioned directly beneath the axial hole in the turntable and connected with a rigid plastic tube allowing free rotation of the turntable while creating a path for the pressure pulse. Due to the multiple turns and area changes in the path, significant pulse distortion would be expected. Figure 2 is a plot of the output of a 1/8 inch B&K microphone positioned 0.06 inch directly above the hole for several input voltage pulse amplitudes and zero rotational speed. The pressure (ordinate) is based on the microphone calibration but due to the size disparity between the probe and the hole, it underestimates the actual amplitude by an undetermined amount. Time is measured from the leading edge of the input voltage pulse to the horn-driver. Each curve is the ensemble average of 256 instantaneous traces to reduce background noise. The pressure signal was also band-pass filtered in the range $63 < f < 10000$ Hz in order to further enhance the signal-to-noise ratio. As can be seen, the negative (i.e., suction) pulse consists of a primary component followed by lower amplitude secondary components presumably due to reflections within the complicated pulse path. The issue of initial disturbance amplitude is important, not only to avoid by-pass transition phenomena, but also to provide validation data for linear stability theory calculations where infinitesimal initial disturbance amplitudes are implicit. In this regard, input pulse amplitudes as low as -0.5-volt were observed to generate disturbances that clearly develop into faint wave-like structures. These structures, however, were difficult to photograph and did not appear, from a cursory investigation, to differ greatly in character from the stronger, more easily photographed - 2.5-volt pulse structures. There is also the consideration that the impulse was input at a radius where the Reynolds number (for stationary roughness) was subcritical and may decay in its early life. Only that remnant surviving though the minimum critical Reynolds number would then be subject to amplification. The higher amplification input pulse was beneficial in this regard. The -2.5-volt pulse was therefore chosen for detailed study as a compromise between low-disturbance input amplitude and observable effect, without appreciably compromising the essential character of the linear process under study.

3. Experimental Procedure

The experimental procedure used a smoke wire and electronic flash to photograph the wave packet at a succession of instantaneous developmental stages as it travelled from the orifice towards the outer edge of the disk. The wire was stretched across the center of the disk at a height of 0.055 inch or approximately 0.8δ. The smoke wire heating current was triggered by the passage of a photoelectrically detected timing tab on the disk (timing mark). After allowing one or more revolutions of the disk for the oil to begin vaporizing sufficiently, the suction pulse was triggered at a preselected time after the timing mark, T_p, by means of an electronic delay circuit. After another delay, T_f, the photographic flash was fired, capturing the current state of the smoke on film through an open-shutter, 150 mm lens positioned approximately 20 inches above the disk surface. By manipulating T_p and T_f, any desired region of the flowfield could be visualized at any desired time. A line-source flash tube was used in a parabolic reflector with a slit mask to provide a light sheet parallel to the disk. The film was 4×5 inch Kodak Tri-X pushed to 800 ASA in Kodak D-19 developer.

The disturbance due to the smoke wire itself was negligible. This was shown by the undistorted smoke streaklines emanating from the wire, as well as by cursory investigation with a hot-wire probe which showed, in general, that disturbances due to a detached roughness within the disk boundary layer, but fixed in the laboratory reference frame, produce disturbances that are highly damped. Occasionally, the smoke wire would vibrate due to its impulse heating and spring/mass characteristics. The consequence of this vibration was visible as azimuthal waves in the smoke streaklines parallel to the smoke wire but they apparently did not couple to any discrete phenomena associated with the wave packet.

4. Results

Photographic data are shown in Figures 3(a-f). Due to the limited persistence of the smoke filaments and problems with obtaining uniform exposure, only a portion of the disk is shown in each case. The photos are all to the same scale and in correct physical placement relative to the smoke wire (coincident with line OW) and the center of the disk (O). The line, OF, locates the orifice relative to the fixed smoke wire at the time the flash was fired (i.e., the angle at which the photo was obtained). The dashed line, OP, locates the orifice at the time the suction pulse arrived there, taking into account the acoustic transit time from the speaker to the orifice (1.5 millisec). The angle between lines OP and OF is β. The arc at r_h represents the physical trajectory of the orifice. The actual orifice is too small to be visible in the photographs and is approximately at the center of the white region (wax filler material) into which the orifice was drilled.

One of the most evident features in the photographs is the clear trail emanating from the orifice. It was shown experimentally in Wilkinson and Malik (1985) that a discrete roughness element fixed on the disk will generate a stationary crossflow vortex. Based on examination of Figure 3, it appears that a surface depression (i.e, the orifice) also acts as a stationary disturbance and generates a trailing crossflow vortex. Interpretation of smoke patterns must, in general, be approached with caution since only those portions of structures marked with smoke are made visible. In this case, the smoke streaklines form a downward-spiralling sheet (due to the mean axial velocity component) which intersects the vortex structure over a finite z-coordinate extent. If the vortex structure is skewed in the z-coordinate direction over its length, as is almost certainly the case due to the shear flow, the smoke streaklines would not necessarily intersect the vortex along its actual 3D axis, and would yield a distorted image of the structure, particularly in regard to local vortex axis orientation. Smoke photographs at different heights above the surface would be required to properly characterize the structure. Even then, knowledge of the actual streakline trajectories (i.e, the full 3D trajectory as opposed to the planar projection shown in the photograph) would be required in order to obtain quantitative structural data. The photographs in Figure 3, however, show the high sensitivity of the technique and provide an invaluable qualitative record of the boundary layer response to the orifice (i.e., surface depression) and acoustic impulse disturbances. The travelling disturbance is initially seen clearly in Figure 3b as a small, roughly rectangular dark region just to the right of line OF. The disturbance was most likely present in Figure 4a, but is still over the white wax plug and not visible. The disturbance gradually develops into the clearly defined waves in Figures 3 d, e, and f. The aforesaid precautions regarding interpretation of the smoke patterns also apply to the travelling disturbance; however, certain features can still be discerned. The first is that the travelling disturbance follows the upstream side of the stationary vortex. This suggests that the same or similar mechanisms are responsible for each phenomenon. Secondly, waves can be characterized by a specific wave orientation, wavelength, and propagation speed. All of these variables could, in principle, be accurately measured with a more extensive, more closely spaced data set defining any skewness of the phenomena with respect to the smoke streaklines. No attempt is made at this time to quantify these features. A third, more subtle, feature is that the photographs seem to indicate that the travelling disturbance may grow faster than the stationary disturbance. The evidence for this is that the travelling disturbance seems to show greater evolution over its trajectory path than the stationary mode. The photographs obviously contain no direct intensity information and detailed hot wire data must be obtained for confirmation of the relative growth rates.

5. Conclusions

This experiment has shown that a small orifice flush with the surface of a rotating disk acts as the nucleation site for stationary crossflow vortices in the same fashion as attached 3D roughness elements in earlier investigations. It has also been shown that a single, low-amplitude, short-duration suction pulse though the same orifice generates a travelling wave pattern which follows a path roughly along that of the stationary vortex pattern. The smoke wire flow-visualization technique has been shown to be effective in the very thin rotating-disk boundary layer. However, quantitative data derived from the flow visualization photographs must be approached carefully due to possible skewness of the structure with respect to the smoke streaklines.

The emergence of travelling waves towards the outer region of the disk is very interesting. It appears that there are unstable modes that travel with greater amplification than the stationary ones. Maybe transition on the disk and on swept leading edges is controlled more by these travelling waves than by the stationary set. Detailed comparison of hot-wire measurements with linear theory will be well worth pursuing.

References

[1] Gaster, M. and Grant, I., 1975, *An Experimental Investigation of the Formation and Development of a Wave Packet in a Laminar Boundary Layer*, Proc. R. Soc. Lond. A., 347.

[2] Mack, L. M., 1985, *The Wave Pattern Produced by a Point Source on a Rotating Disk*, AIAA Paper 83-0490.

[3] Malik, M. R., 1986, *The Neutral Curve for Stationary Disturbances in Rotating Disk Flow*, J. Fluid Mech., Vol. 164.

[4] Malik, M. R., Wilkinson, S. P., and Orszag, S. A., 1981, *Instability and Transition in Rotating Disk Flow*, AIAA J., Vol. 19, No. 9, September.

[5] Waitz, I. A. and Wilkinson, S. P., 1988, *Rotating Disk Transition Due to Isolated Roughness with Intense Acoustic Irradiation*, AIAA Paper 88-3761, First National Fluid Dynamics Congress, Cincinnati, OH, July.

[6] Wilkinson, S. P. and Malik, M. R., 1983, *Stability Experiments in Rotating Disk Flow*, AIAA Paper 83-1760, AIAA 16th Fluid and Plasma Dynamics Conference, Danvers, MA, July.

[7] Wilkinson, S. P. and Malik, M. R., 1985, *Stability Experiments in the Flow over a Rotating Disk*, AIAA J., Vol. 23, No. 4, April.

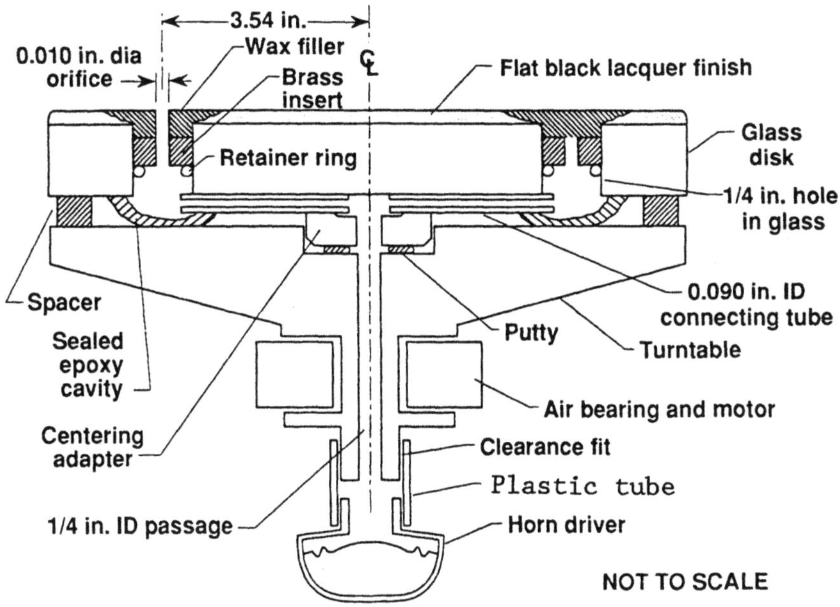

Figure 1. Cross-Sectional Schematic of Rotating Disk Apparatus.

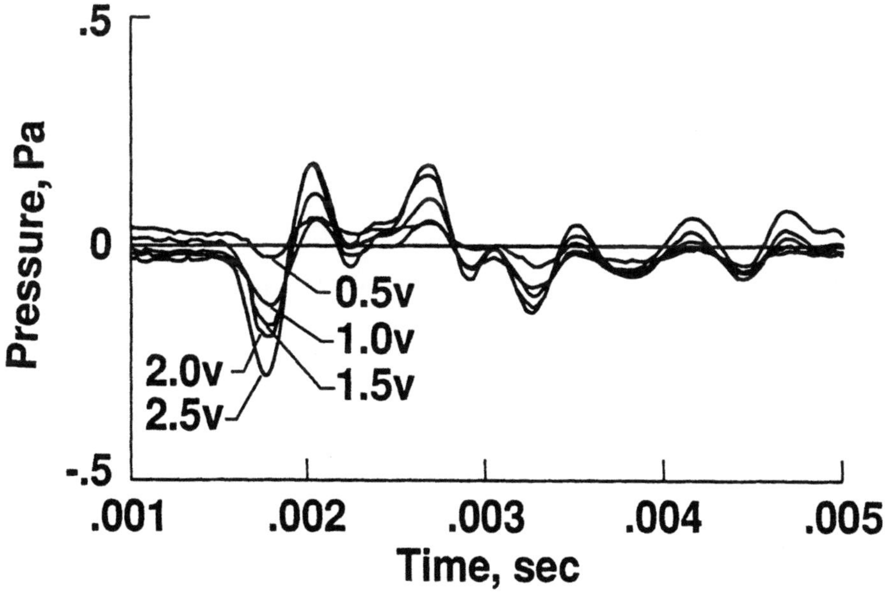

Figure 2. Pressure Pulse Measured Directly above Orifice for Various Amplifier Input Pulse Amplitudes (disk still).

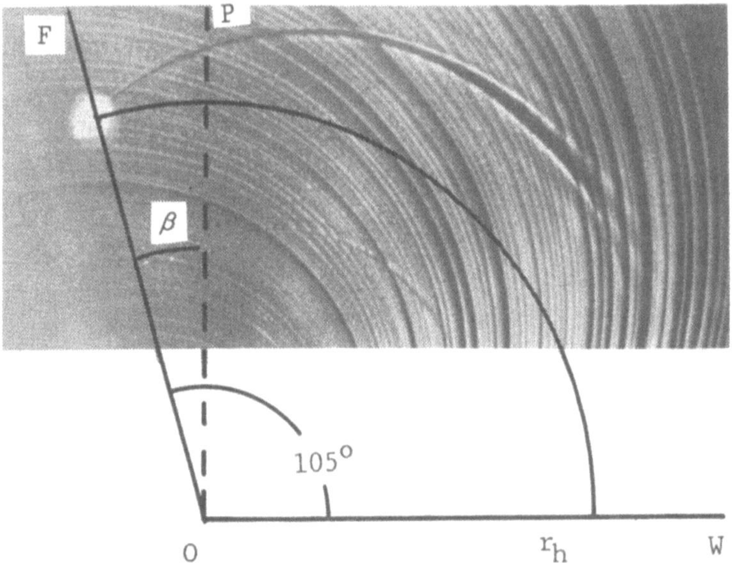

Figure 3(a-f). Smoke Streaklines. (OW smoke wire; OF, flash fired; OP, pulse arrives at orifice.) $\beta = 15^{\circ}$

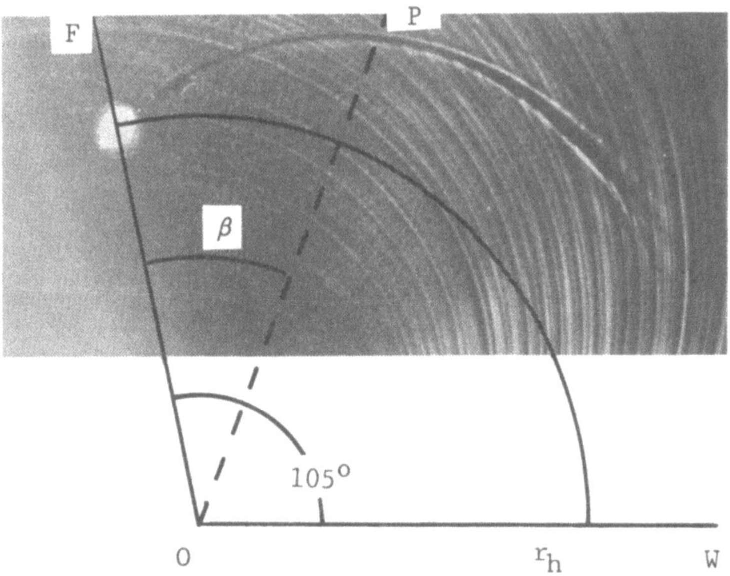

Figure 3b. $\beta = 35^{\circ}$.

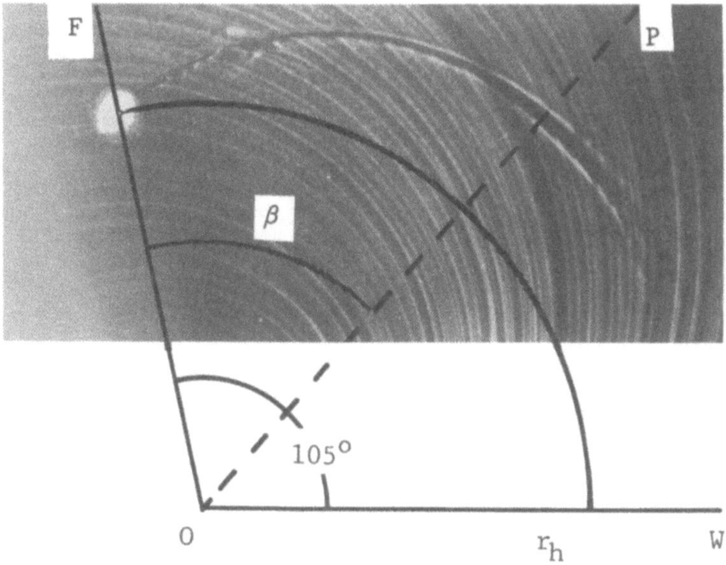

Figure 3c. $\beta = 45^{\circ}$.

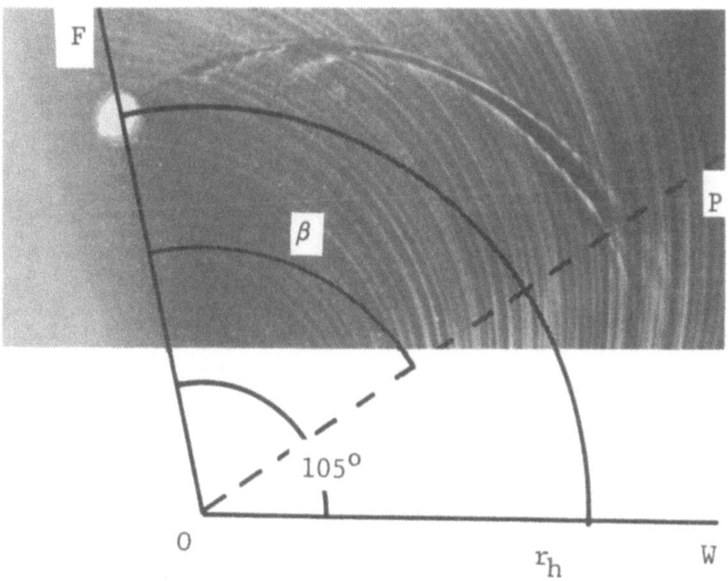

Figure 3d. $\beta = 69^{\circ}$.

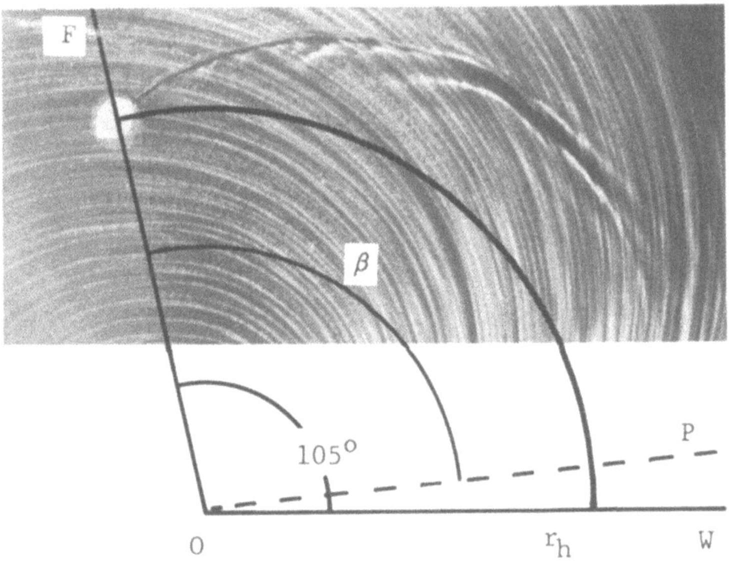

Figure 3e. $\beta = 97^o$.

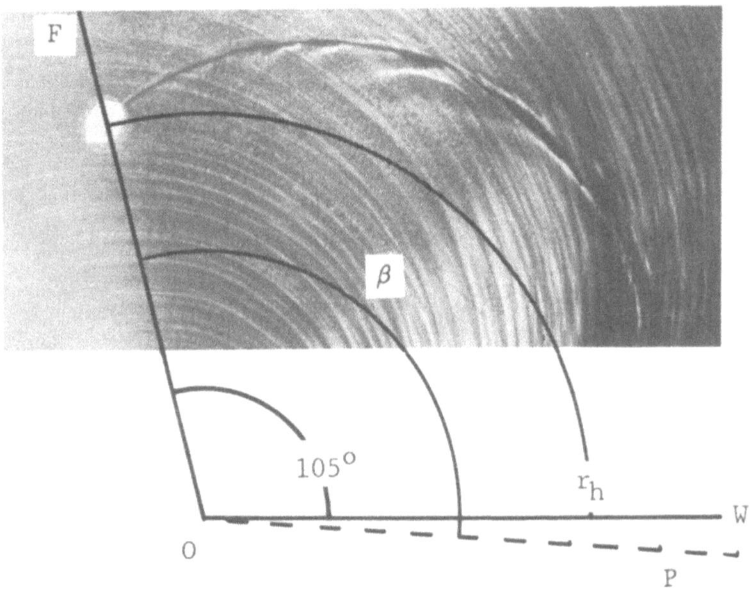

Figure 3f. $\beta = 108^o$.

TRANSITION CONTROL

Mohamed Gad-el-Hak

Department of Aerospace and Mechanical Engineering
University of Notre Dame
Notre Dame, IN 46556

1. Introduction

The ability to actively or passively manipulate a boundary layer to delay or advance laminar-to-turbulent transition is of immense technological importance. The present article is an overview of available or contemplated techniques to control transition. The treatment is pedagogical, making the material accessible to newcomers to the field. Emphasis is placed on two-dimensional, incompressible flows, although more complex situations will be briefly mentioned.

The governing equations are developed in the following section. In Section 3, the routes to transition are mapped. Elements of the linear stability theory are recalled in Section 4. The four basic stability modifiers, wall motion, suction, shaping, and wall heat transfer are discussed in Section 5. This is followed by Section 6 on wave cancellation. Section 7 summarizes available methods to advance transition. Finally, brief concluding remarks are given in Section 8.

2. Governing Equations

The principles of conservation of mass, momentum, and energy govern all fluid motions. In general, a set of partial, nonlinear differential equation expresses these principles, and together with appropriate boundary and initial conditions constitute a well-posed problem. It is of course beyond the scope of this article to derive these equations and the reader is referred to any advanced textbook in Fluid Dynamics (Landau and Lifshitz, 1963; Batchelor, 1967; Hinze, 1975; Kays and Crawford, 1980; Panton, 1984). The equations will be first recalled in as general a form as possible. This approach will become particularly useful when discussing surface heating/cooling (viscosity varies spatially) and other nonconventional situations.

In Cartesian tensor notation, the equation of conservation of mass reads:

$$\frac{\partial p}{\partial t} + \frac{\partial}{\partial x_i}(\rho U_i) = 0, \tag{2.1}$$

and Newton's second law is:

$$\rho\left(\frac{\partial U_i}{\partial t} + U_j\frac{\partial U_i}{\partial x_j}\right) = \frac{\partial}{\partial x_j}\tau_{ji} + F_i, \tag{2.2}$$

where U_i is an instantaneous velocity component, ρ is the density, F_i is the body force per unit volume, and τ_{ji} is the second-order stress tensor which must be related to the rate of deformation tensor in order to reduce the number of unknowns to be equal to the number of equations. The independent variables are time t and the three spatial coordinates x_1, x_2, and x_3 (or $x, y,$ and z). For a Newtonian fluid, a linear relation between the stress and the rate of strain is assumed. If this fluid is further assumed to be isotropic and to have a negligible bulk viscosity (Stokes's hypothesis), its constitutive relation reads:

$$\tau_{ji} = -P\delta_{ji} + \mu\left(\frac{\partial U_i}{\partial x_j} + \frac{\partial U_j}{\partial x_i} - \frac{2}{3}\frac{\partial U_k}{\partial x_k}\delta_{ji}\right), \tag{2.3}$$

where P is the hydrostatic pressure, μ is the dynamic coefficient of viscosity, and δ_{ji} is the unit second-order tensor (Kronecker delta). While (2.3) is valid for air and water under most practical conditions, different constitutive relations must be sought for non-Newtonian fluids such as polymer solutions.

For a Newtonian fluid, (2.3) is substituted in (2.2) to yield the momentum equation:

$$\rho\left(\frac{\partial U_i}{\partial t} + U_j\frac{\partial U_i}{\partial x_j}\right) = -\frac{\partial P}{\partial x_i} + \frac{\partial}{\partial x_j}\left[\mu\left(\frac{\partial U_i}{\partial x_j} + \frac{\partial U_j}{\partial x_i}\right)\right]$$
$$-\frac{2}{3}\frac{\partial}{\partial x_i}\left(\mu\frac{\partial U_k}{\partial x_k}\right) + F_i. \tag{2.4}$$

For a compressible, Newtonian fluid, (2.1) and (2.4) must be complemented by an equation of state and the energy equation to form six equations for the six unknowns U_i, ρ, P and T, where $T(x_i, t)$ is the temperature field.

If the flow is incompressible, then $\frac{\partial U_k}{\partial x_k} = 0$, density is assumed given, and (2.4) reads:

$$\rho\left(\frac{\partial U_i}{\partial t} + U_j\frac{\partial U_i}{\partial x_j}\right) = -\frac{\partial P}{\partial x_j} + \frac{\partial}{\partial x_j}\left[\mu\left(\frac{\partial U_i}{\partial x_j} + \frac{\partial U_j}{\partial x_i}\right)\right] + F_i. \tag{2.5}$$

Note that in (2.5), μ has not been assumed constant, a useful generalization when surface heating/cooling is used.

The familiar Navier-Stokes equation is obtained from (2.2) by assuming a Newtonian, incompressible, constant-viscosity fluid:

$$\rho\left(\frac{\partial U_i}{\partial t}+U_j\frac{\partial U_i}{\partial x_j}\right)=-\frac{\partial P}{\partial x_i}+\mu\frac{\partial^2 U_i}{\partial x_j \partial x_j}+F_i. \tag{2.6}$$

In this case, the momentum and continuity equations form four equations for the four unknowns U_i and P.

For external flows at high Reynolds number, viscous forces are confined to a relatively thin layer along the surface of a body, although this layer's thickness increases in the downstream direction. Outside the boundary layer, the flow could be computed using the potential flow theory. Within the viscous region, the classical boundary-layer approximations apply (Rosenhead, 1963; Schlichting, 1979). Ignoring body forces, the continuity equation and the streamwise and normal momentum equations along a two-dimensional (or axisymmetric) surface of small curvature are obtained from (2.1) and (2.4), respectively:

$$\frac{\partial \rho}{\partial t}+\frac{\partial}{\partial x_1}(\rho U_1)+\frac{\partial}{\partial x_2}(\rho U_2)=0, \tag{2.7}$$

$$\rho\left(\frac{\partial U_1}{\partial t}+U_1\frac{\partial U_1}{\partial x_1}+U_2\frac{\partial U_1}{\partial x_2}\right)=-\frac{\partial P}{\partial x_1}+\frac{\partial}{\partial x_2}\left(\mu\frac{\partial U_1}{\partial x_2}\right), \tag{2.8}$$

$$-\frac{\rho}{R}U_1^2=-\frac{\partial P}{\partial x_2}. \tag{2.9}$$

These equations are valid for variable properties ρ and μ. In (2.7) through (2.9), x_1 is in the main-flow direction along the body, x_2 is normal to the surface, and R is the radius of curvature of the two-dimensional surface or the radius of revolution of the axisymmetric body. Equation (2.9) gives the pressure gradient required to balance the centrifugal effect of flow round a curved wall (Milne-Thomson, 1968). For a plane wall ($R \to \infty$) and within the boundary-layer approximation, the pressure is constant in the normal direction and its value is determined by the inviscid flow at the outer edge of the boundary layer.

For steady flow, the above equations can be integrated in the x_2 direction resulting in the momentum integral equation:

$$\frac{C_f}{2}=\frac{d\delta_\theta}{dx_1}+\delta_\theta\left[\left(2+\frac{\delta^*}{\delta_\theta}\right)\frac{1}{U_\infty}\frac{dU_\infty}{dx_1}+\frac{1}{\rho_\infty}\frac{d\rho_\infty}{dx_1}+\frac{1}{R}\frac{dR}{dx_1}\right]-\frac{\rho_0 v_0}{\rho_\infty U_\infty}. \tag{2.10}$$

In (2.10),

$$C_f = \text{local skin} - \text{friction coefficient} \equiv \frac{\tau_0}{\frac{1}{2}\rho_\infty U_\infty^2}, \tag{2.11}$$

$$\delta^* = \text{displacement thickness} \equiv \int_0^\infty \left(1 - \frac{\rho U_1}{\rho_\infty U_\infty}\right) dx_2, \tag{2.12}$$

$$\delta_\theta = \text{momentum thickness} \equiv \int_0^\infty \frac{\rho U_1}{\rho_\infty U_\infty}\left(1 - \frac{U_1}{U_\infty}\right) dx_2, \tag{2.13}$$

ρ_∞ and U_∞ are the density and velocity outside the boundary layer, respectively, ρ_0 and v_0 are the density and normal velocity of fluid injected through the surface, and τ_0 is the shear stress at the wall.

Since the skin-friction coefficient in the momentum integral equation is defined in terms of the shear stress and not in terms of the velocity gradient at the wall, (2.10) is, in fact, valid for both laminar and turbulent flows as well as for both Newtonian and non-Newtonian fluids; the only assumptions being that the boundary-layer flow is steady and two-dimensional in the mean. For an incompressible fluid, $\rho = \rho_\infty =$ constant. In case of a turbulent flow, the mean streamwise velocity, $\overline{U_1}(x_1, x_2)$, is used in the definition of δ^* and δ_θ. For a Newtonian fluid,

$$\tau_0 = \mu \frac{\partial U_1}{\partial x_2}\bigg|_{\text{wall}} \tag{2.14}$$

3. Transition Delay

Delaying laminar-to-turbulent transition of a boundary layer has many obvious advantages. Depending on the Reynolds number, the skin-friction drag in the laminar state can be as much as an order of magnitude less than that in the turbulent condition (Figure 1). For an aircraft or an underwater body, the reduced drag means longer range, reduced fuel cost/volume, or increased speed. Flow-induced noise results from the pressure fluctuations in the turbulent boundary layer and, hence, is virtually nonexistent in the laminar case. Reducing the boundary layer noise is crucial to the proper operation of an underwater sonar. On the other hand, turbulence is an efficient mixer and rates of mass, momentum and heat transfer are much lower in the laminar state, so early transition may be sought

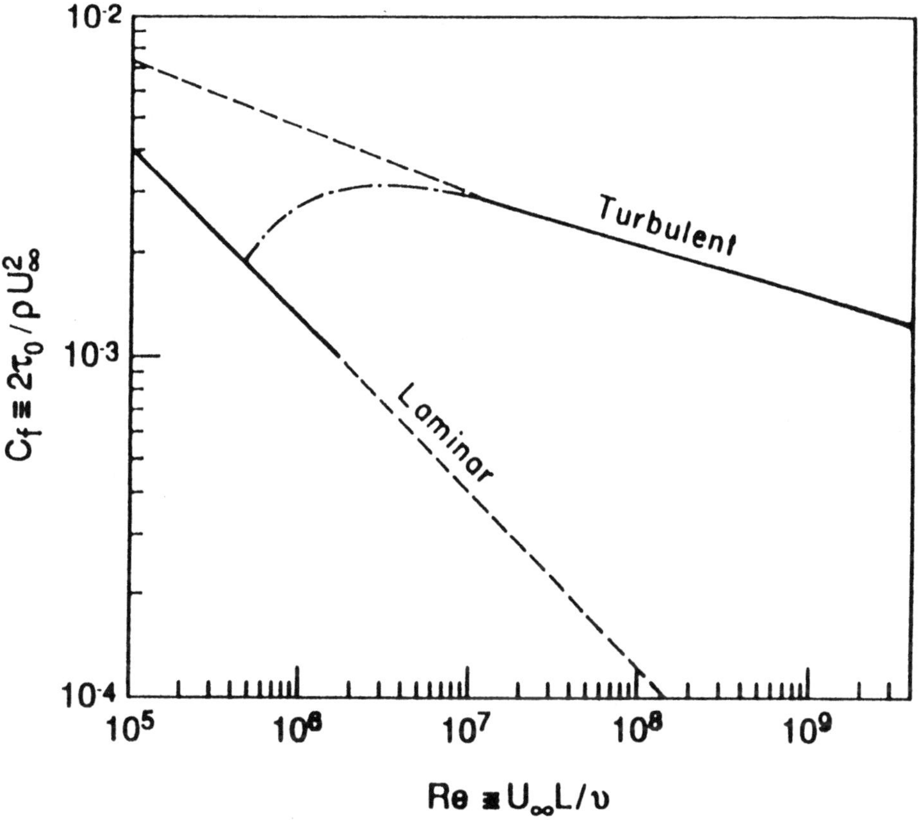

Figure 1. Skin - Friction Coefficient for a Smooth Flat Plate at Zero Incidence to a Uniform, Incompressible Flow.

in some applications as, for example, when enhanced heat transfer rates are desired or when rapid mixing is needed.

The routes to transition are many; some are more understood than others. On a semi-finite, smooth flat plate placed in a clean, uniform, incompressible flow with as little external disturbances as possible, the laminar flow (somewhat downstream of the leading edge) is in the form of a Blasius profile and exists at low enough Reynolds number, typically $R_x < 6 \times 10^4$. This laminar shear layer is, however, unstable to small perturbations that invariably exist in any flow. Squire's (1933) theorem shows that two-dimensional traveling waves (Tollmien-Schlichting waves) are the most dangerous for incompressible-flow instability and become unstable when the Reynolds number exceeds a critical value. However, as soon as the T-S waves are amplified, gain a certain amplitude and nonlinear effects take place, three-dimensional disturbances can non longer be excluded (Itoh, 1987). Following the linear step, the originally two-dimensional waves inherently acquire a nearly periodic spanwise modulation and three-dimensional structures evolve as a result of a secondary instability. Hairpin vortices develop, perhaps due to a tertiary instability, and finally breakdown to turbulence occurs (Klebanoff, et al., 1962). Not surprisingly, these nonlinear processes are far less understood than the linear step. The linear amplification step is, however, the slowest of the successive multiple steps in the transition process and, hence, factors that affect the linear amplification determine the magnitude of the transition Reynolds number.

Reshotko (1976; 1985; 1987) asserts that transition is a consequence of the nonlinear response of the laminar shear layer (a very complicated oscillator) to random forcing disturbances that result from freestream turbulence, radiated sound, surface roughness, surface vibrations, or combination of these environmental factors. If the initial disturbances are small, transition Reynolds number depends upon the nature and spectrum of these disturbances, their signature and excitation of the normal modes in the boundary layer (receptivity; see, for example, Morkovin, 1969; Goldstein and Hultgren, 1989), and the linear amplification of the growing normal modes. Once wave interaction and nonlinear processes set in, transition is quickly completed. If the initial disturbance levels are large enough, the relatively slow linear amplification step mentioned above is bypassed (Morkovin, 1984; 1988) and transition can occur at much lower Reynolds numbers. In fact, a sufficiently violent disturbance,

$u_{\mathrm{rms}}/U_\infty \sim 10\%$, can cause transition of a laminar boundary layer to advance to the position upstream of which perturbations of all wave numbers decay (Klebanoff, et al., 1955).

Other routes to transition include the Taylor-Görtler vortices forming on a concave surface as a set of counter-rotating, streamwise standing eddies (Stuart, 1963; DiPrima and Swinney, 1985), the cross-flow instabilities occurring on a swept wing or a rotating disk (Reed and Saric, 1987; 1989), and any other situation where an instability will occur as a result of disturbing the equilibrium of body forces, inertia and surface forces (Drazin and Reid, 1981). Once again, these instabilities will generally occur at Reynolds numbers lower than the critical Reynolds number for the growth of Tollmien-Schlichting waves.

To delay transition to as far downstream position as possible, the following steps may be taken. First, since factors that affect the linear amplification of Tollmien-Schlichting waves determine the magnitude of the transition Reynolds number, these waves may be either inhibited or cancelled. In the former method of control, the growth of the linear disturbance is minimized using any or a combination of the so-called stability modifiers which alter the shape of the velocity profile. These include increased length of favorable pressure gradient, wall transpiration, wall motion, and surface heating/cooling. Wave cancellation of the growing perturbation is accomplished through exploiting but not altering the stability characteristics of the flow. Secondly, the forcing disturbances in the environment in which the laminar shear layer develops may be reduced. This is accomplished by using smooth surfaces, reducing the freestream turbulence and the radiated sound, minimizing body vibration, and ensuring a particulate-free incoming flow or, in case of a contaminated environment such as the ocean, using a particle-defense mechanism. Practically achieved surface smoothness and levels of radiated noise place an upper limit for unit Reynolds number required for a successful laminar flow control system. For aircraft, this typically translates into a requirement for high altitude operation (above 10 Km). Thirdly, one may provide a flow where other kind of instabilities, e.g., Taylor-Görtler vortices or cross-flow instabilities, will not occur or at least will not grow at a rapid rate. This is done by avoiding as much as possible concave surfaces or concave streamlines, minimizing the sweep on lifting surfaces, etc. In the next section, the elements of linear stability theory are briefly recalled, again for the benefits of newcomers to the field. This is followed by a discussion of the first of the three steps outlined above, stability modifiers and wave cancellation methods.

4. Linear Stability Theory

The linear stability of a laminar flow is governed by the Orr-Sommerfeld equation and appropriate boundary conditions:

$$(U_1 - c)(\phi'' - \alpha^2\phi) - U_1''\phi = \frac{i}{\alpha \mathrm{Re}}(\phi'''' - 2\alpha^2\phi'' + \alpha^4\phi). \qquad (4.1)$$

This fourth-order, linear, ordinary differential equation is derived from the Navier-Stokes equation (2.6)[1] by assuming a two-dimensional small disturbance superimposed upon a steady, unidirectional mean flow, $U_1(x_2)$. The stream function for the perturbation is given by:

$$\Psi(x_1, x_2, t) = \phi(x_2)\mathrm{e}^{i\alpha(x_1 - ct)}, \qquad (4.2)$$

where α is the wave number, c is the wave speed, ϕ is the amplitude function of the fluctuation, the superscript $'$ denotes the derivative with respect to x_2, and Re is the Reynolds number based upon the freestream velocity and the boundary layer thickness. The Orr-Sommerfeld equation has been nondimensionalized using the boundary layer thickness as a length scale and the freestream velocity as a velocity scale.

Either the temporal or the spatial growth of instability waves is considered as an eigenvalue problem. In the former case, a disturbance oscillates in space but either grows or decays exponentially with time. A complex eigenvalue $c = c_r + ic_i$ is determined for each pair of values of the real parameter α and the Reynolds number. The real part of c is the phase velocity of the prescribed disturbance and the sign of the imaginary part determines whether the wave is temporarily amplified ($c_i > 0$) or temporarily damped ($c_i < 0$). The more realistic spatial stability problem involves disturbances that oscillate in time but either grow or decay exponentially with downstream distance. In this case, a complex eigenvalue $\alpha = \alpha_r + i\alpha_i$ is

[1] Note that (2.6) is the equation of motion for a Newtonian, incompressible, constant-viscosity fluid. If viscosity is not constant, additional terms containing the first and second derivatives of viscosity with respect to x_2 result and one obtains the so-called modified Orr-Sommerfeld equation (see, for example, Wazzan, et al., 1968).

determined for each pair of values for the real circular frequency αc and the Reynolds number. The real part of α is the wave number and the sign of α_i determines whether the wave is spatially amplified ($\alpha_i < 0$) or spatially damped ($\alpha_i > 0$). In either case, the major difficulty in numerically integrating the Orr-Sommerfeld equation lies in the fact that it is highly stiff and unstable (Lin, 1945), which makes the use of conventional numerical schemes virtually impossible. Explicit numerical methods with a step size that is commensurate with the global behavior of the solution cannot be used to integrate (4.1) because the numerical instabilities that characterize this ordinary differential equation. The papers by Orszag (1971) and Scott and Watts (1977) give examples of special codes to handle stiff eigenvalue problems.

In a spatially growing boundary layer, transition to turbulence involves the integrated effect of the growth of unstable waves as they are convected downstream by the mean flow. If A represents the amplitude of a wave, then its local spatial growth is given by:

$$\frac{d|A|}{dx_1} = -\alpha_i |A|. \tag{4.3}$$

The net growth of a wave with a given real frequency, from the location x_1^0, where it first becomes unstable, to the location x_1^{00}, where transition to turbulence occurs, is given by:

$$\ln\left[\frac{|A|}{|A_0|}\right] = -\int_{x_1^0}^{x_1^{00}} \alpha_i dx_1 = n. \tag{4.4}$$

Empirically, a modal amplification n of between 8 and 11 is found at transition to turbulence (e.g., Smith and Gamberoni, 1956; Van Ingen, 1956; Jaffe, et al. 1970). Despite the apparent success of the amplitude-ratio method to predict transition, Reshotko (1976) maintains that the method is defective in principle and perhaps also in practice. Transition location depends strongly on the freestream turbulence levels and other environmental factors. Moreover, nonlinear effects that are physically significant in the transition process cannot be accounted for by procedures based on linear stability theory. Notwithstanding these drawbacks, the $\mathbf{e}^n$ rule provides a practical scheme for estimating the effectiveness of a transition- delay method of control (Bushnell, et al., 1988; Bushnell, 1989).

The Reynolds number below which perturbations of all wave numbers decay is termed the critical Reynolds number or the limit of stability. For a given velocity profile, $U_1(x_2)$, the critical Reynolds number and the rate of growth of perturbations depends strongly on the shape of the velocity profile. A profile with an inflectional point ($\partial^2 U_1/\partial x_2^2 = 0$) above the wall provides a necessary and sufficient condition for inviscid instability (Rayleigh, 1880; Tollmien, 1935). Such profiles must have a positive curvature at $x_2 = 0$, since $\partial^2 U_1/\partial x_2^2$ is negative at a large distance from the wall (Figure 2). Even when viscous effects are included (right-hand side of the Orr-Sommerfeld equation $\neq 0$), a velocity profile becomes more stable as its second derivative near the wall becomes negative, $[\partial^2 U_1/\partial x_2^2]_0 < 0$. The profile is then said to be more full, having a smaller ratio of displacement thickness to momentum thickness than, for example, an inflectional velocity profile. In the former case, the critical Reynolds number is increased, the range of amplified frequencies is diminished and the amplification rate of unstable waves is reduced.

5. Stability Modifiers

Stability modifiers are those methods of laminar flow control which alter the shape of the velocity profile to minimize the linear growth of unstable waves. For a boundary layer flow along a two-dimensional surface of small curvature, the streamwise momentum equation (2.8) at the wall ($x_2 = 0$) reads:

$$\left[\rho\frac{\partial U_1}{\partial t}\right]_0 + \left[\rho U_1\frac{\partial U_1}{\partial x_1}\right]_0 + \left[\rho U_2\frac{\partial U_1}{\partial x_2}\right]_0 + \frac{dP}{dx_1} - \frac{d\mu}{dT}\left[\frac{\partial T}{\partial x_2}\frac{\partial U_1}{\partial x_2}\right]_0 = \left[\mu\frac{\partial^2 U_1}{\partial x_2^2}\right]_0, \tag{5.1}$$

where the subscript $[]_0$ indicates flow quantities computed at the wall and T is the temperature field. Equation (5.1) is valid for a fluid with variable properties ρ and μ. For a two-dimensional laminar boundary layer, vorticity is only in the spanwise direction and is given by $\Omega_3 = -\partial U_1/\partial x_2$. The right-hand side of (5.1), therefore, represents the flux of vorticity at the surface, as demonstrated by Lighthill (1963). Any of the terms on the left-hand side of (5.1) can affect the sign of the second derivative of the velocity profile (or the direction of the vorticity flux) at the wall and, hence, the flow stability. Stability modifiers do just that and include wall motion

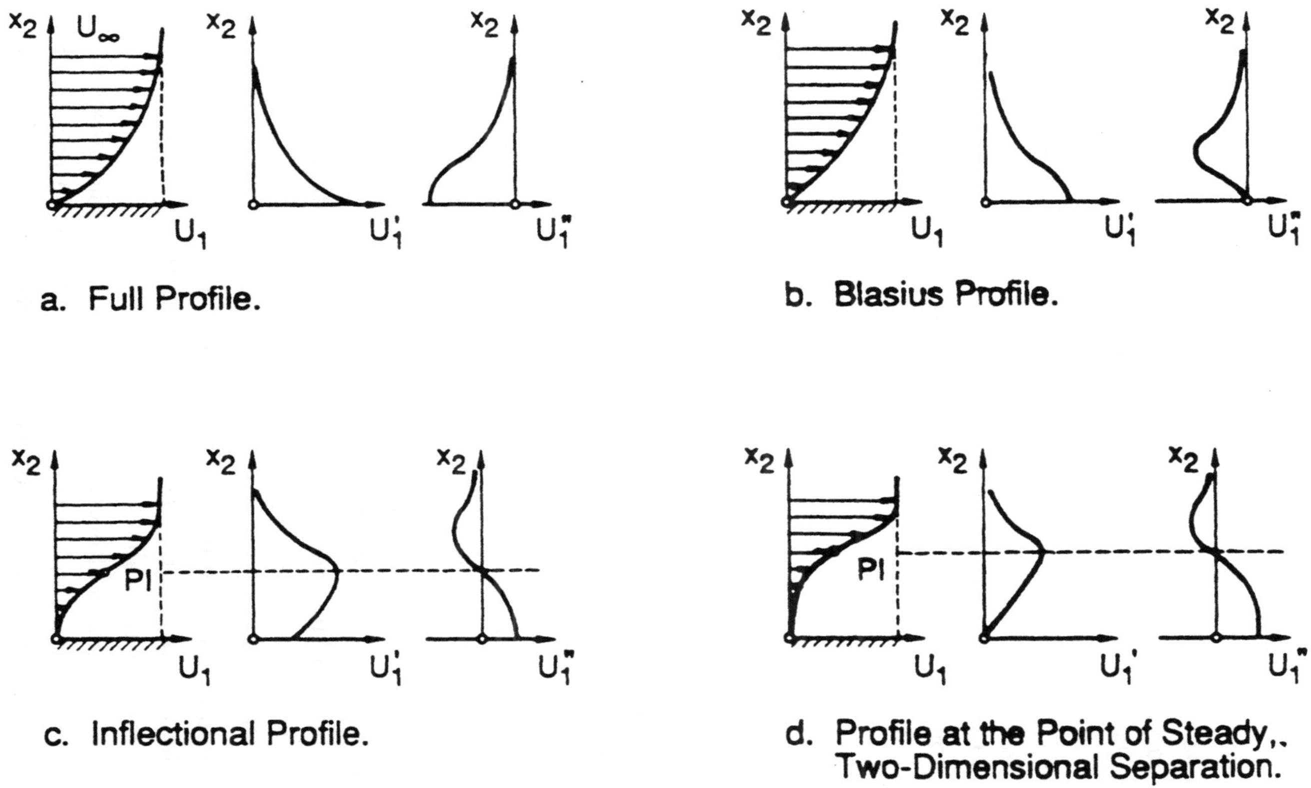

a. Full Profile.

b. Blasius Profile.

c. Inflectional Profile.

d. Profile at the Point of Steady,. Two-Dimensional Separation.

Figure 2. Normal Velocity Profile in a Boundary Layer.

such that the first, second, or third term on the left-hand side of (5.1) is negative, wall suction ($[U_2]_0 < 0$), favorable pressure gradient ($dP/dx_1 < 0$), surface cooling in gases ($d\mu/dT > 0$; $[\partial T/\partial x_2]_0 > 0$), or surface heating in liquids ($d\mu/dT < 0$; $\partial T/\partial x_2]_0 < 0$). Any one or a combination of these methods will cause the curvature of the velocity profile at the wall to become more negative and, hence, increase the lower critical Reynolds number and reduce the spatial or the temporal amplification rates of unstable waves.

Boundary layers which are stabilized by extending the region of favorable pressure gradient are known as natural laminar flow (NLF), while the other methods to modify the stability of the shear flow are termed laminar flow control (LFC). It is clear from (5.1) that the effects of these methods are additive. The term hybrid laminar flow control normally refers to the combination of NLF and one of the LFC techniques.

5.1. Wall Motion

Wall motion can be generated by either actively driving the surface or by using a flexible coating whose modulus or rigidity is low enough so that surface waves are generated under the influence of the stress field in the fluid. In the former case, the wall motion is precisely controlled and can be made to affect the shape of the velocity profile in a desired manner. This method of control is, however, quite impractical and is used mainly to provide controlled experiments to determine what type of wave motion is required to achieve a given result. The more practical passive flexible walls can be broadly classified into two categories: truly compliant coatings which have extremely small damping and modulus of elasticity and can therefore respond with little phase leg to the boundary layer flow; and the more readily available resonant walls in which vibration modes in the solid, acting as an active vibration damper, are excited by the flow-disturbance forcing function (Bushnell, et al., 1977). The flow stabilization in this case may be a result of altering the phase relation in the viscous region rather than changing the curvature of the velocity profile at the wall.

Theoretical studies of boundary-layer stability in the presence of a flexible wall started in the early 1960's, stimulated in part by the pioneering experimental work of Krämer (1960). In both the classical work (e.g., Benjamin, 1960; Landahl, 1962; Kaplan, 1964) and the more recent research (e.g., Carpenter and Garrad, 1985; Willis, 1986; Yeo and Dowling, 1987), steady-state stability theory has been used with an assumed velocity profile which is not allowed to change as the wall moves. It is clear, however, that the wall displacement will induce a traveling pressure signal which in turn will modulate the mean velocity profile. Either a quasi-steady stability analysis of the modulated flow or a true time-dependent calculations would be preferable to conventional linear-stability theory, but neither has been attempted yet due to the obvious complexities of the problem. Notwithstanding the shortcoming of present stability calculations, Willis (1986) obtained a very impressive agreement with his carefully conducted experiments. His results are depicted in Figure 3. Eigenvalue calculations were performed to predict the amplification factors for a range of modal frequencies. The flexible coating was a silicon-rubber/silicon-oil mix covered by a thin latex rubber skin stretched across the surface. The experiments were conducted in a water towing tank using a flat plate, and controlled, harmonic, two-dimensional disturbances were introduced upstream of the compliant surface. As indicated in Figure 3, a reduction of the wave growth by an order of magnitude is feasible when this rather simple coating is used, almost eliminating transition due to Tollmien-Schlichting type of instability. The flexible wall itself is, however, susceptible to other kinds of instability and care must be taken to ensure that these surface waves will not grow to an amplitude that will promote transition through a roughness-like effect (see the recent review article by Riley, et al., 1988).

Although the original Kramer's (1960) experiments were discredited up until a few years ago, new theoretical and experimental evidence confirm Kramer's results (Carpenter and Garrad, 1985). Passive flexible coatings with density the order of the fluid density appear to be capable of considerable transition postponement. The density requirement makes this method of control suitable for water applications only. Transitional Reynolds numbers (based on distance from the lading edge) that are 5-10 times those for a rigid surface seem to be readily achievable with a simple method that does not require energy expenditure, slots, ducts, or internal equipment of any kind.

5.2 Suction

A second method for postponing transition is the application of

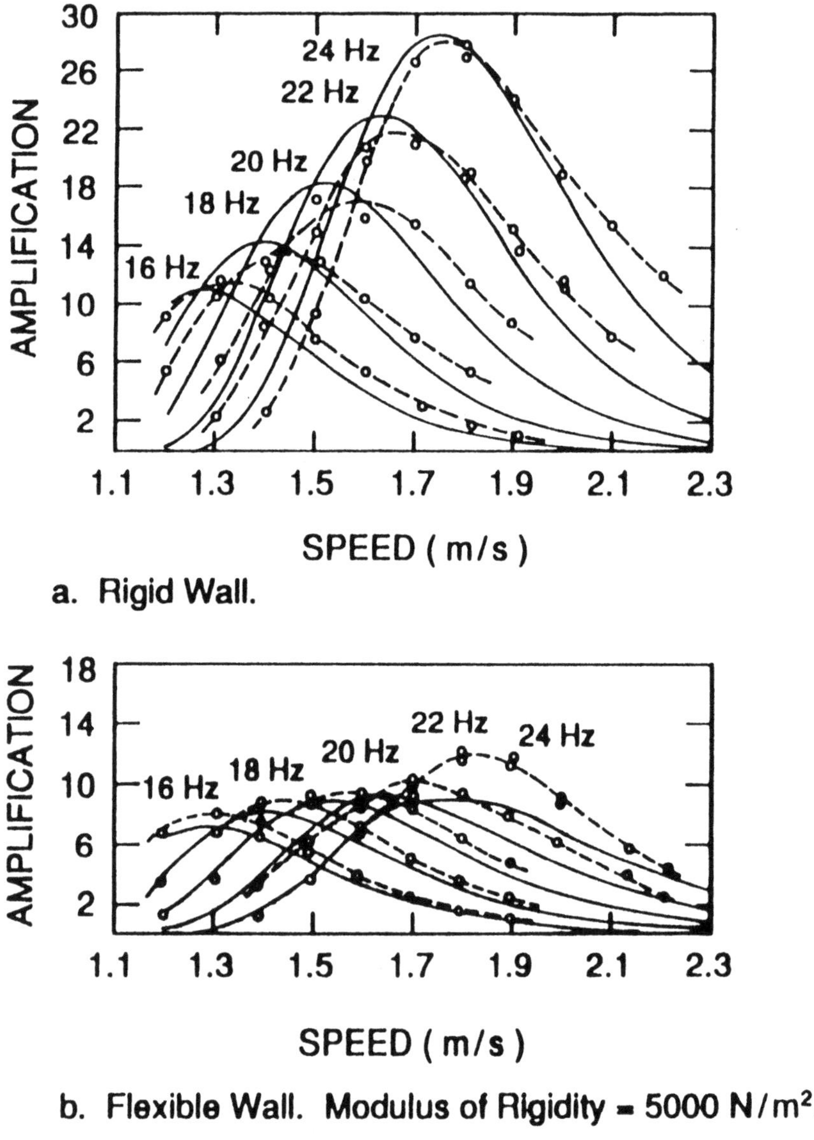

Figure 3. Growth Curves of Artificially Introduced Tollmien - Schlichting Waves. Solid Lines are Theoretical Prediction and Circles are Experimental Data. (from Willis, 1986).

wall suction. As seen from (5.1), small amounts of fluid withdrawn from the near-wall region of the boundary layer change the curvature of the velocity profile at the wall and can dramatically alter the stability characteristics of the boundary layer, so that the critical Reynolds number based on thickness may never be reached.

Although laminar flow can be maintained to extremely high Reynolds number provided that enough fluid is sucked away, the goal is to accomplish transition delay with the minimum suction flow rate. Not only will this reduce the power necessary to drive the suction pump but also the momentum loss due to suction, and hence the skin friction, is minimized. This latter point can easily be seen from the momentum integral equation. Rewriting (2.10) for an incompressible flow ($\rho_0 = \rho_\infty =$ constant) over a flat plate ($dU_\infty/dx_1 = 0$; $dR/dx_1 = 0$) with uniform suction through the wall (v_0 negative), the equation reads:

$$\frac{C_f}{2} = \frac{d\delta_\theta}{dx_1} + \frac{|v_0|}{U_\infty}. \tag{5.2}$$

The second term on the right-hand side is the suction coefficient, C_q, and although withdrawing the fluid through the wall leads to a decrease in the rate of growth of the momentum thickness, C_f increases directly with C_q. Fluid withdrawn through the wall has to come from outside the boundary layer where the streamwise momentum per unit mass is at the relatively high level of U_∞. The second term is proportional to the rate of momentum loss due to withdrawing a mass per unit time and area of $\rho|v_0|$. Note that this term does not exist for pipe flows because of the mass flow constraint. Hence, this momentum penalty is not paid for channel flows with wall transpiration, an important distinction between internal and external flows.

Although Prandtl (1904) used suction to prevent flow separation from the surface of a cylinder near the beginning of this century, the first experimental demonstration that boundary-layer transition can be delayed by withdrawing near-wall fluid did not take place until about four decades later. Holstein (1940), Ackeret, et al. (1941), Ras and Ackeret (1941), and Pfenninger (1946) used carefully shaped, single and multiple suction slits to demonstrate the decrease in drag associated with delaying transition. Braslow, et al. (1951) used continuous suction through a porous wall to maintain a laminar flow on an airfoil to chord Reynolds number of 2.0×10^7. Raspet (1952) conducted unique, noise-free experiments that confirmed the large decrease in drag when suction is applied through the wings of a sailplane. In the early 1960's, test flights of two X-21 aircrafts (modified U. S. Air Force B-66's) indicated the feasibility of maintaining a laminar flow on a swept wing to chord Reynolds number as high as 4.7×10^7 (Whites, et al., 1966). The wing surfaces contained many thin and closely spaced spanwise suction slots, and the total airplane drag was reduced by 20% as compared to the no suction case.

Although discrete suction slots were used first, because of the unavailability of suitable porous surfaces in the early 1940's, the theoretical treatment of the problem is considerably simplified by assuming continuous suction through a porous wall where the characteristic pore size is much smaller than a boundary layer thickness. In fact, the case of a uniform suction from a flat plate at zero incidence is an exact solution of the Navier-Stokes equation (2.6). Assuming weak enough suction that the potential flow outside the boundary layer is unaffected by the loss of mass at the wall (sink effects), the asymptotic velocity profile in the viscous region is exponential and has a negative curvature at the wall:

$$U_1(x_2) = U_\infty[1 - \exp(-|v_0|x_2/\nu)]. \tag{5.3}$$

The displacement thickness has the *constant* value $\delta^* = \nu/|v_0|$, where ν is the kinematic viscosity and $|v_0|$ is the absolute value of the normal velocity at the wall. In this case, (5.2) reads:

$$C_f = 2C_q. \tag{5.4}$$

Bussman and Münz (1942) computed the critical Reynolds number for the above asymptotic velocity profile to be $R_{\delta^*} \equiv U_\infty \delta^*/\nu = 70,000$. From the value of δ^* given above, the flow is stable to all small disturbances if $C_q \equiv |v_0|/U_\infty > 1.4 \times 10^{-5}$. The amplification rate of unstable disturbances for the asymptotic profile is an order of magnitude less than that for the Blasius boundary layer (Pretsch, 1942). This treatment ignores the development distance from the leading edge needed to reach the asymptotic state. When this is included into the computation, a higher $C_q(1.18 \times 10^{-4})$ is required to ensure stability (Iglisch, 1944; Ulrich, 1944). Wuest (1961) presented a summary of transpiration boundary layer computations up to the early 1960's.

The more complicated analysis for the stability of a boundary layer with suction through discrete spanwise strips was only carried out satisfactorily very recently. Reed and Nayfeh (1986) conducted a numerical-perturbation analysis of a linearized, triple-deck, closed-form basic state of a flat plate boundary layer with suction through a finite number or spanwise porous strips. Their results were compared to interacting boundary layer calculations (Ragab and Nayfeh, 1980) as well as to the carefully conducted experiments of Reynolds and Saric (1986). Suction applied through discrete strips can be as effective as suction applied continuously over a much longer streamwise length. Reed and Nayfeh suggested a scheme for optimizing the strip configuration. Their results showed that suction should be concentrated nearer the leading edge (branch I of the neutral stability curve) when disturbances are still small in amplitude.

Suction may be applied through porous surfaces, perforated plates, or carefully machined slots. It is of course structurally impossible to make the whole surface of an aircraft's wing or the like out of porous material and often strips of sintered bronze or steel are used. A relatively inexpensive woven stainless steel, Dynapore, is now available and provides some structural support (Reynolds and Saric, 1986). Superior surface smoothness and rigidity are obtained by drilling microholes in titanium using the recently developed electro-beam technology. The lower requirement for a pressure drop in the case of a perforated plate translates directly into pumping-power saving. However, outflow problems may result from regions of the wing having strong pressure gradients (Saric and Reed, 1986). Outflow in the aft region of a suction strip can cause large destabilizing effects and local three-dimensionality.

While structurally a surface with multiple slits is more rigid than a porous surface, slots are more expensive to fabricate accurately. Moreover, the higher mass flow rates associated with them may result in high Reynolds number instabilities such as separation and backflow, which adversely affect the stability of the basic flow. The rule of thumb is that the Reynolds number based on slot width (or hole diameter in the case of a perforated plate) and the local suction velocity should be kept below 10 to avoid adverse effects on the boundary layer stability, although Saric and Reed (1986) claim a hole Reynolds number an order of magnitude higher than that without destabilization of the basic flow.

Delaying a transition using suction is a mature technology, where most of the remaining problems are in the maintainability and reliability of suction surfaces and the optimization of suction rate and distribution. To protect the delicate suction surfaces on the wing of an aircraft from insect impacts and ice formation at low altitudes, special leading edge systems are used (Wagner and Fischer, 1984; Wagner, et al., 1984). Suction is less suited for underwater vehicles because of the abundance of suspended ocean particulate that can clog the suction surface as well as destabilize the boundary layer.

5.3. Shaping

The third method of control to delay laminar-to-turbulent transition is perhaps the simplest and involves the use of suitably shaped bodies to manipulate the pressure distribution. In (5.1), the pressure gradient term can affect the sign of the curvature of the velocity profile at the wall and, hence, change the stability characteristics of the boundary layer. According to the calculations of Schlichting and Ulrich (1940), the critical Reynolds number based on displacement thickness and freestream velocity changes from about 100 to 10,000 as a suitably nondimensionalized pressure gradient (the shape factor Λ) varies from $\Lambda = -6$ (adverse) to $\Lambda = +6$ (favorable). Moreover, for the case of a favorable pressure gradient, no unstable waves exist at infinite Reynolds number. In contrast, the upper branch of the neutral stability curve in the case of an adverse pressure distribution tends to a non-zero asymptote so that a finite region of wavelengths at which disturbances are always amplified remains even as $Re \rightarrow \infty$.

Streamlining a body to prevent separation and reduce form drag is quite an old art, but the stabilization of a boundary layer by pushing the longitudinal location of the pressure minimum to as far back as possible dates back to the 1930's and led to the successful development of the NACA 6-Series NLF airfoils. Newer, low-Reynolds-number lifting surfaces used in sailplanes, low-speed drones and executive business jets have their maximum thickness point far aft of the leading edge. The recent success of the Voyager's nine-day, unrefueled flight around the world was due in part to a wing design employing natural laminar flow to approximately 50% chord. Application of NLF technology to underwater vehicles is feasible but somewhat more limited (Granville, 1979).

The favorable pressure gradient extends to the longitudinal location of the pressure minimum. Beyond this point, the adverse pressure gradient becomes steeper and steeper as the peak suction is moved further aft. For an airfoil, the desired shift in the point of minimum pressure can only be attained in a certain narrow range of angles of incidence. Depending on the shape, angle of attack, Reynolds number, surface roughness, and other factors, the boundary layer either becomes turbulent shortly after the point of minimum pressure or separates first and then undergoes transition. One of the design goals of NLF is to maintain attached flow in the adverse pressure gradient region and some method of separation control may have to be used there.

Factors that limit the utility of NLF include crossflow instabilities and leading edge contamination on swept wings, insect and other particulate debris, high unit Reynolds numbers at lower cruise altitudes, and performance degradation at higher angles of attack due to the necessarily small leading edge radius of NLF airfoils. Reductions of surface waviness and smoothness of modern production wings, special leading edge systems to prevent insect impacts and ice formation, higher cruise altitudes of newer airplanes, and higher Mach numbers all favor the application of NLF (Runyan and Steers, 1980). To paraphrase a recent statement by Holmes (1988), an NLF airfoil is no longer as finicky as Morris the Cat. It is true that a boundary layer that is kept laminar to extremely high Reynolds numbers is very sensitive to environmental factors such as roughness, freestream turbulence, radiated sound, etc. However, the flow is durable and reliable within certain conservative design corridors which must be maintained by the skillful designer and eventual operator of the vehicle. Current research concentrates on understanding the achievability and maintainability of natural laminar flow, expanding the practical applications of NLF technology, and extending the design methodology to supersonic aviation (Bushnell and Malik, 1988; Bushnell, 1989).

5.4. Wall Heating/Cooling

The last of the stability modifiers is the addition or removal of heat from a surface, which causes the viscosity to vary with distance from the wall. In general, viscosity increases with temperature for gases, while the opposite is true for liquids. Thus, if heat is removed from the surface of a body moving in air, the fifth term on the left-hand side of (5.1) is negative. In that case, the velocity gradient near the wall increases and the velocity profile become fuller and more stable. The term containing the viscosity derivative will also be negative if the surface of a body moving in water is heated. With heating in water or cooling in air, the critical Reynolds number is increased, the range of amplified frequencies is diminished and the amplification rate of unstable waves is reduced. Substantial delay of transition is feasible with a surface that is only a few degrees hotter (in water) or colder (in air) than the freestream.

The first indirect evidence of this phenomenon was the observation that the drag of a flat plate placed in a wind tunnel increases by a large amount when the plate is heated (Linke, 1942). Both Frick and McCullough (1942) and Liepmann and Fila (1947) showed that the transition location of a flat-plate boundary layer in air at low subsonic speeds is moved forward as a result of surface heating. The stability calculations of Lees (1947) confirmed these experiments and, moreover, showed that cooling has the expected opposite effects. The critical Reynolds number based on distance from the leading edge increases from 10^5 to 10^7 when the wall of a flat plate placed in an air stream is cooled to 70% of the absolute ambient temperature. Even a modest cooling of the wall to $0.95T_\infty$ results in doubling of the critical Reynolds number (Kachanov, et al., 1974). With cooling, the range of amplified frequencies is diminished and the growth rate of T-S waves is reduced resulting in a substantial increase in transition Reynolds number. These same trends were dramatically confirmed in subsonic and supersonic flights[2] by Dougherty and Fisher (1980) who studied the transition on an airborne cone over the Mach number range of 0.55 - 2.0. They reported a transition Reynolds number that varied approximately as T_0^{-7}, where T_0 is the wall temperature. For aircraft, this method of transition delay is feasible only for a vehicle which uses a cryo-fuel such as liquid hydrogen or liquid methane. In that case, a sizeable heat sink is readily available. The idea being that the fuel is used to cool the major aerodynamic surfaces of the aircraft as it flows from the fuel tanks to the engines. Reshotko (1979) examined the prospects for the method and concluded that, particularly for a hydrogen-fueled aircraft, substantial drag reductions are feasible. His engineering calculations indicated that the weight of the fuel saved is well in excess of the weight of the required cooling system.

[2]In hypersonic flows, a different mode of stability, Mack's Second Mode, dominates the transition process and cooling, in fact, promotes earlier transition to turbulence.

The above effects are pronounced in water flows due to the larger Prandtl number (good thermal coupling) and the stronger dependence of viscosity on temperature.[3] In a typical low-speed situation, a surface heating of 1°C in water has approximately the same effect on the curvature of the velocity profile at the wall as a surface cooling of 20°C in air or a suction coefficient (in air or water) of 0.0003 (Liepmann, et al., 1982). Wazzan, et al. (1968; 1970) used a modified fourth-order Orr-Sommerfeld equation combined with the $\mathbf{e}^9$ method of Smith (1957) and confirmed that wall heating can produce large increases in the transition Reynolds number of water boundary layers. They predicted a transition Reynolds number, based on a freestream velocity and distance from the leading edge of a flat-plate, as high as 2×10^8 for wall temperatures that are only 40°C above the ambient water temperature.

Lowell and Reshotko (1974) refined Wazzan, et al.'s (1968; 1970) calculations by introducing a coupled sixth-order system of vorticity and energy disturbance equations. The predicted critical Reynolds numbers for wall overheats of up to 2.8°C were confirmed by the experiments of Strazisar, et al. (1977) who measured the growth rates of small disturbances generated by a vibrating ribbon in a heated flat-plate in water. These experiments did not yield data on transition or on stability at higher overheats.

The transition predictions of Wazzan, et al. (1968; 1970) at higher overheats were partially confirmed by the very carefully conducted experiments of Barker and Gile (1981) who used the entrance region of an electrically heated pipe. The displacements thickness was much smaller than the pipe radius and, thus, the boundary-layer development was approximately the same as that of a zero-pressure gradient flat-plate. Barker and Gile reported a transition Reynolds number of 4.7×10^7 for a wall overheat of 8°C. No further increase in $Re_{transition}$ was observed as the wall was heated further, in contradiction to the computations of Wazzan, et al. (1968; 1970). Barker and Gile (1981) investigated possible causes of this discrepancy including buoyancy effects, wall roughness, effects of geometry, flow asymmetries, and suspended particulate matter. Their analysis and numerous related work (e.g., Kosecoff, et al., 1976; Chen, et al.,

---

[3]For water at room temperature, $Pr \simeq 7$ and absolute viscosity is decreased by approximately 2% for each 1°C rise in temperature. For room temperature air, $Pr \simeq 0.7$ and absolute viscosity is decreased by approximately 0.2% for each 1°C drop in temperature.

1979; Hendricks and Ladd, 1983; Lauchle and Gurney, 1984) indicate that increased concentration and size of suspended particulate diminish the stabilizing effect of surface heating until at some point surface heating no longer stabilizes the boundary layer but is in fact a destabilizing influence.

On a heated body of revolution in a high-speed water tunnel, Lauchle and Gurney (1984) observed an increase in transition Reynolds number from 4.5×10^6 to 3.6×10^7 for an average overheat of 25°C. Clearly, surface heating in water can be an extremely effective method of transition delay and, hence, drag reduction for small, high-speed underwater vehicles where the rejected heat from their propulsion system is used to increase the surface temperature along the body length. The detrimental effects of freestream particulate alluded to above are, however, a major obstacle at present for a practical implementation of this method of control. Suspended particulate having a wide-band concentration spectra are abundant in the oceans and "particle-defense" mechanisms must be sought before using any of the transition delay methods in a contaminated environment.

In addition to surface heating, several other techniques are available to lower the near-wall viscosity ($\partial\mu/\partial x_2 > 0$) in a water boundary layer and, thus, favorably affect the stability of the flow. These include film boiling, cavitation, sublimation, chemical reaction, or wall injection of a gas or lower-viscosity liquid. Finally, a shear-thinning additive could be introduced into the boundary layer. Since the shear increases as the wall is approached, the effective viscosity of the non-Newtonian fluid decreases there and $\partial\mu/\partial x_2$ becomes positive.

6. Wave Cancellation

An alternative approach to increase the transition Reynolds number of a laminar boundary layer is wave cancellation. If the frequency, orientation and phase angle of the dominant element of the spectrum of growing linear disturbances in the boundary layer is detected, a control system, and appropriately located disturbance generators may then be used to effect a desired cancellation or suppression of the detected disturbances. In this case, the stability characteristics of the boundary are exploited but not altered (Reshotko, 1985). Wave cancellation is feasible only when the disturbances are still relatively small, their growth is governed by a linear equation, and the principle of superposition is still valid.

The first reported use of wave cancellation is that due to Schilz (1965/66). He used a vibrating ribbon to excite a T-S wave on a test plate which had a flexible surface. A unique wall-motion device flush mounted into the plate moved the flexible wall in a transverse, wavelike manner with a variety of frequencies and phase speeds. A significant amount of cancellation results when the flexible wall motion had the opposite phase but the same frequency and phase speed as the T-S wave. Both Milling (1981) and Thomas (1983) used two vibrating wires, one downstream of the other, to generate and later cancel a single frequency T-S wave. Thomas (1983) observed that interaction between the primary disturbance and background excitations prevented complete cancellation of the primary wave. To further study the consequences of wave interactions, Thomas applied the same method of control to eliminate two interacting waves of different frequency. Although the primary waves were behaving linearly, a nonlinear interaction gave rise to a low-amplitude difference frequency that could only be partially reduced and ultimately led to transition. Thomas (1983) concluded that it is not possible to return the flow completely to its undisturbed base state because of wave interactions and that it is perhaps more appropriate to describe this control method as wave superposition rather than wave cancellation.

The same principle of wave superposition could be applied using wall heating/cooling (Liepmann, et al., 1982; Liepmann and Noshenchuck, 1982; Ladd and Hendricks, 1988), plate vibration (Gedney, 1983), compliant wall (McMurray, et al., 1983), or periodic suction/blowing (Biringen, 1984). Liepmann and Nosenchuck (1982) used flush-mounted hot-film probes to sense natural T-S waves in a flat-plate boundary layer in a water tunnel. A feed-forward control loop was then used to synthesize and introduce disturbances of equal amplitude but of opposite phase via flush-mounted wall heaters. Quite recently, Ladd and Hendricks (1988) performed their experiment in a water tunnel on a 9:1 fineness-ratio ellipsoid. Strip heaters were again used to create and actively attenuate T-S waves. They applied digital filtering techniques to synthesize the attenuation signal. The filter was able to actively adapt the attenuation signal to changes in amplitude and frequency of the artificially introduced instability wave with no loss in attenuation downstream. A sample of Ladd and Hendrick's results is shown in Figure 4. Time records

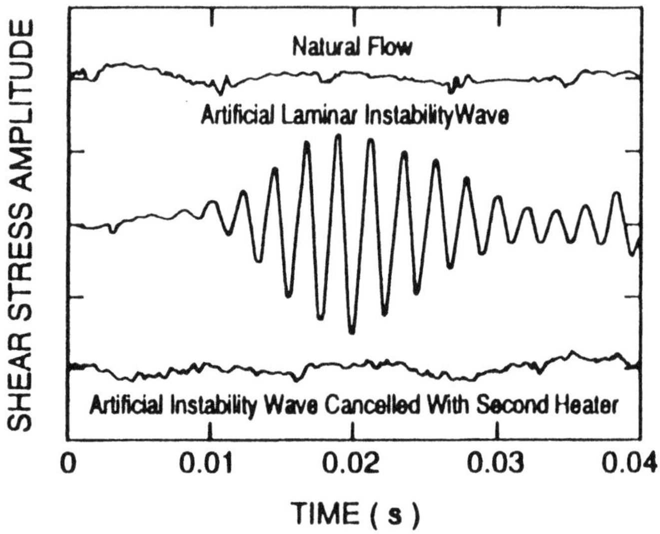

Figure 4. Cancellation of Artificial Instability Waves Using Adaptive Heat. (from Ladd and Hendricks, 1988).

of a flush-mounted hot-film probe are shown for the natural flow, artificially introduced T-S wave using a heating ring, and cancelled wave using downstream heating ring.

The transition delay achieved by active wave cancellation is modest, typically a factor of two or less increase in the transition Reynolds number based on distance from the leading edge. Reshotko (1985) maintains that to achieve significant delay in transition using this technique would require an extensive array of disturbance detectors and generators as well as prohibitively complicated control system that could cancel both the primary and residual disturbance spectra. Significant delay in transition is more readily achieved via the stability modifiers summarized in Section 5.

7. Transition Advance

For some applications, the efficient mixing or transport mechanisms of turbulence may be required. For example, a turbulent boundary layer is in general more resistant to separation than a laminar one; turbulence is used to homogenize fluid mixtures and to accelerate chemical reactions; and turbulent heat exchangers are much more efficient than laminar ones. As an additional example, earlier transition may be required for some low-speed wind or water tunnel testing to simulate high Reynolds number conditions.

For a zero-pressure-gradient boundary layer, transition is completed typically at a Reynolds number based on distance from leading edge of the order of 10^6. The critical Re below which perturbations of all wave numbers decay is about 6×10^4. To advance the transition Reynolds number, one may attempt to lower the critical Re, increase the growth rate of Tollmien-Schlichting waves, or introduce large disturbances that can cause bypass transition. The first two routes involve altering the shape of the velocity profile using wall motion (e.g., transition through deceleration, see Gad-el-Hak, et al., 1984), injection, adverse pressure gradient, or surface heating in gases or cooling in liquids. The third route is much simpler to implement though more difficult to analyze. Morkovin (1984) broadly classifies the large disturbances that can cause bypass transition into steady or unsteady ones originating into the freestream or at the body surface. The most common example is single, multiple or distributed roughness elements placed on the wall. If the roughness characteristic-length is large enough, the disturbance introduced is nonlinear and bypass transition takes place. For a three-dimensional roughness element of height-to-width ratio of one, a transition Reynolds number $R_{\delta^*} \simeq 300$ (below the critical $R_{\delta^*} = 420$ predicted from the linear stability theory) is observed for a roughness Reynolds number, based on its height and the velocity in the undisturbed boundary layer at the height of the element, of about 10^3 (Tani, 1969). Transition occurs at $R_{\delta^*} \simeq 10^3$ for a roughness Reynolds number of about 600. For a smooth surface, transition typically takes place at $R_{\delta^*} \simeq 2.6 \times 10^3$.

Other large disturbances that could lead to early transition include high turbulence levels in the freestream, external acoustic excitations, particulate contamination, and surface vibration. These are often termed environmental tripping. Transition could also be effected by detecting naturally occurring T-S waves and artificially introducing in-phase waves. Transition could be considerably advanced, on demand, using this wave superposition principle.

Early transition could also be achieved by exploiting other routes to turbulence such as Taylor-Görtler or cross-flow vortices. For example, a very mild negative curvature of $(0.003/\delta^*)$ results in the generation of strong streamwise vortices. In this case, transition Reynolds number is lowered from $R_{\delta^*} \simeq 2600$ for the flat-plate case to $R_{\delta^*} \simeq 700$ for the curved surface (Tani, 1969). For high Mach number flows, the general decay in spatial amplification rate of T-S waves makes conventional tripping more difficult as the Mach number increases (Reshotko, 1976). For these flows, trips that generate oblique vorticity waves of appropriate wavelength may be most effective to advance the transition location.

8. Concluding Remarks

This article has attempted to present a unified view of the different control methods available or contemplated to delay or advance transition. Both the science and technology to maintain a laminar boundary layer to a Reynolds number of about 4×10^7 are well established, although some details remain to be worked out. The linear stability theory provides a solid analytical framework, at least for the important first stage of transition. Barring large disturbances in a conventional boundary-layer flow, the linear amplification of Tollmien-Schlichting waves is the slowest of the successive multiple steps in the transition process. Stability modifiers inhibit this linear amplification and, therefore, determine the magnitude of the transition Reynolds number. Shaping to provide extended regions of favorable pressure gradient is the simplest method of control and is well-suited for small underwater vehicles or for the wings of low- or moderate-speed aircraft. Flight tests have demonstrated the feasibility of using suction to maintain a laminar flow on a swept wing to $Re \simeq 4.7 \times 10^7$. The required suction rate is very modest and 20% net drag reduction is possible. Remaining problems are technological in nature and include maintainability and reliability of suction surfaces and further optimization of the suction rate and its distribution. Suction is less suited for underwater vehicles because of the abundance of particulate matters that can clog the suction surface as well as destabilize the boundary layer. For water applications, compliant coatings that increase the transitional Reynolds number by a factor of 5-10 are available in the laboratory but performance in the field is still unknown. This technique is very appealing because of its simplicity and absence of energy requirement. Moderate surface heating also increases the transition Re by an order of magnitude, but a source of rejected heat must be available to achieve net drag reduction. Additionally, a particle-defense mechanism is needed before the technique could successfully be used in the ocean. For futuristic aircraft using cryo-fuel, surface cooling may be a feasible method to delay transition.

Wave cancellation of growing perturbations is accomplished through exploiting but not altering the stability characteristics of the boundary layer. In order to be able to delay or advance transition using this method, the frequency, orientation and phase angle of the dominant element of the spectrum of growing linear disturbances must be detected. A control system together with appropriately located disturbance generators are then used to effect a desired cancellaltion or suppression of the detected disturbance. Wave cancellation is not yet a practical tool for transition control and is feasible only when the growth of stability waves is governed by a linear equation. Transition delay achieved by active wave cancellation is at present modest, typically a factor of two or less in the transition Reynolds number.

References

[1] Ackeret, J., Ras, M., and Pfenninger, W. (1941), "Verhinderung

des Turbulentwerdens einer Grenzschicht durch Absaugung," *Naturwissenschaften 29*, pp. 622-623.

[2] Barker, S. J. and Gile, D. (1981), "Experiments on Heat-Stabilized Laminar Boundary Layers in Water," *J. Fluid Mech. 104*, pp. 139-158.

[3] Batchelor, G. K. (1967), *An Introduction to Fluid Dynamics*, Cambridge University Press, London.

[4] Benjamin, T. B. (1960), "Effects of a Flexible Boundary on Hydrodynamic Stability," *J. Fluid Mech. 9*, pp. 513-532.

[5] Biringen, S. (1984), "Active Control of Transition by Periodic Suction-Blowing," *Phys. Fluids 27*, pp. 1345-1347.

[6] Braslow, A. L., Burrows, D. L., Tetervin, N., and Visconti, F. (1951), "Experimental and Theoretical Studies of Area Suction for the Control of Laminar Boundary Layer," NACA Report No. 1025.

[7] Bushnell, D. M. (1989), "Applications and Suggested Directions of Transition Research," *Fourth Symp. on Numerical and Physical Aspects of Aerodynamic Flows*, Long Beach, CA, January 16-19.

[8] Bushnell, D. M., Hefner, J. N., and Ash, R. L. (1977), "Effect of Compliant Wall Motion on Turbulent Boundary Layers," *Phys. Fluids 20*, pp. S31-S48.

[9] Bushnell, D. M. and Malik, M. R. (1988), "Compressibility Influences on Boundary-Layer Transition," *Symp. on Physics of Compressible Turbulent Mixing*, Princeton, NJ, October 25-27.

[10] Bushnell, D. M., Malik, M. R., and Harvey, W. D. (1988), "Transition Prediction in External Flows via Linear Stability Theory," *Proc. IUTAM Symp. Transsonicum III*, Gottingen, Germany, May 24-27.

[11] Bussmann, K. and Münz, H. (1942), "Die Stabilität der laminaren Reibungsschicht mit Absaugung," *Jharb. Dtsch. Luftfahrtforschung 1*, pp. 36-39.

[12] Carpenter, P. W. and Garrad, A. D. (1985), "The Hydrodynamic Stability of Flow over Kramer-Type Compliant Surfaces. Part 1. Tollmien-Schlichting Instabilities," *J. Fluid Mecho. 155*, pp. 465-510.

[13] Chen, C. P., Goland, Y., and Reshotko, E. (1979), "Generation Rate of Turbulent Patches in the Laminar Boundary Layer of a Submersible," in *Viscous Flow Drag Reduction*, ed. G. R. Hough, AIAA *Progress in Astronautics and Aeronautics 72*, pp. 783-89.

[14] DiPrima, R. C. and Swinney, H. L. (1985), "Instabilities and Transition in Flow between Concentric Rotating Cylinders," in *Hydrodynamic Instabilities and the Transition to Turbulence*, eds. H. L. Swinney and J. P. Gollub, Second Edition, pp. 139-180, Springer-Verlag, Berlin.

[15] Dougherty, N. S. and Fisher, D. F. (1980), "Boundary Layer Transition on a 10-Degree Cone," AIAA Paper No. 80-0154.

[16] Drazin, P. and Reid, W. (1981), *Hydrodynamic Stability*, Cambridge University Press, London.

[17] Frick, C. W. and McCullough, C. B. (1942), "Tests of a Heated Low Drag Airfoil," NACA ARR, December.

[18] Gad-el-Hak, M., Davis, S. H., McMurray, J. T., and Orszag, S. A. (1984), "On the Stability of the Decelerating Boundary Layer," *J. Fluid Mech. 138*, pp. 297-323.

[19] Gad-el-Hak, M. and Ho, C.-M. (1985), "The Pitching Delta Wing," *AIAA J. 23*, pp. 1660-1665.

[20] Gedney, C. J. (1983), "The Cancellation of a Sound-Excited Tollmien-Schlichting Wave with Plate Vibration," *Phys. Fluids 26*, pp. 1158-1160.

[21] Goldstein, M. E. and Hultgren, L. S. (1989), "Boundary-Layer Receptivity to Long-Wave Free-Stream Disturbances," *Ann. Rev. Fluid Mech. 21*, pp. 137-166.

[22] Granville, P. S. (1979), "Drag of Underwater Bodies," in *Hydroballistics Design Handbook*, Vol. 1, pp. 309-341, Naval Sea Systems Command, SEAHAC TR 79-1, Washington, D.C.

[23] Hendricks, E. W. and Ladd, D. M. (1983), "Effect of Surface Roughness on the Delayed Transition on 9:1 Heated Ellipsoid," *AIAA J. 21*, pp. 1406-1409.

[24] Hinze, J. O. (1975), *Turbulence*, Second Edition, McGraw-Hill, New York.

[25] Holmes, B. J. (1988), "NLF Technology is Ready to Go," *Aerospace America 26*, pp. 16-20.

[26] Holstein, H. (1940), "Messungen zur Laminarhaltung der Grenzschicht an einem Flügel," *Lilienthal-Bericht S10*, pp. 17-27.

[27] Iglisch, R. (1944), "Exakte Berechnung der laminaren Reibungsschicht an der längsangeströmten ebenen Platte mit homogener Absaugung," *Schr. Dtsh. Akad. Luftfahrtforschung 8B*, pp. 1-51.

[28] Itoh, N. (1987), "Another Route to the Three-Dimensional Development of Tollmien-Schlichting Waves with Finite Amplitude," *J. Fluid Mech. 181*, pp. 1-16.

[29] Jaffe, N. A., Okamura, T. T., and Smith, A. M. O. (1970), "Determination of Spatial Amplification Factors and Their Application to Predicting Transition," *AIAA J. 8*, pp. 301-308.

[30] Kachanov, Y. S., Koslov, V. V., and Levchenko, V. Ya. (1974), "Experimental Study of the Influence of Cooling on the Stability of Laminar Boundary Layers," *Izvestia Sibirskiogo Otdielenia Ak. Nauk SSSR, Seria Technicheskikh Nauk*, Novosibirsk, No. 8-2, pp. 75-79.

[31] Kaplan, R. E. (1964), "The Stability of Laminar Incompressible Boundary Layers in the Presence of Compliant Boundaries," Sc.D. Thesis, Massachusetts Institute of Technology, Cambridge, MA.

[32] Kays, W. M. and Crawford, M. E. (1980), *Convective Heat and Mass Transfer*, McGraw-Hill, New York.

[33] Klebanoff, P. S., Schubauer, G. B., and Tidstrom, K. D. (1955), "Measurements of the Effect of Two-Dimensional and Three-Dimensional Roughness Elements on Boundary-Layer Transition," *J. Aero. Sci. 22*, pp. 803-804.

[34] Klebanoff, P. S., Tidstrom, K. D., and Sargent, L. M. (1962), "The Three-Dimensional Nature of Boundary Layer Instability," *J. Fluid Mech. 12*, pp. 1-34.

[35] Kosecoff, M. A., Ko, D. R. S., and Merkle, C. L. (1976), "An Analytical Study of the Effect of Surface Roughness on the Stability of a Heated Water Boundary Layer," Dynamics Technology, Inc., Final Report PDT 76-131, Torrance, CA.

[36] Krämer, M. O. (1960), "Boundary Layer Stabilization by Distributing Damping," *J. Am. Soc. Naval Engrs. 72*, pp. 25-33.

[37] Ladd, D. M. and Hendricks, E. W. (1988), "Active Control of 2-D Instability Waves on an Axisymmetric Body," *Exp. Fluids 6*, pp. 69-70.

[38] Landahl, M. T. (1962), "On the Stability of a Laminar Incompressible Boundary Layer Over a Flexible Surface," *J. Fluid Mech. 13*, pp. 609-632.

[39] Landau, L. D. and Lifshitz, E. M. (1963), *Fluid Mechanics*, translated from the Russian, Pergamon Press, Oxford.

[40] Lauchle, G. C. and Gurney, G. B. (1984), "Laminar Boundary-Layer Transition on a Heated Underwater Body," *J. Fluid Mech. 144*, pp. 79-101.

[41] Lees, L. (1947), "The Stability of the Laminar Boundary Layer in Compressible Fluid," NACA Report No. 876.

[42] Liepmann, H. W., Brown, G. L., and Nosenchuck, D. M. (1982), "Control of Laminar Instability Waves Using a New Technique," *J. Fluid Mech. 118*, pp. 187-200.

[43] Liepmann, H. W. and Fila, G. H. (1947), "Investigations of Effects of Surface Temperature and Single Roughness Elements on Boundary Layer Transition," NACA Report No. 890.

[44] Liepmann, H. W. and Nosenchuck, D. M. (1982), "Active Control of Laminar-Turbulent Transition," *J. Fluid Mech. 118*, pp. 201-204.

[45] Lighthill, M. J. (1963), "Introduction - Boundary Layer Theory," in *Laminar Boundary Layers*, ed. L. Rosenhead, pp. 46-113, Clarendon Press, Oxford.

[46] Lin, C. C. (1945), "On the Stability of Two-Dimensional Parallel Flows," Parts I, II, and III, *Q. Appl. Maths. 3*, pp. 117-142, 218-234, 277-301.

[47] Linke, W. (1942), "Über den Strömungswiderstand einer beheizten ebenen Platte," *Luftfahrtforschung 19*, pp. 157-160.

[48] Lowell, R. L. and Reshotko, E. (1974), "Numerical Study of the Stability of a Heated Water Boundary Layer," Case Western University, Report No. FTAS/TR-73-93, Cleveland, OH.

[49] McMurray, J. T., Metcalfe, R. W., and Riley, J. J. (1983), "Direct Numerical Simulations of Active Stabilization of Boundary Layer Flows," *Proc. Eighth Biennial Symp. on Turbulence*, ed. J. L. Zakin and G. K. Patterson, Paper No. 36, University Missouri, Rolla.

[50] Milling, R. W. (1981), "Tollmien-Schlichting Wave Cancellation," *Phys. Fluids 24*, pp. 979-981.

[51] Milne-Thomson, L. M. (1968), *Theoretical Hydrodynamics*, Fifth Edition, Macmillan, London.

[52] Morkovin, M. V. (1969), "Critical Evaluation of Transition from Laminar to Turbulent Shear Layers with Emphasis on Hypersonically Traveling Bodies," Air Force Flight Dynamics Laboratory Report No. AFFDL-TR-68-149, Wright-Patterson AFB, OH.

[53] Morkovin, M. V. (1984), "Bypass Transition to Turbulence and Research Desiderata," in *Transition in Turbines Symposium*, NASA CP-2386.

[54] Morkovin, M. V. (1988), "Recent Insights into Instability and Transition to Turbulence in Open-Flow Systems," AIAA Paper No. 88-3675.

[55] Orszag, S. A. (1971), "Accurate Solution of the Orr-Sommerfeld Stability Equation," *J. Fluid Mech. 50*, pp. 689-703.

[56] Panton, R. L. (1984), *Incompressible Flow*, Wiley-Interscience, New York.

[57] Pfenninger, W. (1946), "Untersuchungen über Reibungsverminderung an Tragflügeln, insbesondere mit Hilfe von Grenzschichtabsaugung," Reports of the Inst. of Aerodynamics, ETH Zürich, No. 13.

[58] Prandtl, L. (1904), "Über Flüssigkeitsbewegung bei sehr kleiner Reibung," *Proc. Third Int. Math. Congr.*, pp. 484-491, Heidelberg, Germany.

[59] Pretsch, J. (1942), "Umschlagbeginn und Absaugung," *Jahrb. Dtsch. Luftfahrtforschung 1*, pp. 54-71.

[60] Ragab, S. A. and Nayfeh, A. H. (1980), "A Comparison of the Second-Order Triple-Deck Theory and Interacting Boundary Layers for Incompressible Flows Past a Hump," AIAA Paper No. 80-0072.

[61] Ras, M. and Ackeret, J. (1941), "Über Verhinderung der Grenzachicht-Turbulenz durch Absaugung," *Helv. Phys. Acta 14*, p. 323.

[62] Raspet, A. (1952), "Boundary-Layer Studies on a Sailplane," *Aeronaut. Eng. Rev. 11*, pp. 52-60.

[63] Rayleigh, F. (1880), "On the Stability, or Instability, of Certain Fluid Motions," *Proc. London Math. Soc. 11*, pp. 57-70.

[64] Reed, H. L. and Nayfeh, A. H. (1986), "Numerical-Perturbation Technique for Stability of Flat-Plate Boundary Layers with Suction," *AIAA J. 24*, pp. 208-214.

[65] Reed, H. L. and Saric, W. S. (1987), "Stability and Transition of Three-Dimensional Flows," *Proc. 10th U. S. National Cong. of App. Mech.*, ed. J. P. Lamb, pp. 457-468, ASME, New York.

[66] Reed, H. L. and Saric, W. S. (1989), "Stability of Three-Dimensional Boundary Layers," *Ann. Rev. Fluid Mech. 21*, pp. 235- 284.

[67] Reshotko, E. (1976), "Boundary-Layer Stability and Transition," *Ann. Rev. Fluid Mech. 8*, pp. 311-349.

[68] Reshotko, E. (1979), "Drag Reduction by Cooling in Hydrogen-Fueled Aircraft," *J. Aircraft 16*, pp. 584-590.

[69] Reshotko, E. (1985), "Control of Boundary Layer Transition," AIAA Paper No. 85-0562.

[70] Reshotko, E. (1987), "Stability and Transition, How Much Do We Know?" *Proc. 10th U. S. National Cong. of App. Mech.*, ed. J. P. Lamb, pp. 421-434, ASME, New York.

[71] Reynolds, G. A. and Saric, W. S. (1986), "Experiments on the Stability of the Flat-Plate Boundary Layer with Suction," *AIAA J. 24*, pp. 202-207.

[72] Riley, J. J., Gad-el-Hak, M., and Metcalfe, R. W. (1988), "Compliant Coatings," *Ann. Rev. Fluid Mech. 20*, pp. 393-420.

[73] Rosenhead, L. (1963), *Laminar Boundary Layers*, Clarendon Press, Oxford.

[74] Runyan, L. J. and Steers, L. L. (1980), "Boundary Layer Stability Analysis of a Natural Laminar Flow Glove on the F-111 TACT Airplane," in *Viscous Flow Drag Reduction*, ed. G. R. Hough, AIAA *Progress in Astronautics and Aeronautics 72*, pp. 17-32.

[75] Saric, W. S. and Reed, H. L. (1986), "Effect of Suction and Weak Mass Injection on Boundary-Layer Transition," *AIAA J. 24*, pp. 383-389.

[76] Schilz, W. (1965/66), "Experimentelle Untersuchungen zur Akustischen Beeinflussung der Strömungsgrenzachicht in Luft," *Acustica 16*, pp. 208-223.

[77] Schlichting, H. (1979), *Boundary-Layer Theory*, Seventh Edition, McGraw-Hill, New York.

[78] Schlichting, H. and Ulrich, A. (1940), "Zur Berechnung des Umschlages Laminar-Turbulent," *Jahrb. Dtsch. Luftfahrtforschung 1*, pp. 8-35.

[79] Scott, M. R. and Watts, H. A. (1977), "Computational Solution of Linear Two-Point Boundary Value Problems via Orthonormalization," *J. Numerical Analysis 14*, pp. 40-70.

[80] Smith, A. M. O. (1957), "Transition, Pressure Gradient, and Stability Theory," *Actes IX Congrès International de Mécanique Appliquée*, Vol. 4, pp. 234-244, Université de Bruxelles, Belgique.

[81] Smith, A. M. O. and Gamberoni, N. (1956), "Transition, Pressure Gradient and Stability Theory," Rep. ES-26388, Douglas Aircraft Company, El Segundo, CA.

[82] Squire, H. B. (1933), "On the Stability for Three-Dimensional Disturbances of Viscous Fluid Flow between Parallel Walls," *Proc. R. Soc. Lond. A 142*, pp. 621-628.

[83] Strazisar, A. J., Reshotko, E., and Prahl, J. M. (1977), "Experimental Study of the Stability of Heated Laminar Boundary Layers in Water," *J. Fluid Mech. 83*, pp. 225-247.

[84] Stuart, J. T. (1963), "Hydrodynamic Stability," in *Laminar Boundary Layer Theory*, ed. L. Rosehead, pp. 492-579, Clarendon Press, Oxford.

[85] Tani, I. (1969), "Boundary-Layer Transition," *Ann. Rev. Fluid Mech. 1*, pp. 169-196.

[86] Thomas, A. S. W. (1983), "The Control of Boundary-Layer Transition using a Wave Superposition Principle," *J. Fluid Mech. 137*, pp. 233-250.

[87] Tollmien, W. (1935), "Ein allgemeines Kriterium der Instabilität laminarer Geschwindigkeitsverteilungen," Nachr. Wiss. Fachgruppe, Göttingen, *Math. Phys. K1 1*, pp. 79-114.

[88] Ulrich, A. (1944), "Theoretische Untersuchungen über die Widerstandsersparnis durch Laminarhaltung mit Absaugung," *Schriften Dtsch. Akad. Luftfahrtforschung 8B*, p. 53.

[89] Van Ingen, J. L. (1956), "A Suggested Semiempirical Method for the Calculation of the Boundary-Layer Transition Region," Rep. V.T.H. 74, Department of Aero. and Eng., Institute of Technology, Delft, Holland.

[90] Wagner, R. D. and Fischer, M. C. (1984), "Fresh Attack on Laminar Flow," *Aerospace America 22*, pp. 72-76.

[91] Wagner, R. D., Maddalon, D. V., and Fischer, M. C. (1984), "Technology Development for Laminar Boundary Control on Subsonic Transport Aircraft," AGARD CP-365, Paper No. 16.

[92] Wazzan, A. R., Okamura, T. T., and Smith, A. M. O. (1968), "Stability of Water Flow over Heated and Cooled Flat Plates," *J. Heat Transfer 90*, pp. 109-114.

[93] Wazzan, A. R., Okamura, T. T., and Smith, A. M. O. (1970), "The Stability and Transition of Heated and Cooled Incompressible Boundary Layers," *Proc. 4th Int. Heat Transfer Conf.*, Vol. 2, ed. U. Grigull and E. Hahne, FC 1.4, Elsevier, New York.

[94] Whites, R. C., Sudderth, R. W., and Wheldon, W. G. (1966), "Laminar Flow Control on the X-21", *Astro. and Aero. 4*, pp. 38-43.

[95] Willis, G. J. K. (1986), "Hydrodynamic Stability of Boundary Layers over Compliant Surfaces," Ph.D. Thesis, University of Exeter, United Kingdom.

[96] Wuest, W. (1961), "Survey of Calculation Methods of Laminar Boundary Layers with Suction in Incompressible Flow," in *Boundary Layer and Flow Control*, Vol. 2, ed. G. V. Lachmann, pp. 771-800, Pergamon Press, New York.

[97] Yeo, K. S. and Dowling, A. P. (1987), "The Stability of Inviscid Flows over Passive Compliant Walls," *J. Fluid Mech. 183*, pp. 265-292.

PERTURBATION AMPLIFICATION IN THE ENTRY REGION OF A TRANSITIONAL PIPE FLOW

Promode R. Bandyopadhyay

Mail Stop 163, NASA Langley Research Center
Hampton, VA 23665-5225

Andrew G. Walton[1]

M. E. Dept., Old Dominion University
Norfolk, VA 23529-0247

ABSTRACT

The forced transition due to the application of large disturbances in the entry region of a pipe flow is considered. The disturbances are not large enough to cause a puff-type transition which is peculiar to a pipe flow; they cause a slug-type transition which is akin to a flat plate turbulence spot. The perturbation amplifications with Reynolds number have been measured at the entrance and also soon after the mean turbulent flow has become fully developed. At the entrance, all dominant frequencies amplify in a nearly universal manner which is exponential in part of the Reynolds number range. Furthermore, the nonlinear process undergoes a change at $4930 \leq \mathrm{Re} \leq 6020$ and again at $8120 \leq \mathrm{Re} \leq 10060$. At the downstream station, an almost saturated amplification rate has been reached and, in the background of a higher turbulence level, the trend is qualitatively similar to that in peak-valley splitting.

1. Introduction

In flat plate transition experiments on the nonlinear stage, disturbances are considered large when they are 1% in amplitude and small when they are 0.3% (Thomas, 1986). Although amplitude does not appear to be the only deciding factor (see below), the formation of arrayed hairpin vortices is commonly attributed to the former and they become staggered in the latter case (Thomas, 1986). It seems that, experimentally the arrayed type was first observed by Hama,

[1]Permanent address: Maths. Dept., University College, London.

et al. (1957) and is sometimes called the K-type (after Klebanoff), while the staggered type was first observed by Knapp, et al. (1966) and is sometimes called the C-type (after Craik) or H-type (after Herbert) depending on whether the spanwise scale is 1.5 times or equal to the Tollmien-Schlichting wave length, respectively (also see Thomson, 1988).

The transition process is still not well understood. For example, in addition to amplitude, the route to transition may also depend on the detailed nature of the disturbances. The channel numerical simulation work of Singer, et al. (1986) shows that the presence of even weak longitudinal vortices can lead to arrayed hairpins where otherwise, there might be the staggered types. There is also a difference in the way the mechanisms are viewed. For example, Thomson (1988) attributes the arrayed and staggered types to fixed and transient disturbances instead of large and small disturbances.

Ultimately, the important question regarding the contemporary research on the later stages of the nonlinear transition process is, how relevant is it to the fully turbulent boundary layer? The preliminary indication is that the staggered type is more relevant. For example, Perry, et al. (1981) have shown that the spread of a turbulence spot involves a domino-like breakdown which is just another name for the staggered array of hairpin vortices. Elsewhere (Bandyopadhyay, 1983a) it has been shown that the characteristic arrowhead angle of a turbulence spot can be arrived at from the longitudinal and spanwise scales of the domino. Furthermore, this relative scaling in the two directions is also observed in fully turbulent boundary layers in a wide range of Reynolds numbers. More recently, it has been possible to calculate the streamwise distribution of the separation intermittency factor rather accurately in a non-statistical approach employing this domino-like character of the spread of a disturbance (Bandyopadhyay, 1989).

Looking at the nonlinear aspects of the transition process in a 'reverse' manner from the fully turbulent side leads to another interesting observation. The forced transition of a flat plate boundary layer due to a circular trip wire shows that such large disturbances indeed, initially, lead to the formation of arrayed hairpins (see flow visualization pictures in Hama, et al. (1957) and Head and Bandyopadhyay (1978)). However, as the discussion in the previous paragraph showed, even if the transitional hairpins are arrayed, during the regeneration of turbulence further downstream, the staggered-type becomes the norm.

Continuing along this 'reverse' approach, an experiment was carried out to further identify the closeness of the fully turbulent flow to that undergoing transition. It was carried out in a large diameter plexiglass pipe (to facilitate flow visualization). Large disturbances were applied at the pipe entrance so that the 'natural' transition Reynolds number (= Re, based on pipe diameter D, and centerline speed U) of 25×10^3 was reduced to a value between 3.2×10^3 and 3.4×10^3. Several different sources of disturbance have been examined. The disturbances are large in the context of flat plate transition research. However, they are not large enough to cause a puff-type transition which is peculiar to pipe flow and occurs at $2000 < \mathrm{Re} < 2700$ (Wygnanski and Champagne, 1973 and Bandyopadhyay, 1983b). In this range, the characteristic centerline puff signatures were also absent in the present experiments. The disturbances are large enough to cause a transition which is sometimes known as the slug-type (Wygnanski and Champagne, 1973); they are similar to turbulence spots and are known to appear at $\mathrm{Re} > 3.2 \times 10^3$ as they also did in the present experiments. So, although carried out in a pipe, the presently studied transition process is expected to be essentially similar to that in a flat plate. The turbulence amplification has been measured at two stations, viz., first, at the pipe entrance at $x/D = 0.45$, where the perturbation due to the trip is interacting with the mean flow and second, at $x/D = 30$, that is, soon after the fully developed turbulent flow has been established which is known to take place at $x/D = 20$.

2. Experiments

The experiments were carried out in a 4 inch diameter and 52.3 foot long plexiglass pipe described in Bandyopadhyay and Weinstein (1988). Figure 1 shows the facility schematically. For the present experiments, a smooth entry section, a 16 mesh wire screen at the entry and a fractional horse power fan at the exit were added.

Four different trips were used to introduce the large disturbances at the pipe entry and the location of their application is shown in Figure 1. The trips are shown schematically in Figure 2 (a to d). The circular ring, the orifice plate and the vortex generator were all of height 6 mm. The saw tooth had a maximum height of 8 mm. The circular ring and the orifice plate are axisymmetric. The vortex generator is three-dimensional and only one is used whereas the sawtooth provided a circumferentially periodic disturbance.

The flow Reynolds number (Re) was monitored with a pitot tube located at the centerline at an x/D of 120. The pitot pressures were read using 1 and 10 Torr head Barocel pressure transducers whose mean values were read with a digital voltmeter. Single hot wire measurements was carried out at x/D of 0.45 and 30. A constant temperature anemometer and a linearizer were used to operate the hot wire. The hot wire was made of platinum coated tungsten and was 5 μm in diameter and 1 mm in length. The hot wires were calibrated in a separate 2.63″ × 6.75″ in cross-section and 21” long wind tunnel. A computer controlled probe traverse was used and distances were measured using an optical encoder. All data acquisition were carried out using a personal computer. A Data Precision Data6000 signal processor was used to digitize the hot wire signal and calculate the spectrum, root-mean-square (rms) and mean (DC) values.

3. Results and Discussion

(a) Mean and turbulence characteristics. The reference mean and turbulence characteristics were first documented to determine the flow quality. The laminar and fully turbulent mean velocity profiles across the pipe were measured at $x/D = 30$ with the hot wire. They are shown in Figures 3 and 4 respectively. The laminar profile is for a clean entry without any trip. In Figure 3, the measurements are compared with the theoretical Hagen-Poiseulle (HP) distribution. The HP profile is supposed to be reached asymptotically and, in practice, a reasonable approximation can be reached only at x/D's of 0(100). Thus, it is surprising that the present measurements are so close to the HP profile. A possible clue to this early development comes from the disturbance intensity measurements in this laminar flow shown in Figure 6. The figure shows that although the flow is laminar (sub-critical), it contains disturbances whose intensities are about 3 to 5% of the centerline velocity. It may be that these disturbances have a predominantly longitudinal vorticity whereby they persist over x/D's 0(10) and enhance viscous mixing across the pipe. It may be worthwhile to verify this explanation numerically. It is an interesting practical aspect that an experimental 'laminar' flow can actually be so disturbed and there is a need to take this into account in theoretical formulations.

The mean velocity measurements across the pipe for the fully developed turbulent condition are shown in Figures 4 and 5 for the circular ring trip case. In Figure 4, the measurements are compared with a 1/6th power law line that the profile is known to follow at the lowest Reynolds numbers where a turbulent flow can be sustained (p. 599 of Schlichting, 1979). The scatter is attributed to the difficulty in the measurements of small velocities (less than 50 cm/s). The wall shear stress and friction velocity (U_τ) were calculated using the 1/6th power law and the mean velocity profile measurements were plotted in the conventional law of the wall form. This is shown in Figure 5 where the measurements are compared with the well-known logarithmic law viz.,

$$\frac{U}{U_\tau} = 5.5 \log\left(\frac{(R-r)U_\tau}{\nu}\right) + 5.4 \tag{1}$$

where R is $D/2$ and r is the radial distance from the pipe center. The measurements follow the log law reasonably well.

At the Reynolds number corresponding to the fully developed turbulent flow case shown in Figures 4 and 5, the distributions of the turbulence intensity across the pipe for two trips, namely, the circular ring and orifice plate, are shown in Figure 6. The turbulence intensities are increased compared to the no-trip laminar case as to be expected.

(b) Perturbation amplification at the fully-developed station. The amplification of the perturbation applied at the pipe entry was measured at the downstream fully developed station ($x/D = 30$) at a near-wall location where the rms level was the highest in the boundary layer in the circular ring case (Figure 6). The variation of the amplification with Reynolds number for the various forcing devices is shown in Figures 7 and 8. In all cases, transition occurs at an Re of 3.2×10^3 to 3.4×10^3. In the subcritical range, the rms level is the same, about 6 to 7%, for all disturbances as well as in the no-trip clean entry case. This may be because the level is determined by the screen located at the entry (Figure 1) and the perturbations due to the trips are damped. A comparison of the circular ring and the vortex generator data in Figure 7 suggests that the maximum amplification level due to the vortex generator is the same as in the circular ring case. It is not clear whether the laminar to turbulent intermittency in the vortex generator case is genuine or it results

from a circumferential wandering of the turbulent patch which is intermittently missed by the hot wire.

In Figure 8, the amplification level of the circular ring and the subcritical cases are compared with that due to the saw tooth trip. The following observations can be made. The saw tooth data falls in a band and the lower level is the same as in the circular ring case. Recall that the maximum amplification in the vortex generator case is also equal to the circular ring level. Since all the amplification levels in the present experiment fall within the saw tooth trend, we surmise that it represents the saturated growth rate.

The upper level in the saw tooth data in Figure 8 shows an interesting maximum followed by a drop which appears to approach the near constant circular ring level. This trend is very similar to Klebanoff, et al.'s (1962), so called, peak valley splitting measurements of arrayed hairpin vortices which has been reproduced in Figure 9.

However, there are three important differences between the saw tooth (Figure 8) and the peak-valley splitting data (Figure 9). First, the former has a higher background disturbance level. Second, in the latter, the transitional length extends over a Reynolds number ratio of about 7 ($(x - x_0)$ from 2 to 14 inches), whereas, the ratio is even less than 2 in the former. In other words, in the saw tooth case, the receptivity has been enhanced. Third, although, receptivity is enhanced in the forced transition case (Figure 8), the rms level approximately doubles whereas, it increases by a factor which is more than 8 in Klebanoff, et al.'s (1962) peak-valley splitting case.

Note that in Figure 9, both the peak and valley levels show a maximum followed by a drop. Also, the difference between the peak and valley levels at a given streamwise station in Figure 9 is of the same order as the band width of the scatter in the saw tooth data in Figure 8. For these reasons, we speculate that the saw tooth data in Figure 8 contains both the peak and valley situations because the longitudinal vortex pairs were not precisely locked circumferentially all the way up to $x/D = 30$. It may be that they were wandering circumferentially at a rate slower than the period over which the rms values were sampled.

(c) Perturbation amplification at the pipe entrance. As mentioned previously, wire trips tend to form arrayed hairpin vortices in transitional Reynolds numbers. So, in view of the striking similarity of the downstream amplification to the peak-valley splitting charac-

teristics, it was necessary to determine the correlation between this downstream station and the pipe entry region where the perturbation interacted with the mean flow.

The spectra of the u-fluctuation were examined across the boundary layer downstream of the circular ring trip at several x/D stations in a range of Reynolds number to determine the location where the first dominant peak appeared. This was judged to be at $x/D = 0.45$ at a distance of $0.09D$ from the pipe surface. Figure 10 shows the variation of the spectrum (whose resolution is 0.98 Hz) with increasing Reynolds number. The dominant frequencies were fairly repeatable. The presence of the vortex shedding became very obvious at Re = 3920 (Figure 10a), but we suspect that it must have started at $3.2 \times 10^3 < \text{Re} < 3.4 \times 10^3$ because the downstream station revealed that to be the transition Reynolds number range.

The amplification of the power in the dominant frequencies with Reynolds number given in Figure 10 are plotted in Figure 11. These two figures indicate that, in the range $4930 \leq \text{Re} \leq 6020$, the amplification rate shows a significant change accompanied by the appearance of several new dominant frequencies. A similar phenomenon seems to occur also at $8120 \leq \text{Re} \leq 10060$ where the appearance of new dominant frequencies is so extensive that the spectrum has become at least twice as wide. This Reynolds number classification may be of value to flow control. Obviously, the nonlinear process is undergoing a change within these two Reynolds number ranges.

The data in Figure 11 has been replotted in Figure 12 normalizing the power values by the maximum reached by the frequency concerned. The data shows a tendency to collapse. Up to $4930 \leq \text{Re} \leq 6020$, the amplification is not large. But, after that, an exponential line is included in Figure 12 to indicate such a trend. In Figure 13, the maximum values reached by the various frequencies are plotted and a systematic behavior is indicated. From Figure 12 it is concluded that, all the dominant frequencies amplify in a similar manner which is exponential in part of the Reynolds number range.

It is of interest to know how the trip vortex shedding frequency is related to the dominant frequencies (see the caption in Figure 11). Consider the forced transition of a flat plate boundary layer due to a circular rod of diameter 16 mm (d) at an Re_d of 160 (Head and Bandyopadhyay, 1978). Flow visualization had showed that the shed hairpin vortex pattern could be frozen up to $x/d = 8$ indicating a Strouhal number St (nd/U where, n is the vortex shedding frequency,

d is rod diameter and U is freestream velocity) of 0.176. This x/d of 8 is close to the x/d of 7.2 of the present experiment (at $x/D = 0.45$). So, the previously described manner in which the present choice of the station $x/D = 0.45$ was made is of significance. Assuming the above mentioned flat plate St to apply in the present case also (Re_d in the pipe experiment is 100), n is estimated to be 6.8 Hz at Re = 3300. However, a frequency of about 7Hz is not so dominant in the spectra in Figure 10. We suspect that this is because at x/d of 7.2 the nonlinear stages have sufficiently advanced. Interestingly, the dominant frequencies in Figure 10 seem to be related to a frequency of 7 to 9 Hz. So, it is suggested that the trip hairpin vortex shedding St acts as a carrier frequency.

4. Conclusions

A large disturbance transition experiment has been carried out in a pipe to identify the similarity of the fully turbulent flow to the highly nonlinear stages of the transition process. The following conclusions have been drawn.

1. The mean flow measurements suggest that the presence of large disturbances in a subcritical laminar flow can enhance the rate of its development.
2. At the pipe entrance, all dominant frequencies amplify in a nearly universal manner which is exponential in part of the Reynolds number range.
3. At the downstream station where the fully turbulent flow is first developed, an almost saturated amplification condition appears to have been reached in the present experiments. Apart from the difference of a high background turbulence level, the Reynolds number trend of the saturated amplification level is qualitatively similar to that in peak-valley splitting.

Acknowledgements

The first author was supported by NASA Contract No. NAS1-18235 and the second author was supported by NAG1-530 during the course of this work. The second author also wishes to thank Professor R. L. Ash for his hospitality.

References

[1] Bandyopadhyay, P. R., 1983a, "Turbulence spot-like features of a boundary layer," Anals. of the New York Acad. of Sciences, pp. 393-395.

[2] Bandyopadhyay, P. R., 1983b, "Aspects of the equilibrium puff in transitional pipe flow," J. Fluid Mech., V163, pp. 439-458.

[3] Bandyopadhyay, P. R., 1989, "A theory of separation intermittency," Bull. Amer. Phys. Soc., V34, No. 10, 42nd Annual Meeting of the Divn. Fl. Dyn. held at Palo Alto, CA.

[4] Bandyopadhyay, P. R. and Weinstein, L. M., 1988, "A simplified oil-film skin-friction meter," in Proc. AIAA/ASME/SIAM/APS 1st National Fl. Dyn. Congress, Part 3, held at Cincinnati, OH, pp. 1487-1499.

[5] Hama, F. R., Long, J. D., and Hegarty, J. C., 1957, "On transition from laminar to turbulent flow," J. Appl. Phys., V28, pp. 388-394.

[6] Head, M. R. and Bandyopadhyay, P., 1978, "Combined flow visualization and hot-wire measurements in turbulent boundary layers," in Workshop on Coherent Structure of Turbulent Boundary Layers, pp. 98-129, Lehigh University.

[7] Klebanoff, P. S., Tidstrom, K. D., and Sargent, L. M., 1962, "The three-dimensional nature of boundary-layer instability," J. Fluid Mech., V12, pp. 1-34.

[8] Knapp, C. F., Roache, P. J., and Mueller, T. J., 1966, "A combined visual and hot-wire anemometer investigation of boundary layer transition," University of Notre Dame, IN Report UNDASTR-866CK.

[9] Perry, A. E., Lim, T. T., and Teh, E. W., 1981, "A visual study of turbulent spots," J. Fluid Mech., V104, p. 387.

[10] Schlichting, H., 1979, *Boundary Layer Theory*, pp. 85, 599, McGraw-Hill, New York.

[11] Singer, B., Reed, H. L., and Ferziger, J. H., 1986, "Investigation of the effects of initial disturbances on plane channel transition," AIAA Paper No. 86-0433.

[12] Thomas, A. S. W., 1986, "Experiments on secondary instabilities in boundary layers," in Proc. of the Ninth U. S. National Congress of App. Mech., University of Texas, Austin, ed. J. P. Lamb, pp. 435-443, Publ. by ASME, New York.

[13] Thomson, K. D., 1988, "Some comments on the later stages of transition from laminar to turbulent flow in the flat plate boundary layer," Aero. J., V92, pp. 309-314.

[14] Wygnanski, I. and Champagne, F. H., 1973, "On transition in a pipe. Part 1. The origin of puffs and slugs and the flow in a turbulent slug," J. Fluid Mech., V59, pp. 281-335.

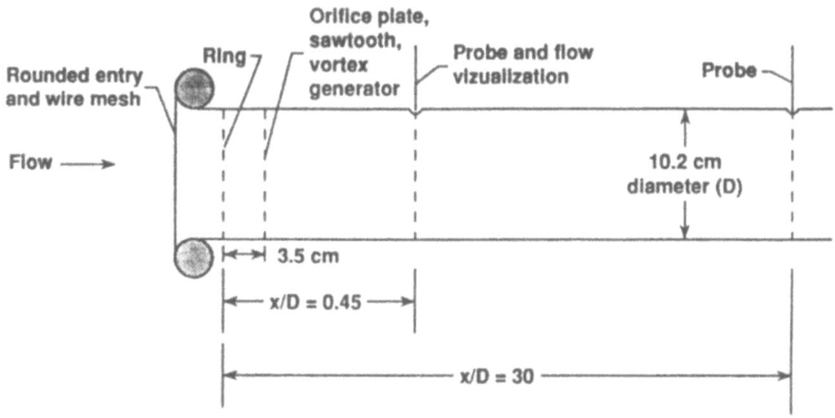

Figure 1. The pipe flow apparatus showing the trip and measurement locations.

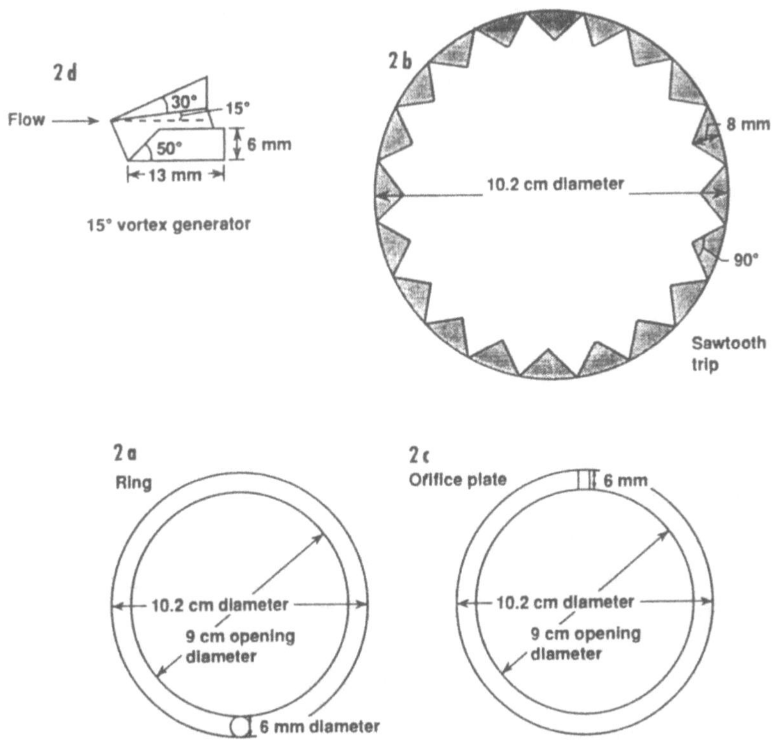

Figure 2. Disturbance generators; (a) circular ring, (b) saw tooth trip, (c) orifice plate, and (d) the 15° vortex generator.

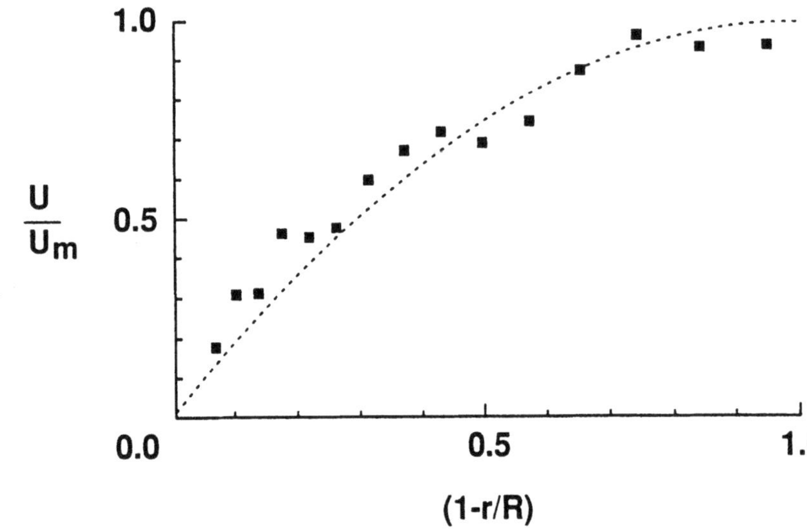

Figure 3. Measurements of the laminar flow velocity profile (symbols) compared with the theoretical Hagen-Poiseulle distribution for fully developed flow (dotted line). Measurements: $x/D = 30$, Re = 3500 and no trip clean entry case.

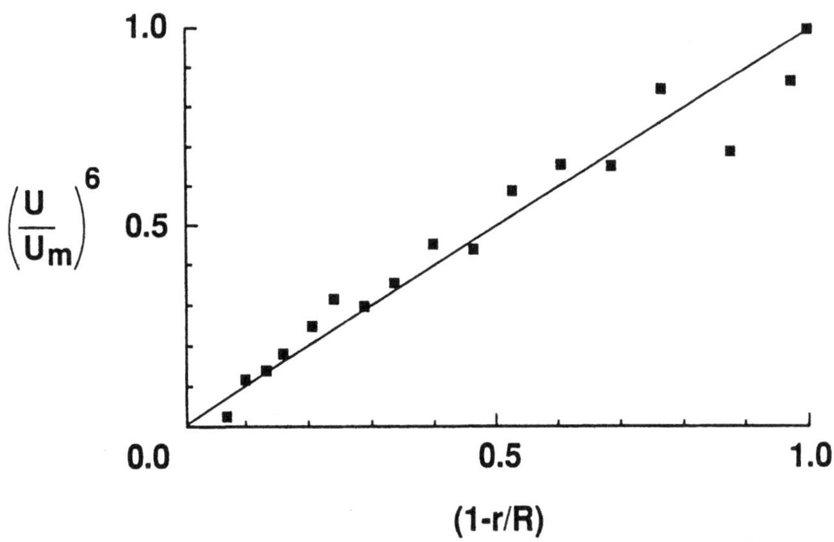

Figure 4. Measurements of the turbulent flow velocity profile (symbols) compared with 1/6th power law distribution (solid line). Measurements: $x/D = 30$, Re = 6000 and circular ring case.

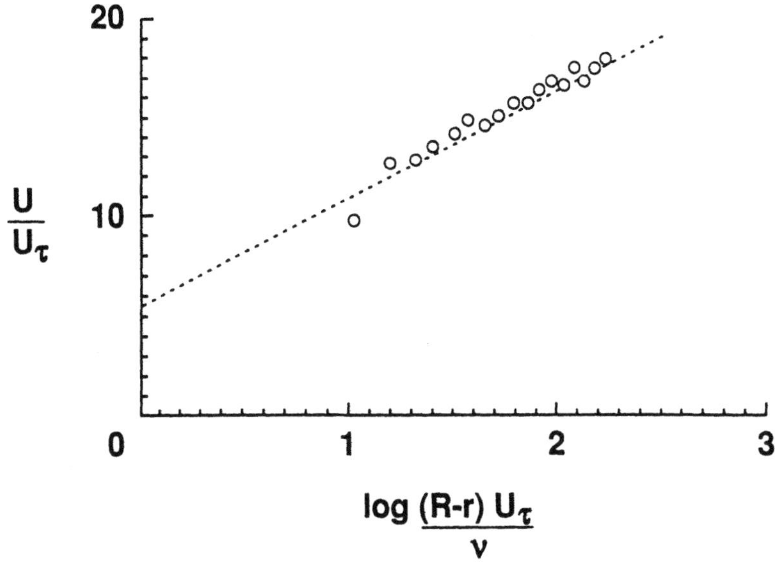

Figure 5. Measurements of the turbulent flow velocity profile (symbols) compared with the log law (dotted line represents Equation 1). Measurements: $x/D = 30$, Re = 6000 and circular ring case.

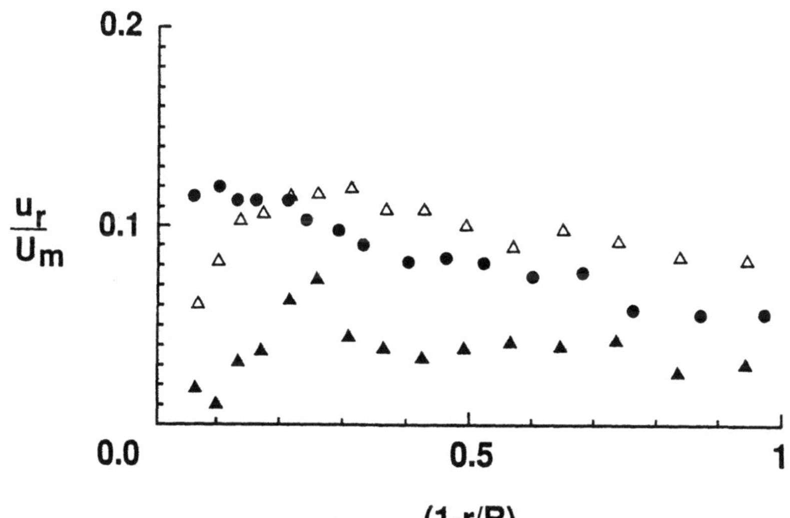

Figure 6. Measurements of the rms distributions of the velocity fluctuations. Filled circle: circular ring at Re = 6000; open triangle: orifice plate at Re = 6000 and filled triangle: no-trip clean entry case at Re = 3500. All measurements at $x/D = 30$.

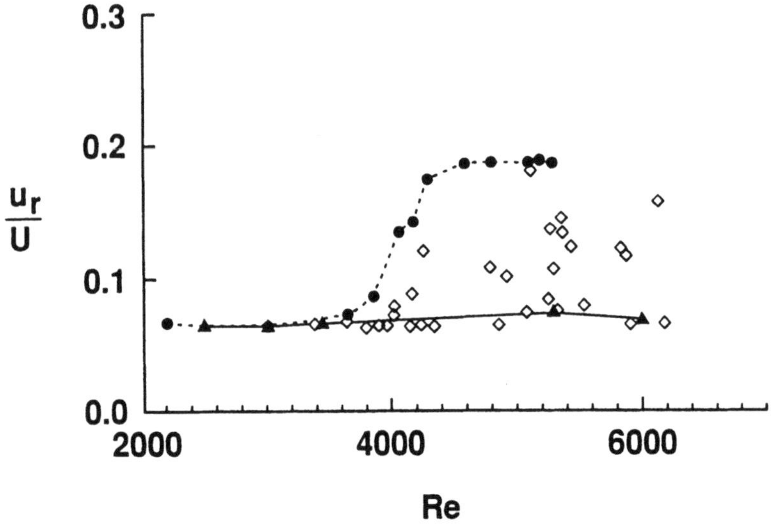

Figure 7. Measurements of the Reynolds number variation of the rms values of the velocity fluctuations at $x/D = 30$ and radially at 0.03D from the bottom pipe surface. Triangle and solid line: no-trip clean entry case; diamond: single vortex generator at the bottom surface, and circle: circular ring.

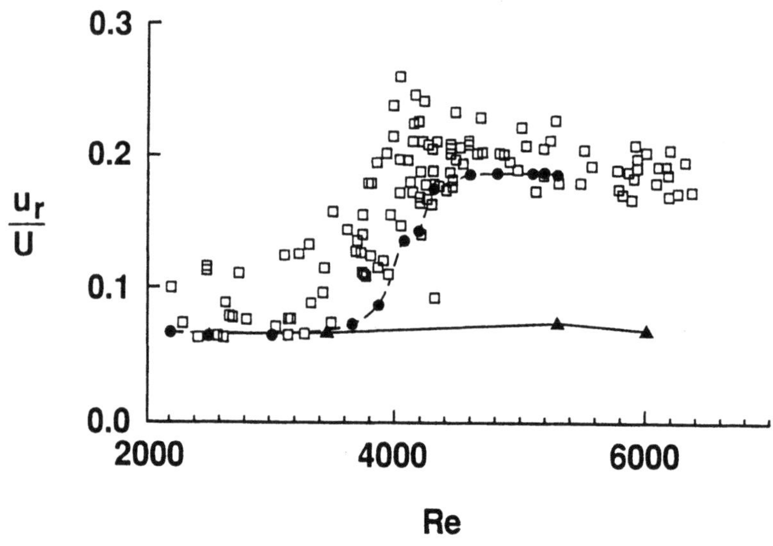

Figure 8. Measurements of the Reynolds number variation of the rms values of the velocity fluctuations at $x/D = 30$ and radially at 0.03D from the bottom pipe surface. Triangle and solid line: no-trip clean entry case; circle: circular ring, and square: saw tooth trip.

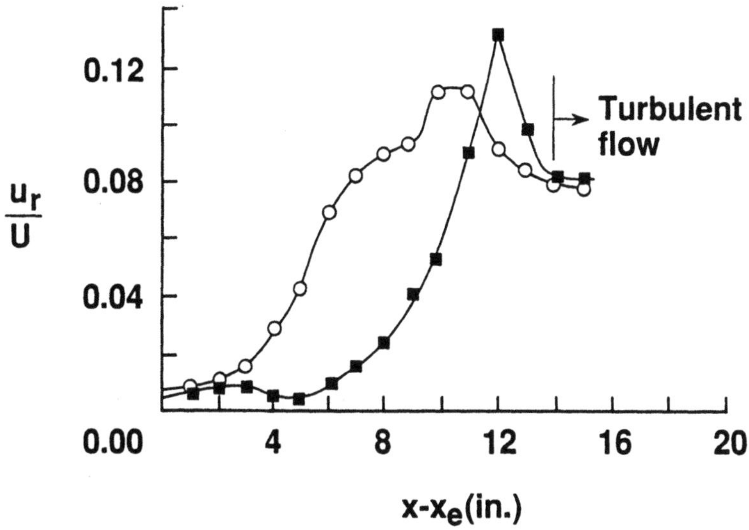

Figure 9. Intensity of u-fluctuation at peak (open circle) and valley (filled square) in a flat plate boundary layer transition, reproduced from Klebanoff, et al. (1962).

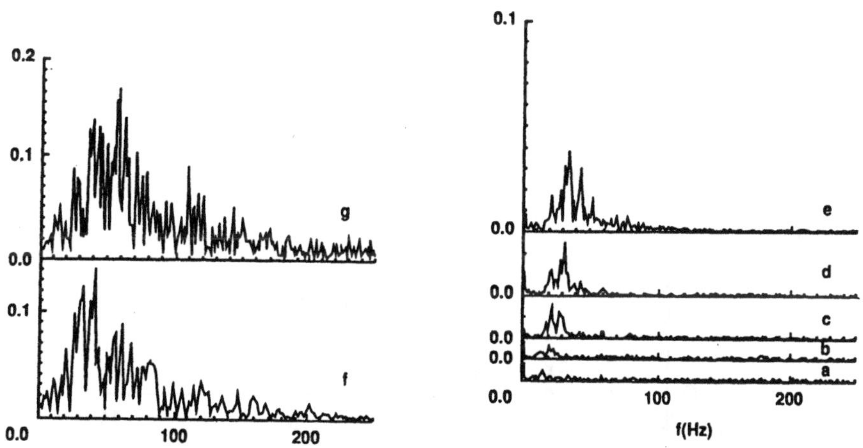

Figure 10. Variation of power spectra of u-fluctuation with Reynolds number near the pipe entrance. Measurement location $x/D = 0.45$ and radially at 0.09D from the pipe bottom surface; circular ring case; Re = 3920 (a), 4930 (b), 6020 (c), 7100 (d), 8120 (e), 10060 (f), and 11640 (g). Note that the plots are staggered along the ordinate which is in arbitrary scale.

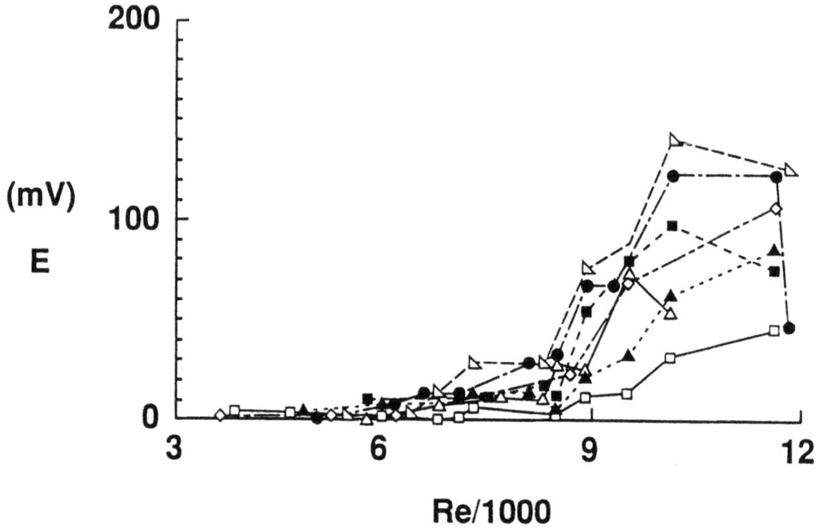

Figure 11. Reynolds number variation of the power in the dominant frequencies shown in Figure 10. The ordinate is in arbitrary scale. Dominant frequencies, f_0 (in Hz): Open square (15.6), filled triangle (22.5), filled square (28.3), filled circle (31.3), diamond (35.2), right triangle (43.0) and open isosceles triangle (51.8).

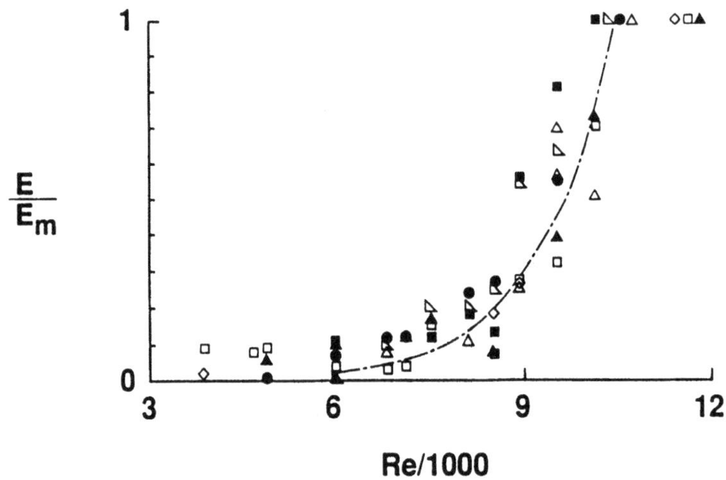

Figure 12. Normalized plot of the amplification of the dominant frequencies in the pipe entry region. Measurement location is the same as in Figure 10. The symbols are the same as in Figure 11. The chain line indicates an exponential amplification rate $E/E_m = 1.3 \times 10^{-4} \exp(0.857\text{Re} \times 10^{-3})$.

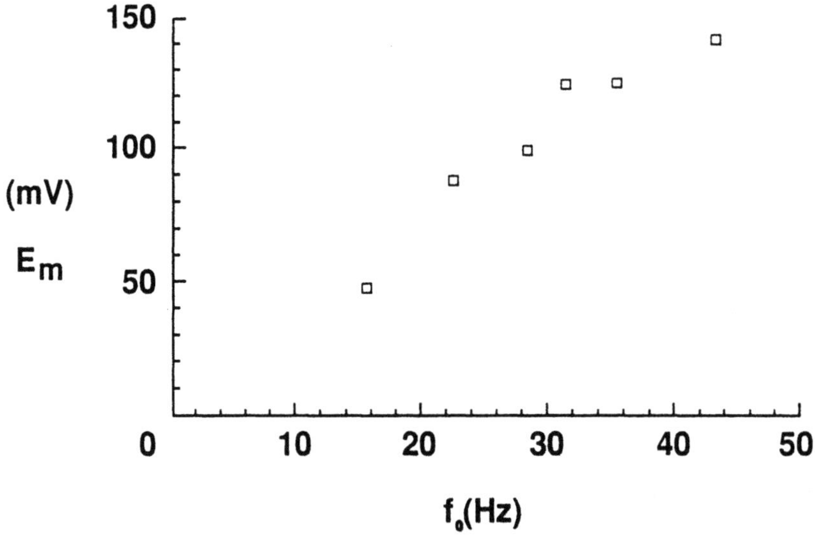

Figure 13. Variation of the maximum power of the dominant frequencies in the pipe entry region shown in Figure 10.

EXPERIMENTAL OBSERVATIONS ON THE RELATIONSHIP BETWEEN STAGNATION REGION FLOW OSCILLATIONS AND EDDY SHEDDING FOR CIRCULAR CYLINDER

S. M. Mangalam and L. R. Kubendran*

Analytical Services & Materials, Inc.
107 Research Dr., Hampton, Virginia 23666

ABSTRACT

Measurements indicate a direct coupling between the dominant frequencies of flow oscillations at the leading-edge stagnation region, laminar-separation unsteadiness, and the vortex shedding. An array of surface mounted, micro-thin hot-films were used to simultaneously determine the state of the boundary layer along the surface of a circular cylinder, and vortex shedding was monitored by a hot wire placed in its wake.

1. Introduction

Multi-element micro-thin hot films have been utilized at NASA Langley Research Center (Holmes et al. 1987, Johnson et al. 1987, Stack et al. 1987, Bobbitt 1988, Mangalam et al. 1988, Johnson & Carraway 1989, and Mangalam et al. 1989) to study boundary-layer separation and transition at speeds ranging from subsonic to low supersonic speeds. They are practically non-intrusive and can be used by themselves or in conjunction with other conventional techniques to simultaneously map the flow field and obtain the spatial as well as temporal flow characteristics. In his comprehensive paper on instability and transition, Morkovin (1988) points out that experimenters, "almost of necessity, must use space-time correlations and coherences between multiple sensors" to better understand the concepts of convective and global instabilities and solve the most vexing problem concerning receptivity of the boundary layer to forced (natural or simulated) disturbances.

*This work was supported in part by the ICASE and the National Aeronautics and Space Administration under NASA Contract No. NAS1-18599.

The actual problem as enunciated above is very complex and time/resource consuming but a simple experiment was conducted during the Instability & Transition Workshop to demonstrate the potential of a multiple hot-film/hot-wire sensor approach to simultaneously determine the state of the boundary layer (laminar, transitional, or turbulent) from the leading to trailing edge of an unswept circular cylinder. Simultaneously, the leading-edge stagnation point, laminar separation, transition, vortex-shedding frequencies and their relation to the oscillations at the front stagnation point were identified for each flow condition.

2. The Test Set-up and Instrumentation

Experiments were conducted on a 6.35 cm diameter circular cylinder (Fig. 1) in the 56 cm x 71 cm low-speed wind-tunnel in the Viscous Flow Branch of NASA Langley Research Center. Tests were carried out in the velocity range of 8 to 37 m/s. Unless otherwise mentioned, the results presented in the next section correspond to the case where the freestream velocity was 29 m/s. The test section had a fairly high turbulence level of about 0.5% but without dominant peaks in the frequency spectra.

The multi-element sensor (Fig. 2) consists of a number of independent Nickel films, electron-beam evaporated on a thin (0.05 mm) polyimide substrate in a straight-line array. Each sensor element consists of a Nickel film 0.9 mm long and 0.15 mm wide with 2-microns thick copper-coated Nickel leads, routed to obtain a practically non-intrusive sensor system. The unheated resistance of each sensor element was a nominal 5 ohms. There were 60 individual hot-film sensors spaced at 2.5 mm intervals. The substrate was bonded to the model with the sensor array placed streamwise along the surface of the circular cylinder model. Since only alternate sensors were used in the present experiment, the sensor angular spacing was 9.2^0, with sensor 1 located at 119.6^0 from the leading edge. A single, straight hot wire mounted slightly off to the side of the cylinder was used for sensing potential flow fluctuations outside the cylinder wake.

The multielement hot-film sensor and the single-wire hot wire were connected to a micro-computer through a bank of constant-temperature anemometers, a multiplexer, signal conditioner, and analog-to-digital converter. The specialized data acquisition and instrumentation system (DAISy) shown in Figure 3 (see Mangalam et al. 1990 for details) allows a software-controlled simultaneous sampling at eight points in the flow field and the multiplexer is used to instantaneously map the field in sets of eight. The maximum sampling rate was 32 kHz per channel.

3. Results and Discussions

Figure 4 presents one of the several sets of simultaneous time history signals acquired from selected sensors on the cylinder and the hot wire in the wake region. For each set, eight channels of signals were simultaneously sampled at a sampling rate of 10 kHz/channel. A total number of 2048 data points were acquired for each channel, resulting in a data time period of 0.2048 sec. Real time analyses of digital data were carried out using the built-in software package.

Figure 4 presents the set containing signals from the sensors 15 through 9 and the hot wire (sensor 16). Even a visual comparison of the raw signals indicates a phase reversal between sensor 15 and 13. Figure 5 gives a composite plot of power spectral density estimates of signals from sensors 13 and 15, their cross-power spectrum, coherence and phase relations. The frequency (Hz) is plotted on x-axis and the power spectral plots have arbitrary y-axis scales. The Phase 1 and Phase 2 plots essentially provide the same information with differing y-axis scales (-180^0 to $+180^0$ for the Phase 1 plot and 0 to 360^0 for the Phase 2 plot).

The results in Figure 5 reveal the presence of a strong peak at a frequency of 90 Hz, excellent coherence, and a 180^0 phase shift at this frequency. The stagnation point was identified by the presence of a phase reversal between sensors 15 and 13 (for details of this technique see Mangalam et al. 1989). The power spectra for these signals have very little energy content beyond about 300 Hz. The hot-wire signals in the wake region have excellent coherence with either of these two signals. The hot-wire power spectrum (Fig. 6) also has a peak at 90 Hz. The presence of multiple peaks in the power spectrum of signals from the sensor 14 (Fig. 7) suggests that the stagnation point was fluctuating across the sensor. The presence of a strong double peak whenever the stagnation point passed over a gage twice during a cycle was also observed by Dwyer & McCroskey 1973.

As we move downstream of the stagnation point, signals from sensors 12 through 9 are in phase with sensor 13 and the hot wire (Fig. 8). As we move further downstream (Fig. 9; sensors 7 through 1), signals from the sensor 7 are still in phase with the hot wire, but the power spectrum for the signals from sensor 7 has started broadening beyond the peak at 90 Hz (Fig. 10). Compared to signals from sensors 15 through 7, signals from the sensor 6 exhibit an increase in amplitude and a loss in periodicity. The presence of double peaks in these signals during a part of the time interval suggests that the sensor is close to laminar separation and that separation itself is unsteady. There is a sharp drop in amplitude and a complete breakdown in periodicity in signals from the sensor 5 resulting from small, random backflow immediately downstream of laminar separation. Transition to turbulence has probably occurred in the separating shear layer farther downstream (the exact location can be determined by traversing a hot-wire in the separated shear layer but the limited scope of the present test precluded it). Signals from sensors 4 through 1 exhibit properties of a turbulent flow with broader power spectra and significant energy content at higher frequencies (Fig. 11). The set shown in Figure 12 provides a composite picture of all the features described above, i.e., stagnation point location, transition, turbulent flow, and shedding characteristics (sensor 16), obtained simultaneously.

One of the unique features of the present technique is the ability to determine whether transition took place in an attached boundary layer or in a separated shear layer. It is based on the identification of a phase reversal signature to detect boundary-layer separation (see e.g., Mangalam et al. 1988). A phase reversal was observed at 90 Hz between signals from sensors 6 and 4 (the coherence between these signals was also a maximum at this frequency), confirming the presence of laminar separation between these two sensors (Fig. 13). This also suggests that the laminar separation was unsteady with a characteristic dominant frequency of 90 Hz. The mean separation "point" was located at approximately 82^0 from the leading edge. On the basis of these observations, it is clear that transition took place in an unsteady separated shear layer.

The spectral content of signals from all the sensors in the laminar region indicated a dominant frequency at 90 Hz. The spectral content of the signal from the hot wire in the wake also showed a dominant frequency of 90 Hz (Fig. 6). In order to find a relation between these frequencies at other flow conditions, experiments were conducted for a range of Reynolds Numbers (20,000 $< Re_D <$ 160,000; Re_D based on the cylinder diameter, D). In Figure 14, the Strouhal Number (based on the dominant frequency at the stagnation point, the cylinder diameter, and the freestream velocity) has been plotted as a function of Re_D. The Strouhal Number remains nearly constant at 0.2. This is also the value of the Strouhal number for the cylinder based on vortex-shedding frequency (Hoerner 1965). Thus the dominant frequency at the stagnation point appears to be directly coupled to the trailing-edge vortex-shedding frequency and it increases linearly with the freestream velocity. The coupling appeared to be robust and the fairly high turbulence level of 0.5% seemed to have little influence on it.

The influence of bluff-body wake oscillations on the pressure field and the associated feedback mechanism have been the subject of extensive studies (see Morkovin 1988). Such a feedback mechanism becomes very weak if the laminar separation is followed by a turbulent reattachment and a relatively thin wake. Poll (1979, 1985) investigated the attachment line contamination and crossflow instability on a swept 50% thick wing with a cylindrical leading-edge while Saric et al. (1989) have looked at crossflow instability and transition on a swept natural laminar-flow wing. Both the investigations indicated the presence of large amplitude disturbances exceeding 20% of the local mean flow velocity. Mangalam et al. 1987 also observed large amplitude disturbances of the order of 45% of the local mean velocity in their experiment on Gortler vortices on an airfoil. It appears possible that in all these cases the presence of such large amplitude disturbances may have been due to the unsteadiness in the mean flow caused by laminar separation downstream of the test region.

The authors of this paper observed the presence of a dominant frequency at the leading edge in a number of other tests, including a few experiments on large swept wings. The the unanswered question in our opinion is: what is the real source of the observed dominant frequency at the leading edge of streamlined bodies (e.g, swept wings)? Is it present only in a tunnel environment? As pointed out by Morkovin (1988), well designed experiments using space-time correlations and coherences between multiple sensors are needed to resolve such problems.

4. Summary

A direct coupling between the dominant frequencies of flow oscillations at the leading-edge stagnation region and the vortex shedding into the wake of a circular cylinder was observed. Unsteadiness associated with laminar separation also had this frequency.

The dominant frequency of oscillations in the leading-edge region was found to be directly proportional to the free-stream velocity within the range of test conditions.

Since the presence of laminar separation could, via pressure feedback mechanism have significant impact on boundary-layer characteristics at the leading edge, in addition to taking proper measures to insure good leading-edge geometries (such as super ellipses for a flat plate) it may be equally important to ensure appropriate trailing-edge conditions to eliminate laminar separation.

Acknowledgements

The authors wish to acknowledge Messrs. J. B. Anders, D. M. Bushnell, and J. R. Dagenhart for their help in providing the model and wind-tunnel facilities for the experiment. The first author wishes to acknowledge the constant support given by Mr. R. D. Wagner, Head, LFCPO and Dr. R. D. Barnwell, Chief Scientist, NASA LaRC. The first author also wishes to gratefully acknowledge thought-provoking discussions with Prof. Morkovin.

References

[1] Bobbitt, P. J., 1988, *Instrumentation advances for transonic testing.* Transonics Symposium, Theory, Application, and Experiment, NASA CP 3020.

[2] Dwyer, H. A. & McCroskey, W. J., 1973, *Oscillating flow over a cylinder at large Reynolds number*, JFM, vol. 61, 753-767.

[3] Hoerner, S. F., 1965, *Fluid-Dynamic Drag*, Published by the Author, New Jersey.

[4] Holmes, B. J., Carraway, D. L., Manuel, G. S., and Croom, C. C., 1987, *Advanced measurement techniques*, Proc. Research in Natural Laminar Flow and Laminar-Flow Control, NASA CP 2487.

[5] Johnson, C. B., Carraway, D. L., Hopson, P. Jr., and Tran, S. Q., 1987, *Status of a specialized boundary-layer transition detection system for use in the U. S. National Transonic Facility*, Proc. 12th ICIASF, Williamsburg, VA.

[6] Johnson, C. B. and Carraway, D. L., 1989, *A transition detection study at MAch 1.5, 2.0, and 2.5 using a micro-thin hot-film system*, 13th ICIASF, Gottingen, West Germany.

[7] Mangalam, S. M., Dagenhart, J. R., and Meyers, J. F., 1987, *Experimental studies on Gortler vortices*, Proc. Research in Natural Laminar Flow and Laminar-Flow Control, NASA CP 2487.

[8] Mangalam, S. M., Stack, J. P., and Sewall, W. G., 1988, *Simultaneous detection of separation and transition in surface shear layers*, AGARD CP 438, 12.

[9] Mangalam, S. M., Kubendran, L. R., and Balister, R. L., 1989, *An experimental investigation to determine the cause of phase reversal signatures at stagnation and separation*, in preparation.

[10] Mangalam, S. M., Balister, R. L., and Kubendran, L. R., 1990, *An advanced PC-based instrumentation system for flow diagnostics*, Submitted for presentation at the 36th ISA International Instrumentation Symposium, Denver, CO.

[11] Morkovin, M. V., 1988, *Recent insights into instability and transition to turbulence in open-flow systems*, AIAA 88-3675.

[12] Poll, D. I. A., 1979, *Transition in the infinite swept attachment-line boundary layer*, Aero. Q., 30, 607-628.

[13] Poll, D. I. A., 1985, *Some observations of the transition process on the windward face of a long yawed cylinder*, J. Fluid Mech., 150, 329-356.

[14] Saric, W. S., Dagenhart, J. R., and Mousseux, M. C., 1989, *Experiments in swept-wing transition*, 4th Symposium on Numerical and Physical Aspects of Aerodynamic Flows, Long Beach, CA.

[15] Stack, J. P., Mangalam, S. M., and Berry, S. A., 1987, *A unique measurement technique to study laminar-separation bubble characteristics on an airfoil*, AIAA 87-1271.

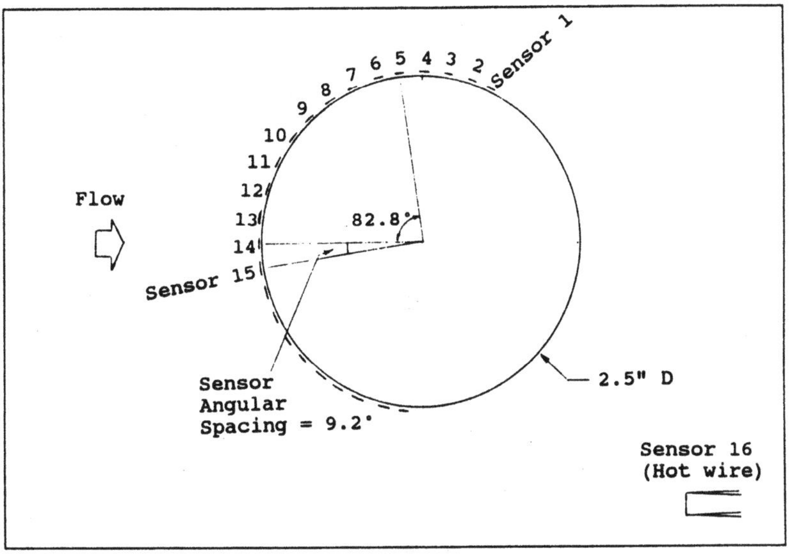

Figure 1. - Schematic of instrumented cylinder and hot wire.

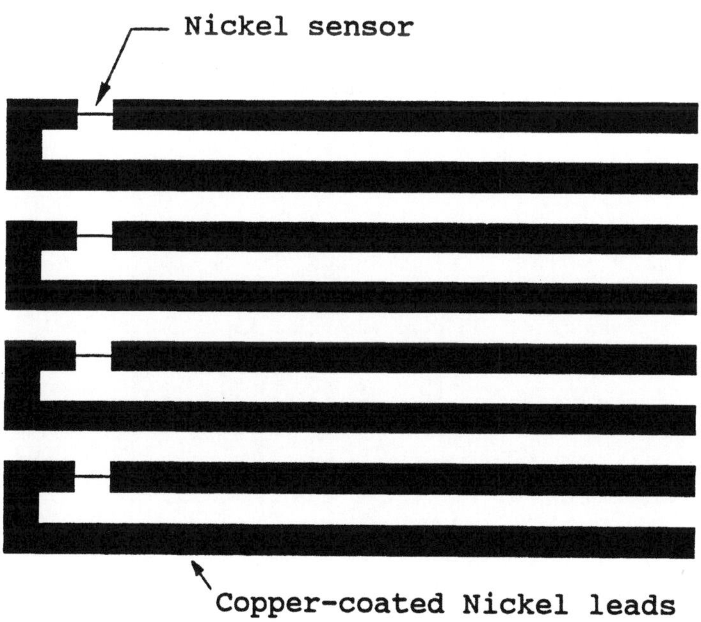

Figure 2. - Multi-element hot-film Sensor.

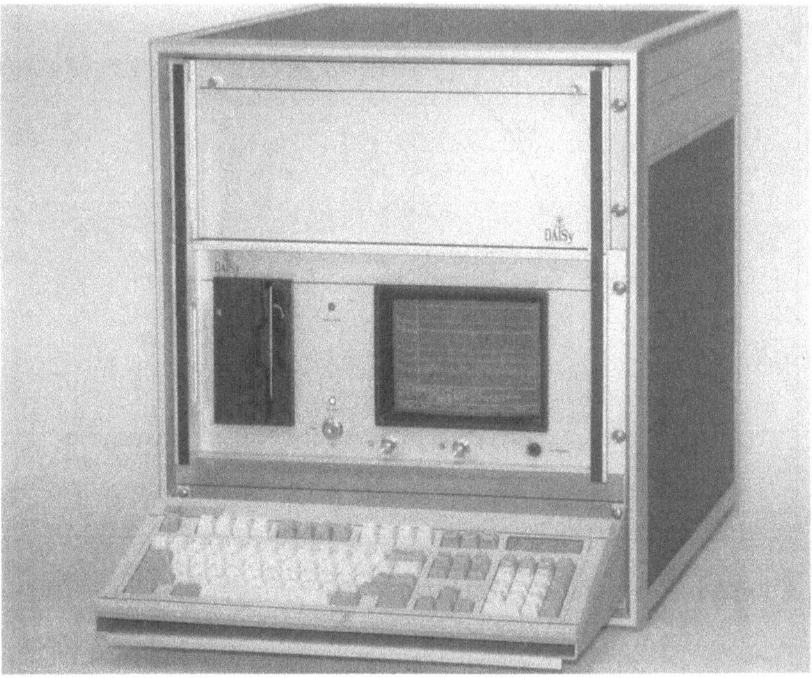

Figure 3. - Data Acquisition and Instrumentation System (DAISy).

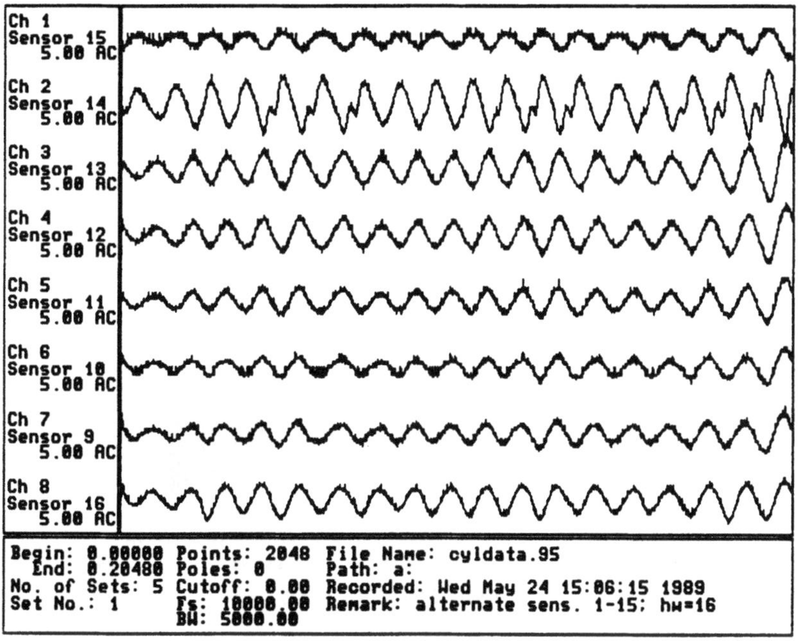

Figure 4. - Simultaneous signals from sensors 9-15 and hot wire.

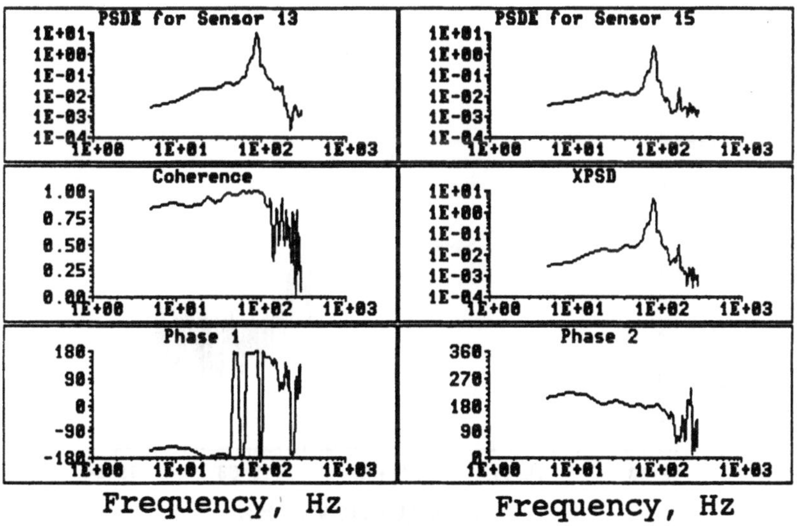

Figure 5. - FFT analysis of signals from sensors 13 and 15.

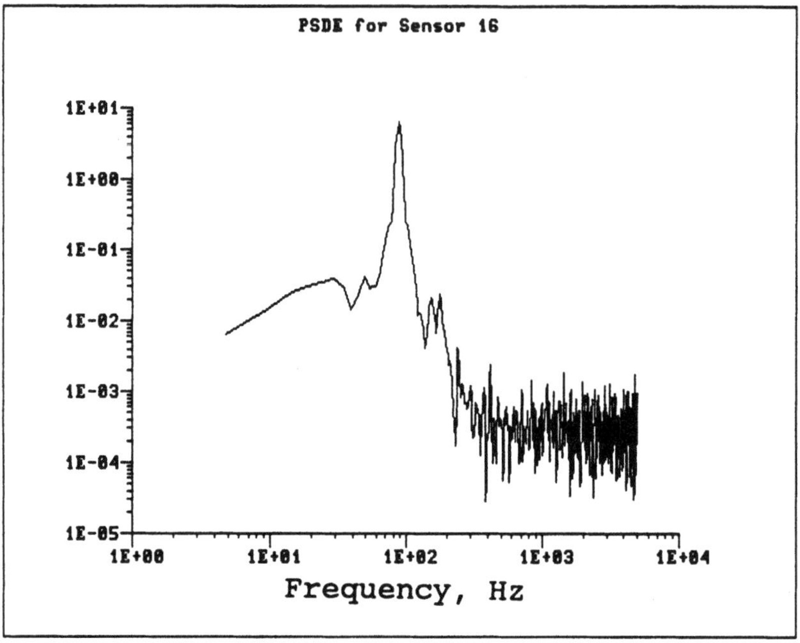

Figure 6. - Power spectrum of signal from hot wire in the wake.

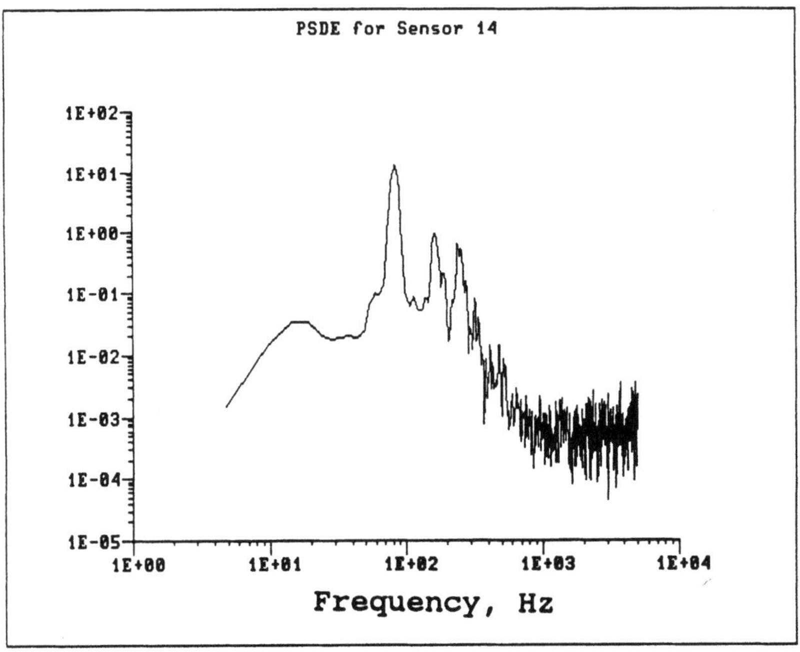

Figure 7. - Power spectrum of signal from sensor 14.

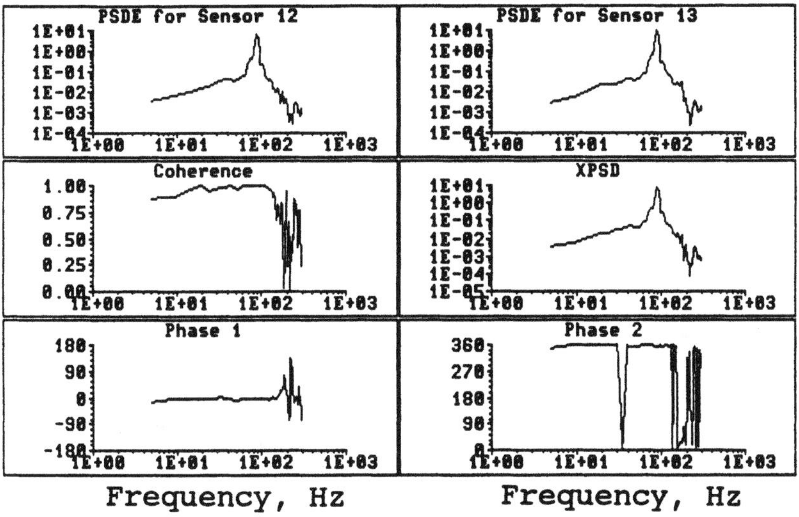

Figure 8. - FFT analysis of signals from sensors 12 and 13.

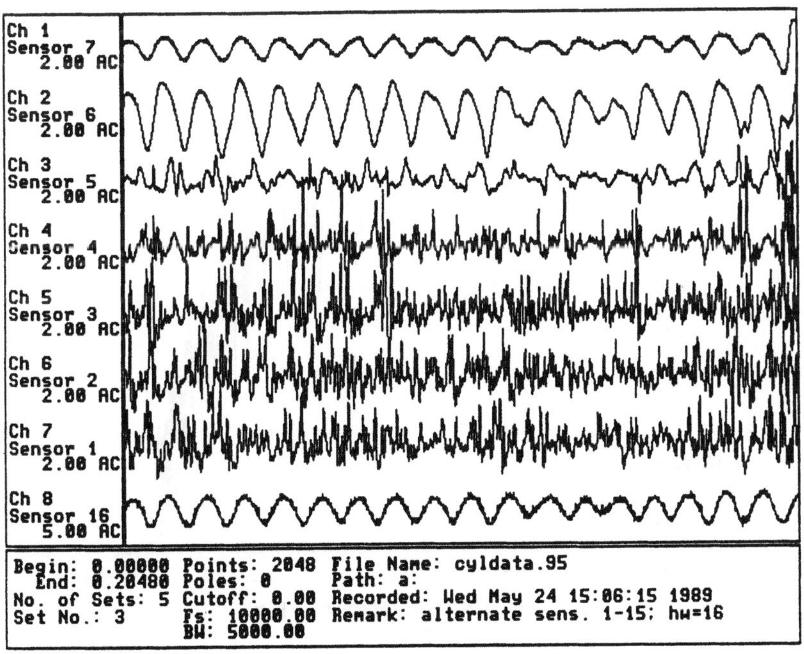

Figure 9. - Simultaneous signals from sensors 1-7 and hot wire.

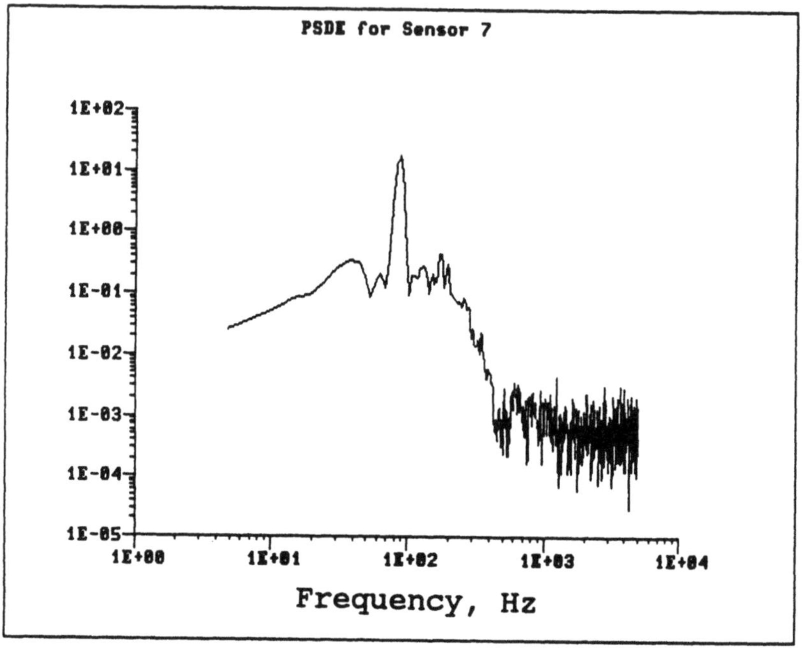

Figure 10. - Power spectrum of signal from sensor 7.

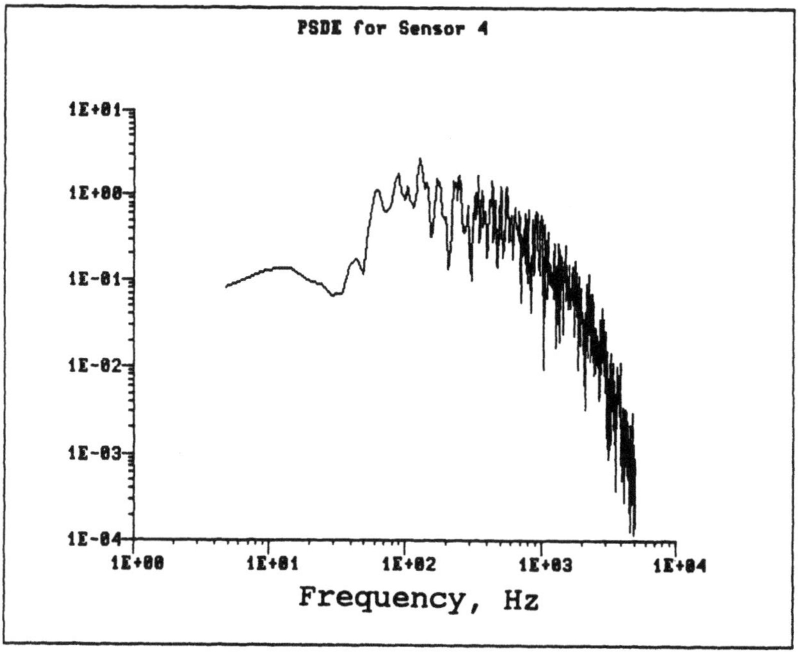

Figure 11. - Power spectrum of signal from sensor 4.

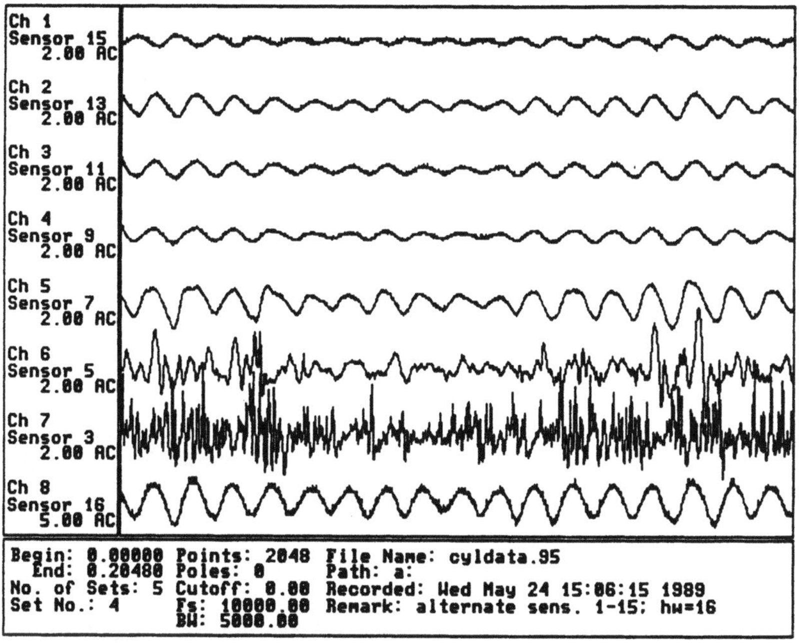

Figure 12. - Simultaneous signals from sensors 3-15 and hot wire.

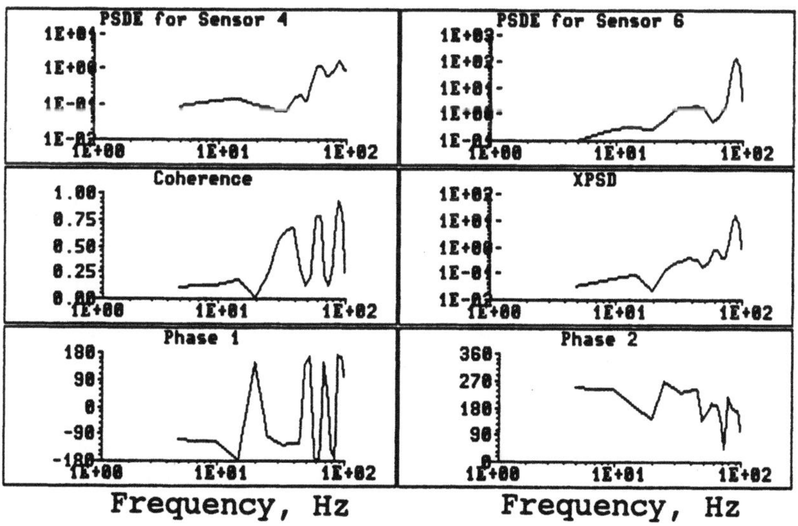

Figure 13. - FFT analysis of signals from sensors 4 and 6.

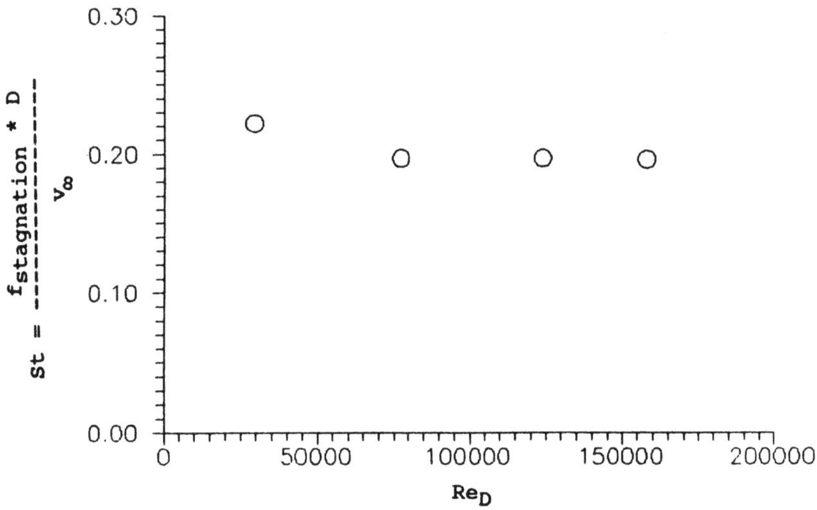

Figure 14. - Strouhal Number based on stagnation frequency as a function of Reynolds Number.

EXPERIMENTS ON A SEPARATION BUBBLE OVER AN EPPLER 387 AIRFOIL AT LOW REYNOLDS NUMBERS USING THIN-FILM ARRAYS

Marc C. Mousseux

Department of Mechanical and Aerospace Engineering
Arizona State University
Tempe, AZ 85287-6106

Ndaona Chokani

Department of Mechanical and Aerospace Engineering
North Carolina State University
Raleigh, NC 27695-7910

John P. Stack and Robert J. McGhee

NASA Langley Research Center
Mail Stop 359
Hampton, VA 23665

ABSTRACT

Preliminary experiments were carried out at the NASA Langley Low-Turbulence Pressure Tunnel on an Eppler 387 airfoil at several angles of attack for Reynolds numbers of 100,000 to 300,000 using surface pressure taps and surface-mounted thin-film arrays. Data are presented for a laminar separation bubble with turbulent reattachment primarily for the case of Reynolds number = 200,000 and angle of attack = 4°. Wavespeeds, wave direction and whether the flow was locally turbulent or laminar were documented and phenomena such as frequency doubling at stagnation regions and wave packets that led to breakdown are presented. There were some efforts to eliminate the bubble over several angles of attack and Reynolds numbers. The pressure and thin-film data along with the Drela-Giles airfoil design/analysis code and the oil-flow visualization results are compared with each other. All methods agree well in demarking the beginning and end of separation as well as showing where transition occurs within the bubble. Several data-reduction techniques were used to calculate wavespeeds and wave direction, and then compared. Spectra were used to check for unstable frequency bands and to document through what region in transition the flow was going.

1. Introduction

Low-Reynolds number airfoils are used in such applications as remotely-piloted vehicles, wind turbines, ultra-light aircraft and propellers. One of the difficulties associated with such airfoils is the presence of separated flow regions on the upper airfoil surface, which are inadequately modeled in the commonly used airfoil design and analysis codes.

The multiple-deck analysis of Smith (1987) provides a simple formula by which the distance between the point of separation and the onset of enhanced instability in a separation bubble may be determined. However, the use of Smith's formula requires that the frequencies, f, be large, and that the detachment distances be small compared to the boundary-layer thickness. Unfortunately, the formula may not be used here because the former criterion is not met in the present experiment. Quantitatively, this is expressed as

$$\mathrm{Re}^{\frac{1}{4}} U/c \ll f \ll \mathrm{Re}^{\frac{1}{2}} U/c.$$

Re is the chord Reynolds number; U and c are the freestream speed and the chord length respectively. In this experiment, the frequencies found are not large enough. The Smith formula has been used with some success when compared to experimental data as stated in the above-mentioned paper.

During the NASA-sponsored Instability and Transition Workshop, experiments were conducted on a low-Reynolds number airfoil in the Low-Turbulence Pressure Tunnel (LTPT). Several items were looked at which related to the laminar separation bubble with turbulent reattachment that occurs over the top surface of the airfoil. The geometry of the separation bubble was mapped out and some transition information and unique phenomena that occurs at stagnated flow are given.

2. Facility and Model

The LTPT is a closed-return, pressurized, continuous-flow facility and detailed information on it is contained in Stainback et al.

(1986). This facility is unique in the fact that the Mach number and Reynolds number may be varied independently by pressurization and fan speed. The turbulence levels in the tunnel for the conditions run in these tests were on the order of $u'_{rms}/U = .06\%$. These measurements were taken using a hotwire with a bandpass of 1-50,000 Hz (McGhee et al. (1988)). The test section is .914 m wide by 2.29 m high by 2.29 m long. The contraction-cone ratio is 17.6:1 and upstream of it is located a heat exchanger which can either cool or heat the flow. The heat-exchanger tubes act as honeycomb with a streamwise dimension of 12.7 cm. Downstream of the tubes 45.7 cm are 9 screens placed 7.6 cm apart with the last screen being 38.1 cm upstream of the contraction cone. The screens are made of .127 mm stainless steel with a mesh size of 15 wires/cm (39 wires/inch).

The Eppler 387 airfoil is a low-Reynolds number airfoil which harbors a separation bubble at certain angles of attack as fully documented by McGhee et al. (1988) and McGhee and Walker (1989). For a Reynolds number of 200,000, a separation bubble occurs over this airfoil for angles of attack varying from $-2°$ to $7°$ and then again for $8.5°$ and above. At $\alpha = 8°$, natural transition occurs with no separation bubble. A large portion of the experiments during the workshop were conducted at a Reynolds number of 200,000, a Mach number of .06, and at $4°$ angle of attack, although $\alpha = 6°$ and Reynolds numbers of 100,000 and 300,000 were also investigated. In the most common configuration, the separation bubble starts at about $x/c = .4$ and turbulent reattachment occurs at about $x/c = .6$. A proposed model of the separation bubble is shown in Figure 1.

Pressure ports on the top and bottom surfaces of the airfoil were used to measure the pressure distribution over the chord. The ports are located about 7.6 cm off of the center span at a spacing of 5% chord starting at $x/c = .1$ out to $x/c = .95$, plus another 10 taps are present from the leading edge to 10% chord. The airfoil is 15.2 cm in chord and 91.4 cm in span.

3. Equipment and Technique

One of the purposes of these tests was to show the versatility of the thin-film array technique in measuring different flow phenomena in situations such as stagnating and recirculating flows. One of these phenomena is phase reversal across separation and reattachment regions which is documented in Stack et al. (1988). The high resolution of the array permits one to acquire accurately pinpoint stagnation lines and even to detect their unsteadiness. A typical thin-film array is shown in Figure 2. The array used here is a scaled-down version of the one shown with a sensor spacing of .127 cm to improve the resolution. The sensor are described in more detail in Sewall et al. (1988). Only every other sensor was used because of the signal analyzer's limited capability to resolve small enough time differences. The sensors were etched onto a .0008 mm Kapton sheet. The sheet was bonded to the airfoil, such that sensors were located 15.2 cm from the center span, and extended from the undersurface of the leading edge over the upper airfoil surface to the undersurface of the trailing edge. The sensors were wrapped around the airfoil, extending onto the undersurface at both the leading and trailing edges, at 15.2 cm from the center span. The airfoil with sensors is shown in Figure 3. The hot-film sensors respond to the surface shear stress, and not to the velocity. Since only the fluctuations in magnitude of the shear stress were of interest, and only correlations of the sensor were required, the amplifiers were not of equal gain nor were the sensors calibrated.

A schematic of the equipment setup is shown in Figure 4. The ac signals were recorded on a Honeywell 101 FM tape. The frequency response of the sensors was limited by the FM-recording equipment to 10 kHz. The tape recorder had less than 3° phase shift between either a pair of odd or even channels, but up to 60° between odd-even pairs of channels for frequencies less than 6 kHz. Thus in the calculation of propagation speeds (described below) only odd or even pairs of channels were used. The signals were post processed using a HP 54470B Signal Analyzer which had a maximum A/D sampling rate of 25.6 kHz. During the post-processing, secondary band-pass filters (DISA TYPE 55D26 Signal Conditioners set at 18dB/oct roll-off) were used to look at specific frequency bands. Also, time traces of the ac portion of the signals were recorded on an oscillograph.

Several methods were used and compared to analyze the signals. Cross-correlations and phase diagrams were relied on heavily to calculate disturbance velocities and directions. Coherence was used to see which frequency bands were correlated and to what extent. The coherence function is a normalized spectrum of the correlation showing which frequencies are correlated. A maximum coherence of unity shows perfect correlation as one would see between two signals coming from the same source. The time shift corresponding to the maximum cross-correlation is used with the distance between the two sensors to calculate wave speed:

$$V_{xc} = \Delta x / \tau$$

where Δx is the spacing between sensors and τ is the time shift corresponding to the maximum correlation. The phase diagram can also be used to calculate velocities as long as the spacing between the two sensors used is less than one wavelength of the particular disturbance under consideration:

$$V_{ph} = 360 \Delta x f / (\Delta \phi + 360 n)$$

where f is the frequency, $\Delta \phi$ is the phase difference between the two channels at the frequency, f, and n is the number of integral wavelengths between the two sensors. If the sensors are less than one wavelength apart, $n = 0$. Spectra were used to show amplified frequencies and the state of the flow, i.e., laminar, turbulent, or transitional.

In general, the way that data were analyzed was by taking two sensors straddling the test point and using those to figure the velocity and the dominant correlated frequencies. The coherence plots show the coherent frequency bands between the two sensors and phase plots show the relative phase of the sensor downstream to that upstream of the test point. The cross-correlation shows the time it takes the disturbance to get from one sensor to the other.

4. Experiment and Results

Since time was a limitation, only one case was run. That was for $\mathrm{Re} = 200,000, \alpha = 4°$, and $U = 18.3$ m/sec. Locations of separation and reattachment, frequency doubling and phase reversal at stagnation regions, and wave packets in the transition region were all observed.

Essentially, all the methods used to determine locations of separation and reattachment were in good agreement. These included oil-flow visualization, pressure-distribution measurements and the thin-film measurements. Also, the Drela-Giles airfoil design/analysis code, which is documented in detail by Drela (1986) and Drela and Giles (1987), was compared to. Basically, the code is a viscous-inviscid interaction code with the ability to calculate across separated regions by operating in an inverse mode. A pressure distribution for Re $= 200,000$ is shown in Figure 5. The laminar separation occurs at approximately $x/c = .4$ and reattachment at about $x/c = .65$. Unfortunately, there is a defective pressure tap at $x/c = .65$. The Drela-Giles code was run for this case and the calculated C_p agreed with the measured pressure distribution over the whole chord and almost exactly away from the discontinuity. Figure 6 shows the shear-stress calculation for this case. The oil flow showed laminar separation to be at $x/c = .4$ and turbulent reattachment at $x/c = .62$ (McGhee et al. (1988)). The sensors located on either side of $x/c = .425$ and $x/c = .616$ show a 180° phase difference as also seen by Stack et al. (1988) in stagnated flow. Also, the time shift in the cross-correlation plots goes to zero at these points.

The velocity estimates using phase measurements and those using correlations agreed well in magnitude for the lower velocities but the agreement was not as good for the higher velocities. The directions agreed, however, so that the large recirculation region in the separation bubble was apparent. The lower velocities were on the order of 9.1-12.2 m/sec which is about .5U to .67U. The A/D frequency of 12.3 kHz limited the accuracy of wavespeed estimates for the very closely spaced sensors. For this reason, alternate sensors were employed to increase proportionally the time shifts to be resolved. However, very fast A/D conversion (at least kHz) is necessary to fully resolve the wavespeed estimates over the entire frequency range. The wavespeeds determined seem reasonable except possibly for very small time shifts which produce wavespeeds greater than the freestream velocity.

The bubble occurs in a region where the flow is already unstable as seen by the pressure plot in Figure 5. The top-most spectrum in Figure 7 is of a sensor located at $x/c = .375$ (ahead of the bubble). It shows an amplified region with a central frequency at 3.5 kHz. Figure 7 shows the spectral evolution across the proposed separation region. Only the relative amplitudes within each plot should be considered since none of the gains in the anemometers are the same nor are they calibrated, i.e., none of the amplitudes can be compared directly from plot to plot. The spectra are taken at a 12.3 kHz sampling rate over 40 m/sec and are averaged 50 times yielding a stationary result. The spectra are shown for $x/c = .375, .392, .408, .425, .442$, and $.458$ from top to bottom respectively. Separation takes place at $x/c = .425$ at which the spectrum shows a much lower-amplitude low-frequency region. The first two spectra show amplified frequencies in the range of 2.5 kHz. The spectrum at $x/c = .408$ (just ahead of separation) shows higher amplitudes both at about 5 kHz and 10 kHz. The spectrum at $x/c = .458$ is typical of spectra further into the bubble but ahead of transition. The frequency doubling across separation can be explained by considering that the shear-stress sensors are only sensitive to the magnitude of the shear. (To calculate direction, one needs at least two sensors along with cross-correlation or phase-difference measurements). When considering an ideal case, if the stagnation line is unsteady such that it oscillates sinusoidally entirely across a given sensor, that sensor will show a rectified sine wave as a time signal. A rectified sine wave has double the frequency of a standard sine wave, i.e, the frequency doubling phenomena at a stagnation region is not physical, but an artifact of the measurement technique, at least in this case.

Figure 8 is typical of the coherence plots in the laminar-recirculation region of the bubble until $x/c = .517$, that is, the high frequencies are completely incoherent. The only coherent frequency range is from 10 Hz to 400 Hz. Figure 8 shows the phase differences between $x/c = .433$ and .45. From $x/c = .517$ to .558 the amplified region is down to 2 kHz versus 4 kHz seen just upstream as in the last plot of Figure 7. Figure 9 shows the spectrum at $x/c = .525$ (note that the frequency scale is again different). This change in the band of amplified waves is accompanied by an increase in the coherence over the amplified frequency range as seen in Figure 10 which shows the coherence between $x/c = .525$ and .541. Figure 10 also shows a higher phase-change slope over the high-frequency range than over the low frequencies. Since V_{ph} is inversely proportional to this slope, the high frequencies are moving slower than the low frequencies. Actually, when one looks at cross correlations, wave packet phenomena can be observed as shown in Figure 11 which is a correlation between $x/c = .525$ and .541. The high frequencies in this plot are close to 1900 Hz and the low frequency is 290 Hz. These lie right in the centers of the coherent frequency bands.

Starting at $x/c = .558$ out to .625, there appear to be higher harmonics growing out of the spectrum as seen in Figure 12. The spectrum slowly fills in until turbulence is reached. Here again, the correlation plots show wave packets which are also seen in the oscillograph traces in Figure 13. Time is increasing to the right and the streamwise direction is from bottom to top in the figure. The traces are for $x/c = .541$ to $.608$ in increments of $x/c = .008$ The small-amplitude high-frequency waves, which are superposed on the lower frequency waves, soon grow large and higher frequencies come into play until almost completely turbulent flow is achieved. When looking at the spectra, the higher-harmonic of the amplified band coming out of the background is easily seen in the oscillograph traces as the high-frequency wave growing in amplitude.

When compared to the geometry of Figure 1, the smaller recirculation regions near the trailing edge of the bubble were not clear. Many of the phase plots for this region showed phase differences for different frequency bands but the coherence between the signals was very low ($< .15$). Figure 14 shows several of these phase plots along with the corresponding coherence plots. The first set is for $x/c = .583$ and it is just upstream of the designated reattachment region. It shows a phase difference of about 90° as seen for the frequency range of 2.5-4. kHz. Then, an abrupt phase change of -180° is seen in the plots of $x/c = .6$. The next two points, $x/c = .616$ and .633, show phase differences of about 90° to 135° after which the last plot shows a phase reminiscent of attached flow. Perhaps the sensors are still not close enough to resolve the geometry in Figure 1 or maybe this geometry does not exist as shown. A possibility is that the reattachment region is unsteady such that a given sensor may meander into and out of the bubble creating completely incoherent signals with the other sensors. This idea may be supported by the observations of a slightly unsteady wake at this Re and α by McGhee et al. (1988).

Cancellation efforts were carried out using different methods to implement the same idea. That is, to trip the boundary layer to earlier turbulence so that separation will either be delayed or eliminated completely. The methods basically consisted of using a piece of adhesive paper cut into strips with a sawtooth pattern on one edge bonded to the surface of the airfoil in various configurations and locations. Two paper thicknesses were used, .127 mm and .254 mm. The best success was with a .254 mm strip that was placed at $x/c = .2$ along the entire span perpendicular to the flow. This virtually terminated the separation bubble at almost all Re and α. Figure 15 shows three C_p plots at $\alpha = 4°$, each for a different Re. The first shows a marked improvement over the no-strip case and the two lower plots show the higher Re cases which seem to be completely cured of separation. This was the case for $\alpha = 6°$ as well.

5. Discussion and Conclusions

Several items were investigated concerning the separation bubble over an Eppler 387 airfoil. Cross-correlations and phase diagrams were used to calculate velocities, spectra were used to check for amplified frequencies, and coherence plots for causality. The cross-correlation and phase calculations for velocity were compared. The hot-film measurements, oil-flow visualization, pressure measurements, and Drela code were compared for finding the leading and trailing edges of the bubble. The following conclusions can be stated.

1. All methods agreed well in showing the beginning and end of the separation bubble.

2. Frequency doubling is not physical, but rather an artifact of the hot-film measurements. That is, this occurs because the hot-films measure only the magnitude of the shear stress and are unable to distinguish the stream direction.

3. The velocity measurements using phase and those using correlations agreed in magnitude for the lower velocities but not as much for the higher velocities. The directions agreed however so that the large recirculation region in the separation bubble was apparent. The signal analyzer was at fault here because the sampling rate and total sampling time could not be varied independently. Therefore, the frequency resolution of the plot as well as the time resolution were inversely proportional so that if the time shift could be resolved properly, the frequency resolution in the correlation would be very low such that one would only be accounting for a few discrete frequencies in the range of interest. This can easily be resolved through the use of a small computer (PC) to acquire the data and process it. In this way, the total sampling time and sampling frequency could be controlled independently.

4. Either the spatial resolution of the sensors is not high enough or the trailing edge of the bubble is unsteady. The phase diagrams are not clear enough for one to conclude that Figure 1 is a correct representation of the bubble geometry.

5. Wave packets were observed in the cross-correlations. The spectra upstream of the bubble show amplified frequencies that

are possibly associated with T-S waves. Within the separated bubble, the velocity profile exhibits an inflection point which gives rise to an inviscid instability; in this region the wave packets may be associated with Rayleigh waves.

Many items still need to be studied further in order to validate some of the observations and conclusions. A computer with fast A/D board should be used to get better quantitative information from the sensors. The other Re cases should be investigated in more detail and, maybe the sensors should be reduced in size even further to really pin down the geometry of the bubble at its trailing edge. Above-surface measurements need to be made to get velocity profiles in the separated region and ahead of this region so that one may infer whether T-S and or Rayleigh waves exist. Also, stability codes such as SALLY should be run to obtain unstable frequency bands both inside and outside of the bubble as well as predictions for transition locations within the bubble.

Acknowledgements

The authors would like to thank J. Ray Dagenhart for his helpful discussions throughout the course of the workshop concerning some of the observations and results. Also, the use of the NASA Langley facilities is greatly appreciated.

References

[1] Drela, M., *Two-Dimensional Aerodynamic Design and Analysis Using the Euler Equations*, Massachusetts Institute of Technology, Gas Turbine Laboratory Report 187 (February 1986).

[2] Drela, M. and Giles, M. B., *Viscous-Inviscid Analysis of Transonic and Low Reynolds Number Airfoils*, AIAA Journal 25, No. 10, 1347-1355 (September 1987).

[3] McGhee, R. J. and Walker, B. S., *Performance Measurements of an Airfoil at Low Reynolds Numbers*, Proposed Paper for Conference on Low Reynolds Number Aerodynamics, University of Notre Dame, IN (June 5-7, 1989).

[4] McGhee, R. J., Walker, B. S., and Millard, B. F., *Experimental Results for the Eppler 387 Airfoil at Low Reynolds Numbers in the Langley Low-Turbulence Pressure Tunnel*, NASA TM 4062 (October 1988).

[5] Sewall, W. G., Stack, J. P., McGhee, R. J., and Mangalam, S. M., *A New Multipoint Thin-Film Diagnostic Technique for Fluid Dynamic Studies*, SAE Technical Paper 881453, Aerospace Technology Conference and Exposition, Anaheim, CA (October 1988).

[6] Smith, F. T., *Break-up in Unsteady Separation*, 1987.

[7] Stack, J. P., Mangalam, S. M., and Berry, S. A., *A Unique Measurement Technique to Study Laminar-Separated Bubble Characteristics on an Airfoil*, AIAA Paper 87-1271, 1987.

[8] Stack, J. P., Mangalam, S. M., and Kalburgi, V., *The Phase Reversal Phenomenon of Flow Separation and Reattachment*, AIAA Paper 88-0408, 1988.

[9] Stainback, P. C. and Owen, F. K., *Dynamic Flow Quality Measurements in the Langley Low Turbulence Pressure Tunnel*, Technical Papers - 13th Aerodynamic Testing Conference, AIAA (1984), 257-265.

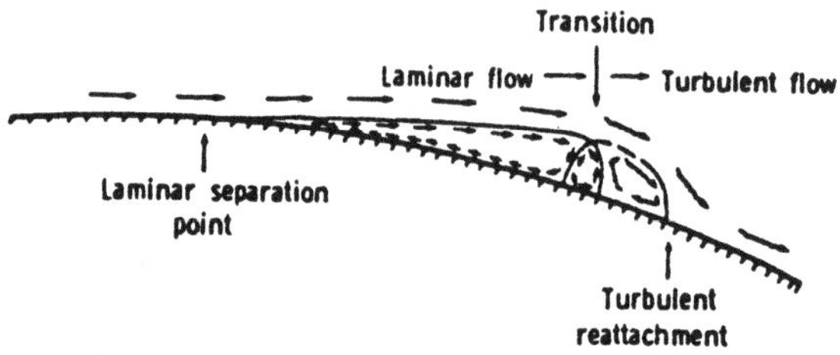

Figure 1. Proposed bubble geometry showing three recirculation regions.

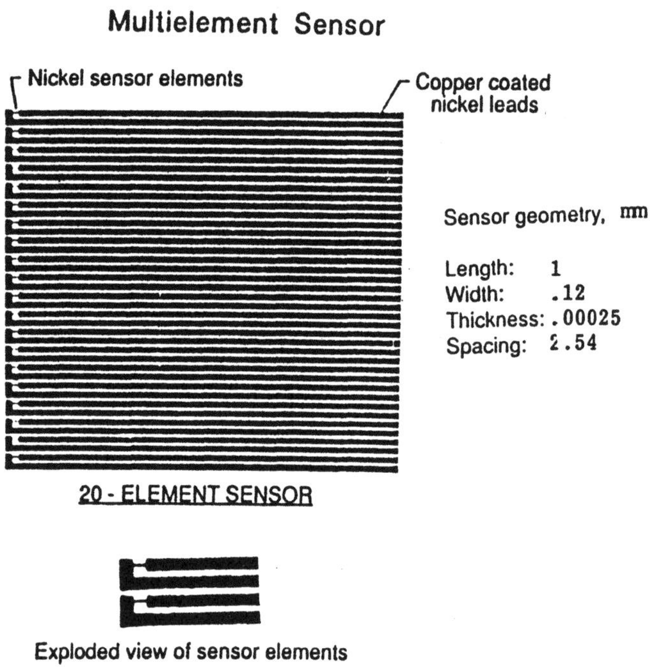

Figure 2. Typical hot-film array.

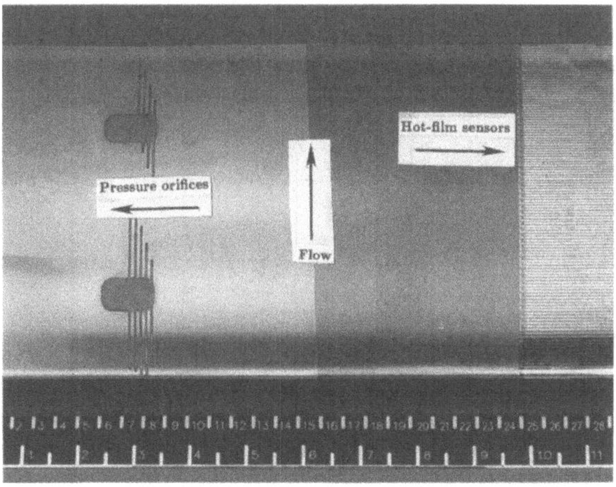

Figure 3. Photograph of top of model showing pressure-tap locations and films.

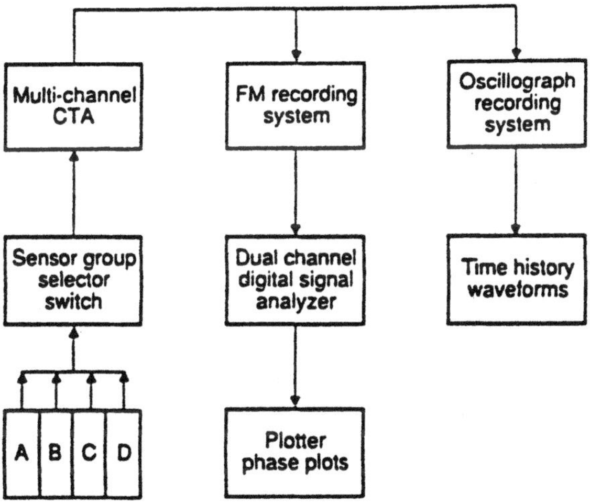

Figure 4. Schematic of equipment setup.

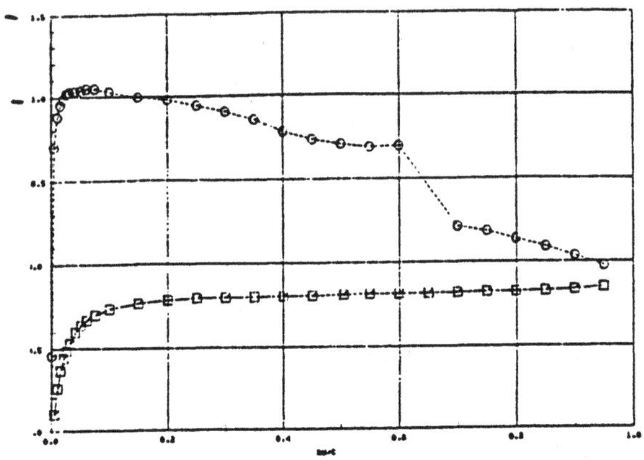

Figure 5. C_p at a=4° and Re=200,000.

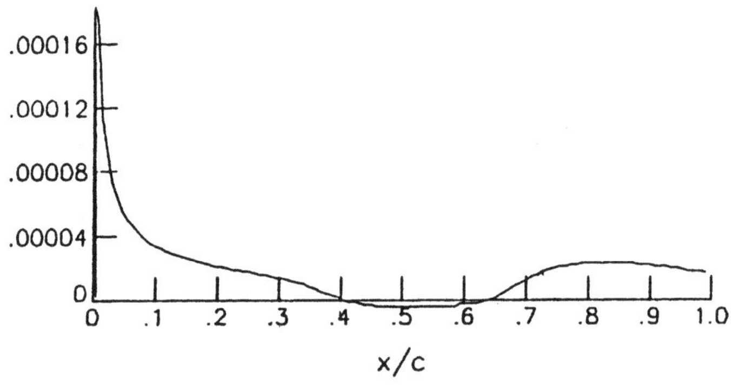

Figure 6. Shear-stress calculation from the Drela code for a=4° and Re=200,000.

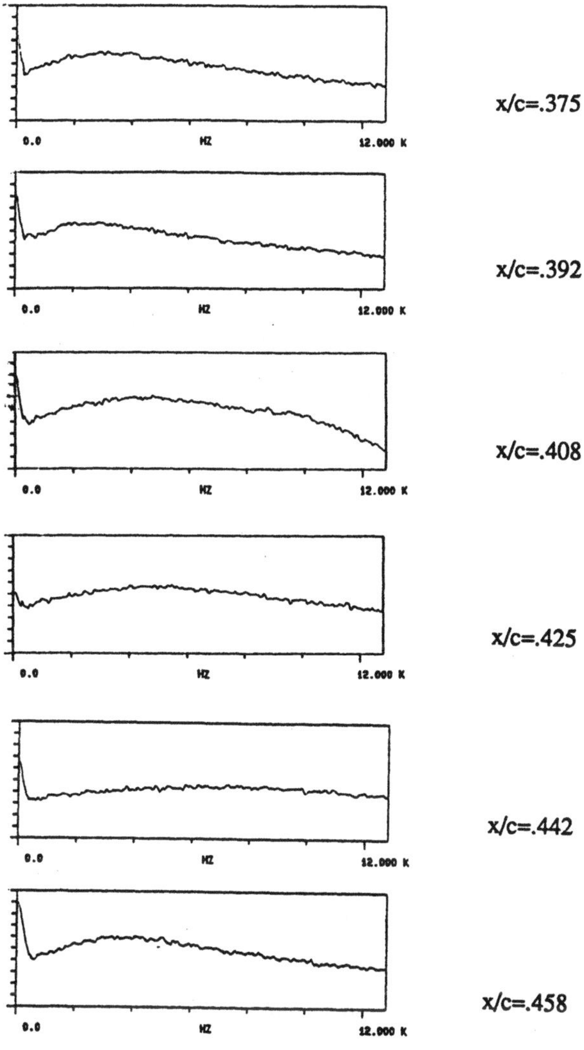

Figure 7 Spectral evolution from x/c=.375 to .45 in increments of x/c=.016.

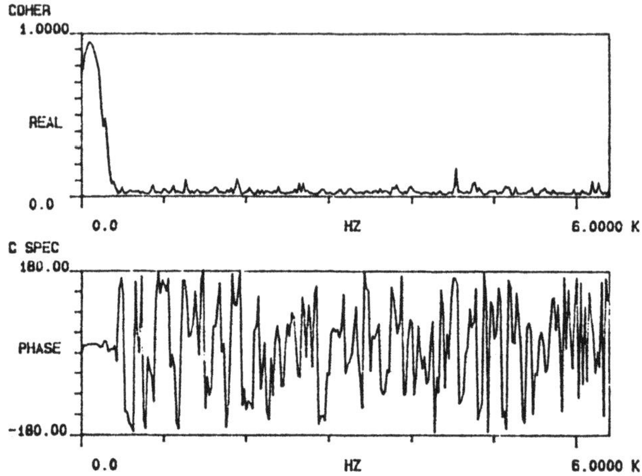

Figure 8. Phase and coherence between x/c=.433 and .45.

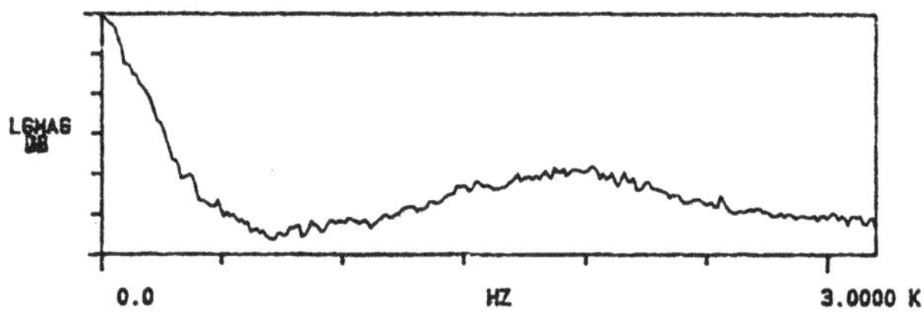

Figure 9. Spectrum at x/c=.525.

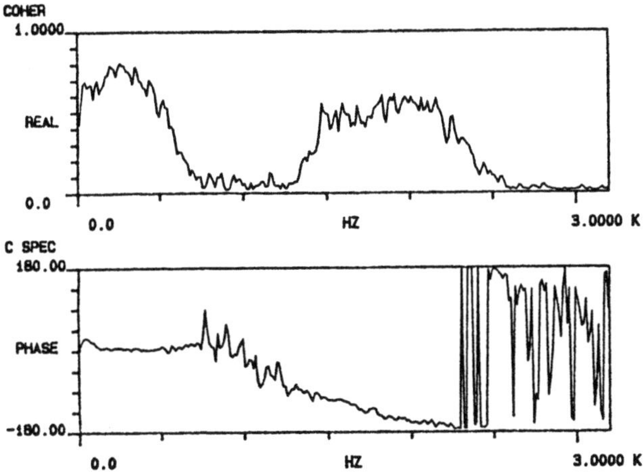

Figure 10. Phase and coherence between x/c=.525 and .541.

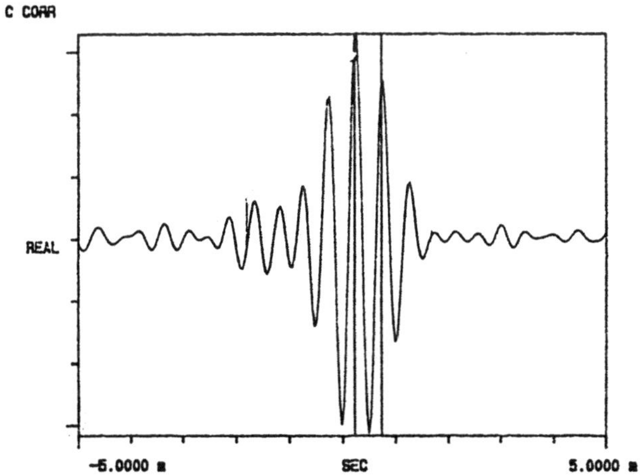

Figure 11. Cross-correlation between x/c=.525 and .541.

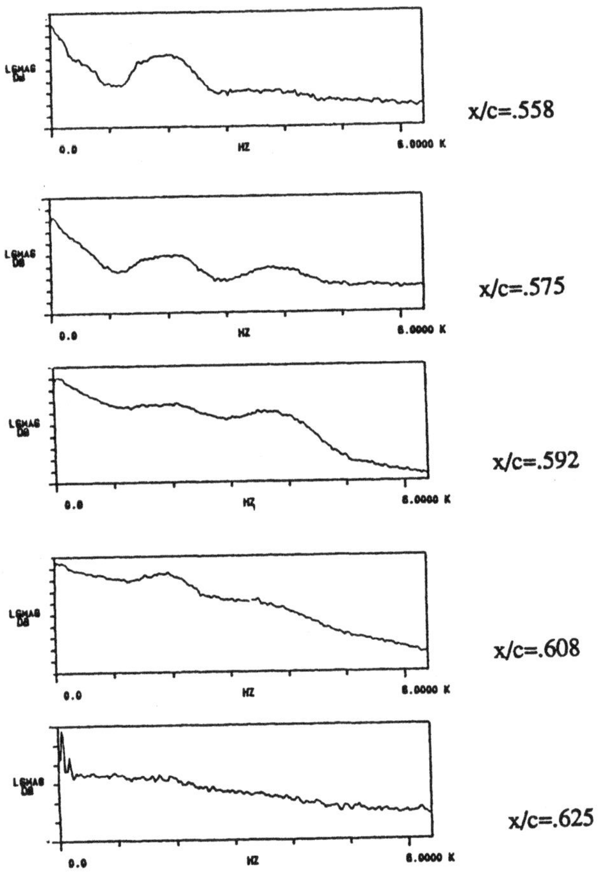

Figure 12. Spectral evolution from x/c=.558 to .625 in increments of x/c=.016.

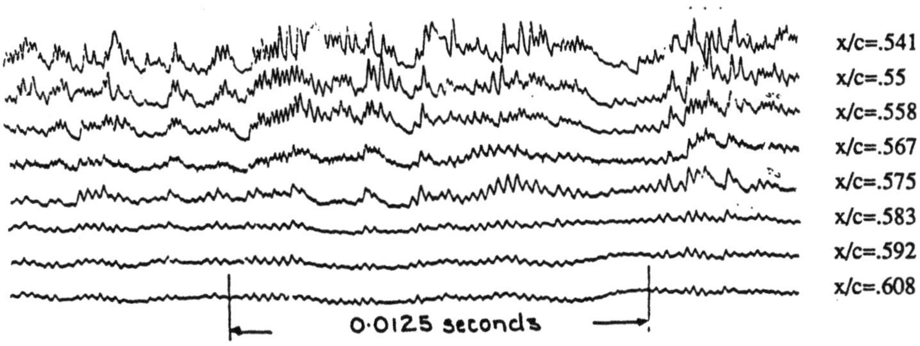

Figure 13. Oscillograph traces for x/c=.55 to .608 in increments of x/c=.008.

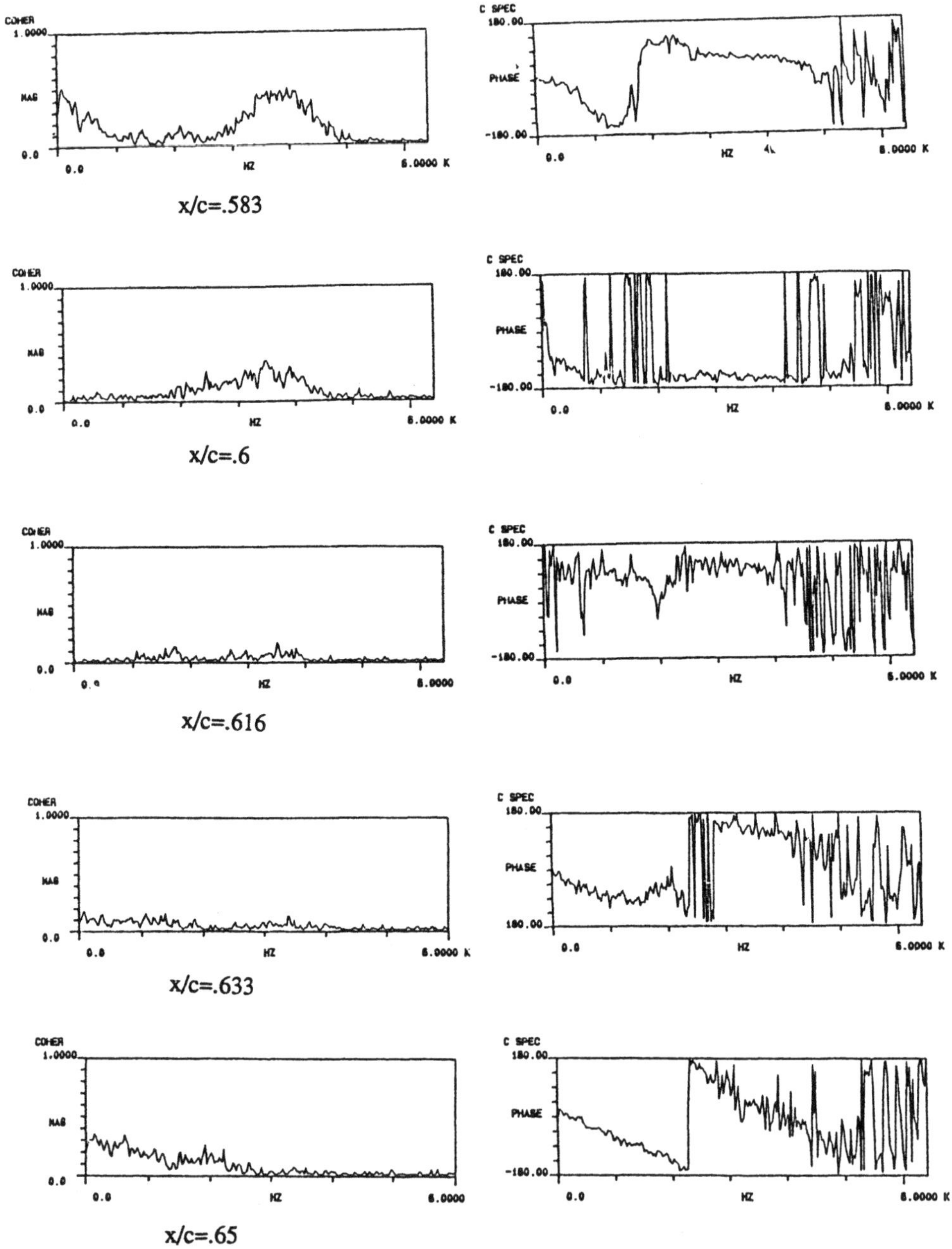

Figure 14. Phase and coherence across x/c=.583 to .65 in increments of x/c=.016.

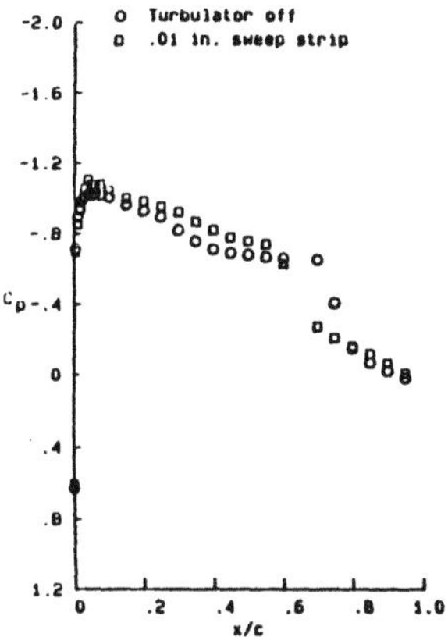

Re=100,000,

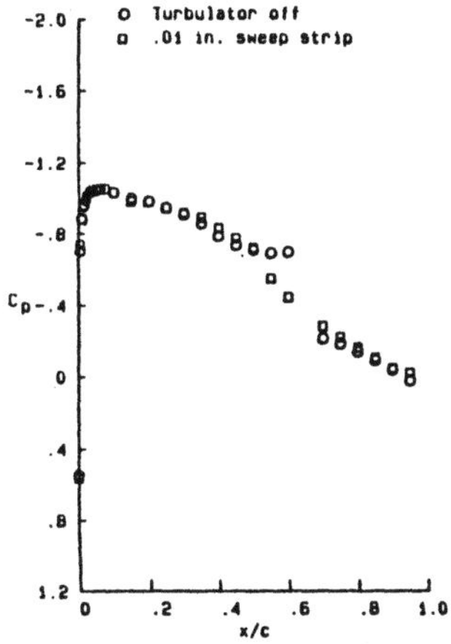

Re=200,000

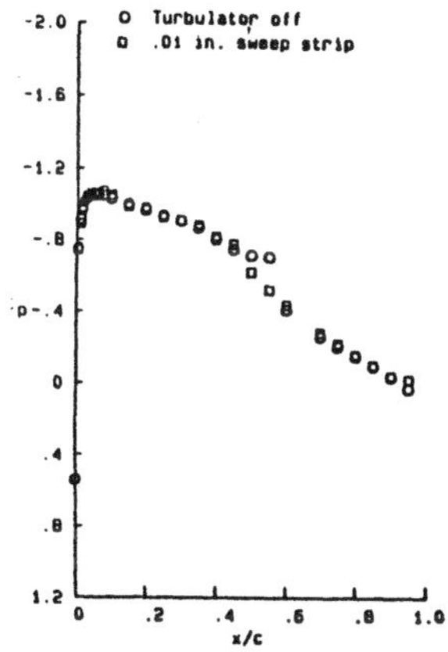

Re=300,000

Figure 15. C_p for alpha = 4°

RECEPTIVITY

GROUP SUMMARY: RECEPTIVITY

T. B. Gatski

NASA Langley Research Center
Hampton, VA 23665

The majority of work conducted during the workshop focused on the analytical or numerical treatment of receptivity problems; however, due to the initiation of an experimental study on the wave-packet phenomenon on a rotating-disk during the workshop, an analytical study on the modulational stability of such flows was begun. The receptivity studies initiated during the workshop were both for the incompressible and compressible regimes as well as for two- and three-dimensional flows.

M. Choudhari and E. Kerschen studied the receptivity due to three-dimesional convected gusts. The gust is essentially a vorticity disturbance which is convected by the free stream. Since weak free stream turbulence can be viewed as a spectrum of time harmonic gusts, it was advantageous to consider three-dimensional disturbances. In addition, the effects of compressibility, where the dominant linear instabilities are often three-dimensional, must also be considered for aeronautical applications.

The receptivity due to a convected gust results from the short scale pressure fluctuations generated by the interaction of the convected gust with a wall hump. This interaction occurs just outside the boundary layer, and is inviscid in nature. Hence, a number of useful deductions regarding the amplitudes of the generated instability waves can be made by simply analyzing the inviscid interaction between the gust and the wall hump. The present study focused on the scattering of a three-dimensional convected gust by a shallow wall roughness element in compressible flow.

This inviscid calculation provides the kernel for a number of interesting problems. Examples are an oblique gust interacting with a two-dimensional wall hump to produce a single three-dimensional instability wave or a convected gust scattered by a three-dimensional wall hump to produce a wedge-shaped pattern of instability waves downstream of the interaction region. This framework can also be easily extended to consider the interaction of the gust with regions of rapid variation in wall suction.

R. Heinrich, T. Gatski, and E. Kerschen examined boundary-layer receptivity to free stream disturbance fields. It has been speculated that this can be the result of distributed mechanisms which cause the receptivity to take place over a streamwise distance of several T-S wavelengths due to the interaction of the streamwise modulation of the free stream disturbance and the otherwise parallel boundary-layer flow. This is distinct to the usual localized receptivity mechanisms which result from the interaction of the long wavelength free stream disturbance with a short scale (local) surface irregularity. Linear triple-deck theory is used to model recent experiments where the effects of distributed receptivity mechanisms are assumed to be present. These experiments are modeled by introducing unsteady sources into the upper deck of the triple-deck structure to appropriately account for the effects of the unsteady pressure gradients in the free stream. This analysis reproduced the essential features of the experimental results, and the resulting solution structure suggested that distributed mechanisms can only cause exponentially small receptivity levels.

A. Cain has performed numerical computations on boundary-layer receptivity to an acoustic disturbance interacting with a wall slot. During the workshop, the asymptotic triple-deck structure of the disturbance field for the compressible boundary-layer flow was examined in collaboration with M. Choudhari and E. Kerschen. The numerical formulation uses fourth-order compact spatial differencing and fourth-order Runge-Kutta time integration. The code is highly vectorized and runs at 2.0×10^{-5} sec/(point-timestep) on a Cray X-MP. The disturbance field is introduced by superposition to the baseflow at the outer boundary. A corresponding experimental effort is also now being initiated.

D. Riahi studied the problem of linear, and weakly nonlinear, modulational primary stability of rotating-disk flow for stationary modes using the method of multiple scales. The system of equations for the disturbances is converted into one integro-differential equation for the vertical component of disturbance velocity. Linear and weakly nonlinear theories are applied to this equation to determine the solutions which represent the linear and nonlinear developments of wave packets. The linear solution indicates that the disturbance spreads out as the slow time, t_s, increases, and, at the center, it grows exponentially, after an initial decrease produced by a t_s^{-1} factor. The nonlinear solution obeys a Landau-type equation. Development of such a solution depends on the precise values of the coefficients of the Landau equation.

BOUNDARY LAYER RECEPTIVITY DUE TO THREE-DIMENSIONAL CONVECTED GUSTS

Meelan Choudhari and E. J. Kerschen

University of Arizona
Tucson, Arizona 85721

ABSTRACT

Boundary layer receptivity refers to the process by which external disturbances generate instability waves in the boundary layer. The inherently three-dimensional nature of both the transition process and the free-stream turbulence, as well as the dominance of three-dimensional instability waves in compressible boundary layers suggest that three-dimensional receptivity problems must be investigated. It has been shown [4] that receptivity to a convected vorticity disturbance occurs through the short scale pressure fluctuations generated by the inviscid interaction of the gust with short scale mean flow gradients. Weak free-stream turbulence can be considered as a spectrum of time harmonic vortical gusts. In this paper, rapid distortion theory has been utilized in conjunction with triple-deck concepts to analyze compressible, fully three-dimensional interactions involving convected free-stream disturbances. The results obtained are applicable in a variety of situations, such as receptivity due to shallow 3-D roughness elements or three-dimensional wall suction distributions. It is found that a three-dimensional instability wave can also be generated by the interaction of a very oblique gust with a two-dimensional mean flow disturbance.

1. Introduction

Conceptually, the phenomenon of boundary layer transition can be separated into three stages. These stages are the receptivity process, the linear amplification of the instability waves, and the nonlinear breakdown to turbulence. The receptivity process refers to the generation of instability waves by external disturbances such as free-stream turbulence and sound waves. The physical mechanism responsible for the coupling between the long wavelength free-stream disturbances and the short wavelength boundary layer instabilities was not understood for a long time.

Recent work (Refs. 1-4) has shown that the wavelength conversion process takes place in regions where the mean boundary layer changes rapidly in the streamwise direction. The regions where the boundary layer is "receptive" can thus be separated into two classes. The first class consists of the region near the leading edge, where the boundary layer is thin and growing rapidly. The second class consists of regions further downstream, where some local feature, such as a wall hump, forces the boundary layer to make a short scale adjustment. This paper considers only the second class of regions.

References (1-4) have examined only the generation of two-dimensional instability waves. However, transition is an inherently three-dimensional process. In compressible flows, even the linear regime of boundary layer instability is dominated by three-dimensional instabilities. Free-stream turbulence is also three-dimensional, and hence, capable of generating three-dimensional instability waves. In this paper, we utilize rapid distortion theory [6,7] in conjunction with triple-deck concepts to analyze instability wave generation by free-stream turbulence.

2. Background

For most wind tunnels and in-flight situations, the free-stream turbulence is weak, and can, therefore, be modelled as a linear superposition of its different Fourier modes. Each of these Fourier modes corresponds to a time harmonic gust (a vortical disturbance) which remains "frozen" in the mean flow.

As noted in the Introduction, localized receptivity occurs due to the interaction of the free-stream disturbance with a region of rapid streamwise variation in the wall boundary condition, such as a wall hump. Due to the short scale, the local mean flow becomes interactive, and is described by the triple-deck structure. The unsteady flow also has a triple-deck structure. If the strength of the mean flow disturbance is assumed to be small enough, both mean and unsteady triple-deck problems can be solved in closed form using Fourier transform methods, and the amplitude of the generated T-S wave is given by the residue of an appropriate pole of the Fourier transform solution.

Receptivity to a convected gust occurs through the short scale pressure fluctuations generated due to the inviscid interaction of the gust and the mean flow gradients in the upper deck outside the boundary layer [4,5]. In contrast, acoustic receptivity occurs through the interaction of mean flow gradients with the Stokes wave generated by the free-stream acoustic wave. This interaction takes place in the lower deck, very close to the wall. These general concepts, although derived originally for the two-dimensional case, are also valid for the case of three-dimensional interactions.

In the following section, we examine the inviscid interaction of a three-dimensional, time harmonic gust with an arbitrary, interactive, localized disturbance to a two-dimensional boundary layer (Fig. 1). The cause of this mean flow disturbance can be any one of the local features, such as a wall hump, suction strip, or shock-boundary layer interaction. This inviscid calculation provides the kernel for a number of interesting applications which are described in Section 4.

3. Theory

The specification of the incident gust is considered first. We assume that the velocity field associated with the incident gust, far upstream of the airfoil, is given by

$$\mathbf{u}_{-\infty}(x,t) = A_g \, \mathbf{e}_g \exp[i(\mathbf{k}_g \cdot \mathbf{x} - k_x U_\infty t)] \quad , \tag{3.1a}$$

where $\mathbf{e}_g = (e_x, e_y, e_z)$, is a unit vector indicating the direction of velocity fluctuations, $\mathbf{k}_g = (k_x, k_y, k_z)$, and

$$\mathbf{e}_g \cdot \mathbf{k}_g = 0 \quad , \tag{3.1b}$$

thus implying that the velocity field is solenoidal. The temporal dependence in (3.1a) reflects the fact that the gust is convected with the free-stream, and hence, the pressure fluctuations vanish to linear order. We assume the nondimensional gust frequency, $k_x L$, to be such that

$$k_x L = O(\varepsilon^{-2}), \; (\varepsilon = (U_\infty L/\nu)^{-1/8}) \quad , \tag{3.2}$$

which is of the same order as the T-S frequency near the lower branch of the two-dimensional neutral stability curve.

Having specified the incident gust, we now analyze the unsteady motion just upstream of the region of the mean flow disturbance (Region I in Fig. 1). The appropriate streamwise length scale in this region corresponds to the gust wavelength. Then in view of Eq. (3.2), the unsteady problem in this region reduces to the interaction of the gust with a flat plate which is infinite in both directions. Since the gust produces an unsteady upwash on the airfoil, the zero normal flow condition is satisfied by adding an irrotational disturbance $\mathbf{u}_0 = \nabla\phi_0$, where ϕ_0 is given by

$$\phi_0 = \frac{A_g\, e_y}{\sqrt{k_x^2 + k_y^2}} \exp\left[i(k_x x + k_z z - k_x U_\infty t) - \sqrt{k_x^2 + k_y^2}\, y \right] \quad (3.3)$$

One may observe that the irrotational component of the motion has the same convected phase as the incident gust and hence, does not produce pressure fluctuations. However, since we are analyzing the inviscid interaction, the horizontal component of the velocity does not vanish near the wall.

We now briefly consider the signature of the gust within the boundary layer. The unsteady motion in the boundary layer region is driven solely by the shearing effect due to the inviscid slip velocity at the edge, since the pressure fluctuations are zero to linear order. At high Reynolds numbers, the viscous effect is very weak and therefore, the unsteady motion is concentrated in the outer part of the boundary layer [8]. In fact, it can be shown that the velocity fluctuations close to the wall are exponentially small.

The upstream solution (in Region I) is modified within Region II by the short scale mean flow gradients. As remarked above, the local motion is governed by the triple-deck equations, and the interaction between the gust and the mean flow gradients occurs within the upper deck. The mean flow disturbance in the upper deck is inviscid and irrotational. The unsteady upper deck is somewhat unconventional in that the flow there is rotational by virtue of the vortical fluctuations associated with the gust. We then use Rapid Distortion Theory [6,7] to solve the upper deck problem.

Rapid Distortion Theory describes the interaction of small amplitude unsteady vortical disturbances with a non-uniform, potential mean flow. It accounts for vorticity distortion by the non-uniform mean flow, but neglects the viscous and nonlinear effects which change a small amplitude convected disturbance much more slowly. The basic equations of Goldstein's Rapid Distortion Theory [6] are summarized below.

The total disturbance velocity and pressure can be written as $\mathbf{u} = \mathbf{v} + \nabla\phi$, and $p = -\rho_m D_m\phi/Dt$ respectively, where $\mathbf{v}$ is a known function of the imposed upstream disturbance, ϕ is a solution to the linear inhomogeneous wave equation

$$\frac{D_m}{Dt}\left(\frac{1}{c_m^2}\frac{D_m\phi}{Dt}\right) - \frac{1}{\rho_m}\nabla\cdot(\rho_m\nabla\phi) = \frac{1}{\rho_m}\nabla\cdot\rho_m\mathbf{v} \quad , \qquad (3.4)$$

D_m/Dt denotes the convective derivative relative to the mean potential flow, and ρ_m, c_m correspond to the mean density and speed of sound respectively. The part $\mathbf{v}$ contains all the vorticity and is convected with the local mean flow. When the far upstream vortical motion corresponds to a single gust, as in (3.1), $\mathbf{v}$ is given by

$$v_i = A_g\, e_g \cdot \partial \mathbf{X}_{la}/\partial x_i \exp[i(\mathbf{k}_g \cdot \mathbf{X}_{la} - k_x U_\infty t)] \quad , \tag{3.5a}$$

where the subscript i denotes the component along coordinate direction i, and $\mathbf{X}_{la}$ is defined by

$$D_m \mathbf{X}_{la}/Dt = 0 \quad (\mathbf{X}_{la} \rightarrow \mathbf{x} \text{ far upstream}) \quad , \tag{3.5b}$$

and hence, represents the Lagrangian coordinates of the mean potential flow. Solving the equations (3.4,5) in a general situation is a formidable task. However, due to the simplifications in the problem at hand, a closed form solution can be obtained in an easy manner.

Presently, we will consider the more general case of a three-dimensional mean flow disturbance. However, one should note that very oblique gusts can also generate three-dimensional instability waves by interacting with two-dimensional mean flow disturbances. This special case will be discussed briefly in the last section of this paper. For now, we assume that the mean velocity in the upper deck is given by

$$\mathbf{U} = (U_\infty, 0, 0) + \delta\, \mathbf{U}_1(\mathbf{X}) \quad , \; (\mathbf{U}_1 = \nabla_{\mathbf{X}}\, \Phi_1(\mathbf{X})) \tag{3.6a}$$

where

$$\mathbf{X} = \frac{\mathbf{x}}{\varepsilon^3 \chi} \; , \; \chi = \mu_0(0)^{1/4}\, U_0'(0)^{1/4} \left| 1 - M_\infty^{\;2} \right|^{-3/8} \rho_0(0)^{-1/2} \tag{3.6b}$$

are the upper deck coordinates. The quantities M_∞, $U_0'(0)$, $\rho_0(0)$, $\mu_0(0)$ denote the local Mach number, and the nondimensional wall shear stress, wall density, and absolute viscosity at the wall conditions respecively, all corresponding to the undisturbed boundary layer profile just upstream of region II. Introducing the factor χ reduces the lower deck problem to a canonical statement independent of the upstream boundary layer profile. It is worth noting that the nondimensional wall viscosity $\mu_0(0)$ is assumed to be given (by experimental measurement or some accurate empirical correlation). It is not necessary to assume the viscosity variation to be governed by any particular law, such as Chapman's law. The quantity δ reflects the strength of the mean flow disturbance, e.g., for a hump of height of $O(\varepsilon^5 hL)$, δ is equal to $\varepsilon^2 h$. The solution for Φ_1 is assumed to have been obtained by solving the appropriate mean flow triple-deck problem.

The linearized nature of the mean flow disturbance in the upper deck allows the local unsteady solution to be linearized about the upstream solution (3.3). In addition, since the triple-deck length scale is much shorter than the gust wavelength, the upstream solution is locally uniform, and the unsteady motion in the upper deck is, in fact, quasi-steady. Hence, one finds,

$$\mathbf{X}_{la} = \mathbf{x} + \delta \mathbf{X}_{la1}(X) \text{ , where } \mathbf{X}_{la1} = \int_{-\infty}^{X} U_1/U_\infty \, dx \text{ , (3.7a)}$$

$$\mathbf{v}_i = A_g \exp(-ik_x U_\infty t) \{ \mathbf{e}_g + \delta \mathbf{e}_g \cdot \partial \mathbf{X}_{la1} / \partial x_i \} \text{ , } \quad (3.7b)$$

$$\mathbf{u} = \mathbf{u}_0 + \delta \nabla_{\mathbf{X}} \phi_1 \text{ , } \phi_1 = \exp(-ik_x U_\infty t) \tilde{\phi}_1(\mathbf{X}) \text{ , } \quad (3.7c)$$

$$p = \delta p_1 \text{ , } p_1 = -\rho_\infty U_\infty \partial\phi_1/\partial X \text{ . } \quad (3.7d)$$

Substituting expansions (3.7) into (3.4,5), one finds that the source term on the right hand side of Eq. (3.4) vanishes to O(δ), and $\tilde{\phi}_1$ satisfies the homogeneous equation,

$$(1-M_\infty{}^2) \frac{\partial^2\tilde{\phi}_1}{\partial X^2} + \frac{\partial^2\tilde{\phi}_1}{\partial Y^2} + \frac{\partial^2\tilde{\phi}_1}{\partial Z^2} = 0 \text{ . } \quad (3.8)$$

The wall boundary condition for $\tilde{\phi}_1$ is decided by the requirement of matching with the main deck. Since these two decks are inviscid, it is convenient to match the normal components of the velocity. It is easy to verify that unsteady motion within the main and lower decks satisfies the homogeneous, linearized triple-deck equations. By matching of the vertical velocities within the main and upper decks, we obtain the required boundary condition for $\tilde{\phi}_1$ at Y = 0,

$$A_g \, \mathbf{e}_g \cdot \int_{-\infty}^{X} \frac{\partial U_1/U_\infty}{\partial Y} dX + \frac{\partial\tilde{\phi}_1}{\partial Y} = -\sqrt{|1-M_\infty{}^2|} \, U_\infty \frac{\partial a_1}{\partial X} \text{ ,(3.9)}$$

where $a_1(X,Z)$ corresponds to the appropriately normalized unsteady perturbation to the boundary layer displacement thickness. Solving the problem (3.8,9) using Fourier transform methods and utilizing (3.7d), we obtain the following interactive relationship between the unsteady pressure p_1 and displacement thickness a_1:

$$\frac{\bar{p}_1(\alpha,\gamma)}{\rho_\infty U_\infty A_g} = e_g \cdot \bar{U}_1(\alpha,0,\gamma)/U_\infty \; \exp(-ik_x U_\infty t)$$

$$+ \frac{\alpha^2 \bar{a}_1}{\sqrt{\gamma^2/|1-M_\infty{}^2| \pm \alpha^2}} , \qquad (3.10)$$

where the plus and minus signs refer to subsonic and supersonic flows respectively, the overbar denotes the Fourier transform with respect to X, Z and α, γ are the wavenumbers in those respective directions. The first term on the right hand side is the particular solution corresponding to the gust-mean flow gradient interaction, while the second term corresponds to the homogeneous upper deck response, and hence, contains the eigensolutions representing the instability waves. The solution for $\bar{p}_1$, and thus $\bar{a}_1$, can be obtained by solving the lower deck problem along with (3.10). This yields

$$\frac{\bar{p}_1}{\rho_\infty U_\infty A_g} = \frac{e_g \cdot \bar{U}_1(\alpha,0,\gamma)}{U_\infty} \frac{(-i\alpha)^{5/3} \sqrt{\gamma^2/(1-M_\infty{}^2) \pm \alpha^2}\; Ai'(\zeta_0)}{\alpha^2(\alpha^2+\gamma^2)\, \Delta}$$

$$\exp(-ik_x U_\infty t) , \qquad (3.11a)$$

where,

$$\Delta \equiv \int_{\zeta_0}^{\infty} Ai(\zeta) d\zeta - \frac{\sqrt{\gamma^2/(1-M_\infty{}^2) \pm \alpha^2}}{(-i\alpha)^{1/3} (\alpha^2+\gamma^2)} Ai'(\zeta_0) , \qquad (3.11b)$$

$$\zeta_0 = - \frac{iS_0}{(-i\alpha)^{2/3}} = \frac{e^{-\frac{i\pi}{6}} S_0}{\alpha^{2/3}} , \qquad (3.11c)$$

$$S_0 = \varepsilon^2 \sigma k_x L , \; \sigma = (\mu_0(0)^{-1/2} \; U_0'(0)^{-3/2} \; (1-M_\infty{}^2)^{-1/4}) , \qquad (3.11d)$$

and the equation $\Delta = 0$ corresponds to the lowest order triple-deck approximation to the T-S dispersion relationship.

The unsteady pressure response, p_1, can now be obtained by inversion of the Fourier transform solution (3.11). This, however, is beyond the scope of the present paper and the reader is referred to [9], where a detailed investigation was carried out for similar Fourier inversion integrals in the low Mach number limit. Here, it is sufficient to note that far downstream of the interaction zone, the unsteady response is dominated by the T-S waves within a wedge shaped region in the X-Z plane. The global features of the T-S wave pattern are relatively independent of the cause of the interaction, and they only depend on the normalized gust frequency S_0. Furthermore, it was shown in ref. [9] that the unsteady fluctuations at any observation point (X,Z) are dominated by the vicinity of a single (complex) wavenumber, say (α^*, γ^*), which depends only on the angular location, Z/X, of the observation point.

4. Discussion

The fact that the unsteady response at each point is dominated by a single mode is a very useful result. It shows that Eq. (3.11) is sufficient to infer (i) the influence of the gust orientation $\mathbf{k}_g$, or equivalently $\mathbf{e}_g$, on the instability amplitude along each ray in the X-Z plane, and (ii) the relative importance of the various types of mean flow disturbances. Both these inferences can be made by evaluating the factor $\mathbf{e}_g \cdot \mathbf{U}_1$ at the locally dominant wavenumber (α^*, γ^*). In view of (3.6a) we can rewrite this factor as

$$\mathbf{e}_g \cdot \mathbf{U}_1 = \left[-i\alpha^* e_x - \sqrt{(1-M_\infty{}^2)\alpha^{*2} + \gamma^{*2}}\; e_y - i\gamma^* e_z \right] \Phi_1(\alpha^*, 0, \gamma^*) \; . \quad (4.1)$$

There is a need for a parametric study using (4.1) to find out (i) and (ii) above.

We now briefly discuss the generation of a three-dimensional instability wave by the interaction of a convected gust with a two-dimensional mean flow disturbance. The essential difference between this case and the case of receptivity due to a three-dimensional mean flow disturbance (examined above) is as follows. In the case examined above, the gust length scale is much larger than the instability wavelength in both streamwise and spanwise directions. It is the mean flow disturbance which provides the required short length scales in both these directions. Since the spatial spectrum of the mean flow disturbance is continuous in both α and γ, the generated instabilites are also broadband in terms of their spatial spectrum.

On the other hand, a convected gust with very oblique wavenumber vector $\mathbf{k}_g$ (specifically, such that $k_z L = O(\gamma_0/\varepsilon^3\chi)$, $k_x L = \varepsilon^{-2}\sigma S_0 = O(k_y L)$, or alternatively, $e_z = O(\varepsilon)$, $e_x = O(e_y) = O(1)$), already has a spanwise wavelength which is of the same order as the instability wavelength. Therefore, for this gust to generate an instability wave, a wavelength reduction mechanism is necessary only in the streamwise direction, and this can be provided by a two-dimensional mean flow disturbance. Much of the qualitative discussion in the previous section, regarding the simplifications in rapid distortion theory, also holds for this case. The unsteady flow in the triple-deck region is now periodic in the spanwise direction as well as in time. The inviscid interaction in the upper deck yields the following interactive relationship

$$\frac{\bar{p}_1(\alpha)}{\rho_\infty U_\infty A_g} = \frac{\mathbf{e}_g \cdot \mathbf{U}_1(\alpha,0)}{U_\infty} \exp[i(\gamma_0 Z - k_x U_\infty t)]$$

$$+ \frac{\alpha^2 \bar{a}_1}{\sqrt{\gamma_0^2/(1-M_\infty{}^2) \pm \alpha^2}} \quad , \tag{4.2}$$

where $\mathbf{U}_1 = (U_1, V_1)$ now. Solving the lower deck problem, we obtain,

$$\frac{\bar{p}_1}{\rho_\infty U_\infty A_g} = \frac{\mathbf{e}_g \cdot \bar{\mathbf{U}}_1(\alpha,0)}{U_\infty} \frac{(-i\alpha)^{5/3} \sqrt{\gamma^2/(1-M_\infty{}^2) \pm \alpha^2}\, Ai'(\zeta_0)}{\alpha^2(\alpha^2+\gamma_0^2)\, \Delta}$$

$$\exp[i(\gamma_0 Z - k_x U_\infty t)] \ , \tag{4.3}$$

and hence, a single T-S wave, corresponding to the pole of the Fourier transform solution (4.3), will be generated. It corresponds to the three-dimensional mode with the same frequency and spanwise wavenumber as the gust, the streamwise wavenumber α_0 being decided by the T-S dispersion relation. Note that the factor $\mathbf{e}_g \cdot \mathbf{U}_1(\alpha)$ can be expressed as

$$\mathbf{e}_g \cdot \mathbf{U}_1(\alpha) = \left[-i\mathbf{e}_x + \mathbf{e}_y \sqrt{|1-M_\infty{}^2|} \right] \alpha \bar{\Phi}_1(\alpha,0) \quad , \tag{4.4a}$$

for the subsonic case, and as

$$\mathbf{e}_g \cdot \mathbf{U}_1(\alpha) = \left[-\mathbf{e}_x + \mathbf{e}_y \sqrt{|1-M_\infty{}^2|} \right] i \alpha \bar{\Phi}_1(\alpha,0) \quad , \tag{4.4b}$$

for the supersonic case. A parametric study investigating the influence of gust orientation and the different mean flow disturbances is required here, too.

The results obtained in this paper can be used to analyze the receptivity in a variety of problems which can occur in LFC applications. Some of these cases, illustrated in Fig. 2, are: interaction of a vortical gust with two and three-dimensional roughness elements, suction strips, periodic array of suction holes, in both subsonic and supersonic flows. Note that this paper considered the receptivity to a single convected gust. However, if the spectrum of the free-stream turbulence is known, the correlation spectra of the generated instability waves can be obtained by using the "coupling coefficients" derived from (3.11) and (4.3).

Finally, some remarks are necessary regarding the importance of receptivity to vortical gusts in supersonic flows. The disturbance environment in supersonic wind tunnels is dominated by acoustic waves radiated from turbulent boundary layers on the tunnel wall, while in supersonic flights, entropy disturbances due to temperature variations in the combustion chamber are likely to be important. The vortical gusts do not make a very significant contribution to oncoming disturbance field. However, supersonic Mach numbers invariably imply the presence of a leading edge shock which preprocesses the incident disturbance field. Passage of acoustic and entropy disturbances through the shock wave generates vorticity fluctuations which are convected to the airfoil surface and can generate instability waves through the mechanism considered in this paper. Hence, for supersonic flows, the importance of receptivity to convected vortical disturbances cannot be ruled out in practice.

In summary, the receptivity to three-dimensional convected gusts due to the presence of local mean flow gradients was analyzed. A general solution, valid for both subsonic and supersonic Mach numbers, as well as for any type of local, interactive mean flow disturbance was obtained by using rapid distortion theory in triple-deck framework. In view of the simplifications in this problem, the results for the inviscid interaction in the incompressible case can also be verified by using Bernoulli's equation.

The results obtained here show the influence of the gust orientation, frequency and the type of mean flow disturbance on the amplitudes of the generated instability waves. Another interesting outcome of this paper is that a three-dimensional instability wave can also be generated by the interaction of a very oblique gust with a two-dimensional mean flow disturbance.

Acknowledgements

The authors acknowledge the support of the ICASE/LaRC Instability and Transition Workshop. Additional support was also provided by the National Science Foundation under grant MEA-8351929 and by the McDonnell Douglas Research Laboratory.

References

1. Goldstein, M.E., "The Evolution of Tollmien-Schlichting Waves near a Leading Edge," *J. Fluid Mech.*, Vol. 127, 1983, pp. 59-81.

2. Goldstein, M.E., "Scattering of Acoustic Waves into Tollmien-Schlichting Waves by Small Streamwise Variations in Surface Geometry," *J. Fluid Mech.*, Vol. 154, 1985, pp. 509-529.

3. Heinrich, R.A., Choudhari, M. and Kerschen, E.J., "A Comparison of Boundary Layer Receptivity Mechanisms," AIAA Paper 88-3758, 1988.

4. Kerschen, E.J., "Boundary Layer Receptivity," AIAA Paper 89-1109, 1989.

5. Kerschen, E.J., "Receptivity of Boundary Layers to Acoustic and Vortical Free-stream Disturbances," Proceedings of ICASE Symposium on Stability and Transition, 1989.

6. Goldstein, M.E., "Unsteady Vortical and Entropic Distortions of Potential Flows Round Arbitrary Obstacles," *J. Fluid Mech.*, Vol. 89, pp. 433-468, 1978.

7. Hunt, J.C.R., "A Theory of Turbulent Flow Round Two-Dimensional Bluff Bodies," *J. Fluid Mech.*, Vol. 61, pp. 625-706, 1973.

8. Rogler H.L. and Reshotko E., "Disturbances in a Boundary Layer Introduced by a Low Intensity Array of Vortices," *SIAM J. Appl. Math.*, Vol. 28, No. 2, pp. 431-462, 1975.

9. Choudhari, Meelan, "Boundary Layer Receptivity in Laminar Flow Control Applications," Ph.D. Dissertation, University of Arizona, 1989.

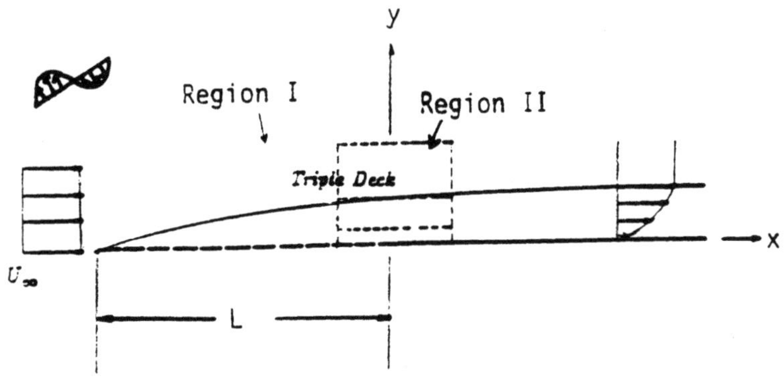

Fig. 1. Different regions in the gust - mean flow disturbance interaction.

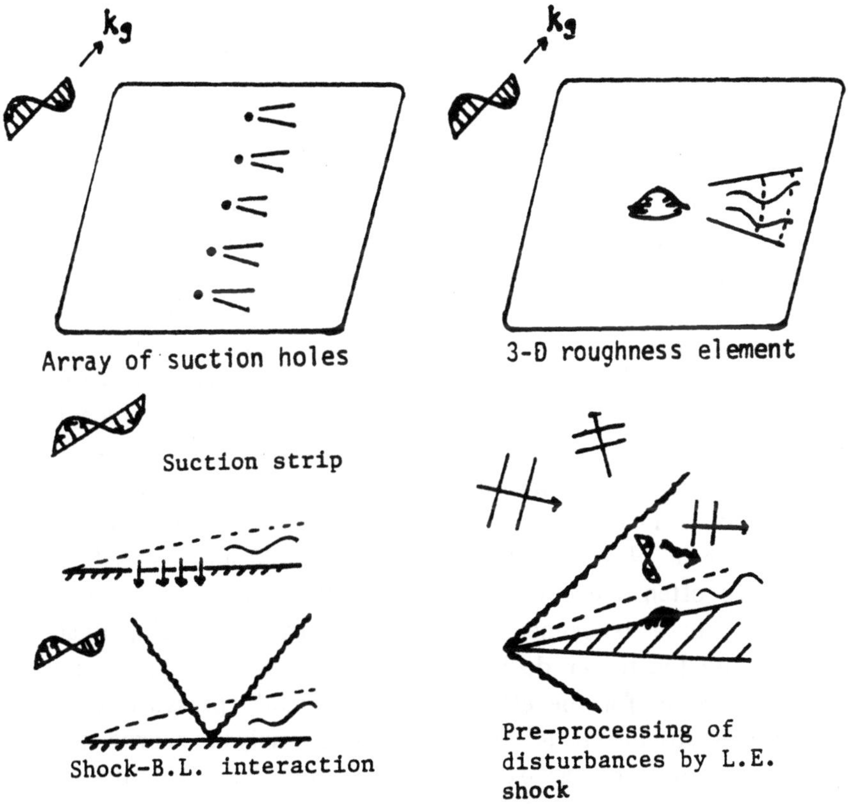

Fig. 2. Various applications of the theory.

BOUNDARY LAYER RECEPTIVITY TO UNSTEADY FREE-STREAM PRESSURE GRADIENTS

Roland A. E. Heinrich

University of Arizona
Tucson, Arizona

Thomas B. Gatski

NASA Langley Research Center
Hampton, Virginia

Edward J. Kerschen

University of Arizona
Tucson, Arizona

ABSTRACT

It has been speculated that boundary layer receptivity to free-stream disturbance fields can be the result of distributed mechanisms. These distributed mechanisms cause the receptivity to take place over streamwise distances of several Tollmien-Schlichting wavelengths, due to the interaction of the streamwise modulation of the free-stream disturbance and the otherwise parallel boundary layer flow. This is distinct to the usual localized receptivity mechanisms which result from the interaction of the long wavelength free-stream disturbance with a short scale (local) surface irregularity. Linear triple-deck theory is used to model the recent experiments of Nishioka and Morkovin (1986) and Kendall (1987) which are often viewed as supporting the distributed receptivity concepts. These experiments are modeled by introducing unsteady sources into the upper deck of the triple-deck structure to appropriately account for the effects of the unsteady pressure gradients in the free-stream. It is shown that this analysis reproduces the essential features of the experimental results. The structure of the solution suggests that distributed mechanisms can only cause exponentially small receptivity levels.

1. Introduction

The means by which particular free-stream disturbances enter the boundary layer and excite instability (Tollmien-Schlichting) waves is known as the "receptivity" problem. The term "receptivity" for this first stage of the laminar turbulent transition process was introduced by Morkovin in 1969, although awareness of the influence of the disturbance environment on the transition process dates back to the 1880s. Over the last few years, substantial progress in our understanding of the receptivity process has been achieved. Recent reviews of boundary layer receptivity research are presented in Goldstein (1989), Kerschen (1989) and Reshotko (1984).

Nishioka and Morkovin (1986) generated instability waves by forcing the boundary layer with weak unsteady pressure fluctuations produced by a local acoustic source just outside the boundary layer. In order to avoid leading edge effects, the boundary layer on the back wall of their test facility was investigated. Three-dimensional Tollmien-Schlichting waves were observed in the boundary layer downstream of the acoustic source. The authors explain the instability wave generation using arguments involving the re-distribution of the vorticity in the boundary layer. This experiment illustrates the detailed level of measurements required to allow comparison with computational and theoretical studies of receptivity. Nishioka and Morkovin also present an illuminating review of the Russian view of receptivity.

Recently, Kendall (1987) has examined the receptivity of the laminar boundary layer to a traveling pressure field in the region far downstream of the leading edge. To generate this pressure field, he rotated two cylinders on a circular path outside the boundary layer (Fig. 1). The axes of the cylinders were parallel to the plate's leading edge and normal to the oncoming flow. Changing the angular speed of this rotorlike assembly, Kendall could vary the speed of the traveling pressure field generated by the cylinders. Each time a cylinder swept by the plate on the lower part of its circular path, a wave packet was generated in the boundary layer. Kendall found significant levels of Tollmien-Schlichting wave generation only if the cylinder speed was close to the propagation speed of the Tollmien-Schlichting waves, $u_{TS} \cong 0.35U_\infty$ (Fig. 2). The boundary layer was not receptive in this downstream region if there was a mismatch in the phase speeds (or wavelengths) of free-stream disturbance and the Tollmien-Schlichting wave.

Over the years, various attempts have been made to develop theoretical models of receptivity. Early attempts were based on the Orr-Sommerfeld equation, which assumes that the mean boundary layer flow is locally parallel. In a review article, Reshotko (1976) emphasized that the major theoretical issue outstanding at that time was the mechanism by which energy was transferred from long wavelength free-stream disturbances to short wavelength instability waves. In recent analyses based on large Reynolds number asymptotics, Goldstein (1983, 1985, 1987) and Heinrich, Choudhari and Kerschen (1988) have shown that receptivity occurs in regions of the boundary layer where the mean flow varies rapidly in the streamwise direction, invalidating the parallel flow assumption of the Orr-Sommerfeld equation. These regions are generally localized, either near the leading edge or in some region further downstream where the wall properties have a rapid local variation. Hence, it seems appropriate to classify these as localized receptivity mechanisms. However, in a number of theoretical and experimental studies, investigators have suggested that distributed receptivity mechanisms are also important. This distributed receptivity is assumed to result from interactions of the free-stream disturbances with a nearly parallel boundary layer over a streamwise distance of several Tollmien-Schlichting wavelengths. The experimental results of Nishioka and Morkovin (1986) and Kendall (1987) are often viewed as supporting the distributed receptivity concept. In this paper we present a rather different interpretation of these experiments. Our view is that the salient features of these experiments are fully explained by the localized receptivity concept. The main distinction between the localized receptivity mechanism considered here, and that considered in previous investigations, is that here the localization is related to the free-stream disturbance rather than to the region of nonparallel mean flow effects.

Golstein (1985) and Kerschen (1989) utilized triple-deck theory to explain localized receptivity mechanisms produced by long wavelength acoustic and vortical free-stream disturbances interacting with a short scale mean flow adjustment due to rapid changes in wall geometry. In this paper, we utilize linear triple-deck theory to examine the response of a locally parallel boundary layer to short scale variations in the free-stream disturbance field. In Section 2, the linear unsteady triple-deck structure is outlined.

Using this structure, a theoretical model of the two-dimensional analog of Nishioka and Morkovin's (1986) experiment is developed in Subsection 3.1, and in Subsection 3.2 a model of Kendall's (1987) rotating rod experiment is presented.

2. General Triple-Deck Structure

As alluded to in the previous section, we use linear triple-deck theory to examine the response of a boundary layer to short scale variations in an unsteady free-stream disturbance field. We consider two-dimensional incompressible flow and assume a locally parallel Blasius mean flow. It is most convenient to outline the structure of the equations in each deck and then, in the next section, to focus on the particular cases of interest. Here we are concerned with an unsteady disturbance flow and hence an appropriate scaling of the disturbance frequency is required. Following Smith (1979), the appropriate scaling in accordance with classical stability theory at the lower branch of the neutral stability curve is $\omega = (\epsilon^{-2})$ with $\epsilon^{-8} = U_\infty \ell/\nu = Re$ and ℓ the downstream distance from the origin of the boundary layer. Hence we introduce

$$\omega\ell/U_\infty = \epsilon^{-2} S_0 \ , \quad S_0 = O(1) \ . \tag{2.1a,b}$$

The streamwise spatial scale $X = \epsilon^{-3}x^*/\ell$ and the cross-stream scales in upper deck ($\hat{y} = \epsilon^{-3}y^*/\ell$), main deck ($y = \epsilon^{-4}y^*/\ell$) and lower deck ($Y = \epsilon^{-5}y^*/\ell$) follow standard triple-deck scaling. The ansatz for the flow variables in each deck are given by

upper deck: $u = 1 + \epsilon^2\beta\hat{u}_2 \ , \ v = \epsilon^2\beta\hat{v}_2 \ , \ p = \epsilon^2\beta\hat{p}_2$ (2.2a)
main deck: $u = U_0(y) + \epsilon\beta\tilde{u}_1 \ , \ v = \epsilon^2\beta\tilde{v}_1 \ , \ p = \epsilon^2\beta p_0(X)$ (2.2b)
lower deck: $u = \epsilon[\lambda_0 Y + \beta U_1] \ , \ v = \epsilon^3\beta V_1 \ , \ p = \epsilon^2\beta p_0(X)$ (2.2c)

where $\beta << 1$ is the strength of the small free-stream forcing and λ_0 denotes the wall shear stress. These expansions are substituted into the Navier-Stokes equations. We then linearize in β and keep only the leading terms in each deck. The results in each deck will be considered in turn.

It is probably best to consider the main deck first. The equations in the main deck are found to be quasi-steady and inviscid. The solution is given by

$$\tilde{u}_1 = a_1(X)\, U_0'(y) \tag{2.3a}$$

$$\tilde{v}_1 = -a_1'(X)\, U_0(y) \tag{2.3b}$$

where the prime denotes differentiation with respect to the indicated dependent variable, and $a_1(X)$ is the modification to the displacement thickness.

The lower deck equations reduce to the lineraized unsteady boundary layer equation which can be written as

$$\frac{\partial^2 w}{\partial Y^2} + iS_0 w - \lambda_0 Y \frac{\partial w}{\partial X} = 0 \tag{2.4a}$$

where

$$w = \frac{\partial U_1}{\partial Y} \ . \tag{2.4b}$$

A solution to this equation can be obtained in Fourier space. Defining

$$F(k) = \int_{-\infty}^{\infty} f(x)\, e^{ikx}\, dx \tag{2.5a}$$

$$f(x) = \frac{1}{2\pi} \int_{-\infty}^{\infty} F(k)\, e^{-ikx}\, dx \tag{2.5b}$$

and applying this transform to (2.4) reduces the equation to Airy's equation

$$W_{zz} - z\,W = 0 \tag{2.6a}$$

where we introduced

$$z = Y\, e^{-i\pi/6} (k\lambda_0)^{1/3} - i\, S_0\, e^{i\pi/3} (k\lambda_0)^{-2/3} \ . \tag{2.6b}$$

Since the solution is required to decay as $k \to \infty$, it is appropriate to choose

$$W = C_1(k)\, Ai(z) \tag{2.7}$$

where $Ai(z)$ denotes the Airy function and the branch cut is choosen to lie along the negative imaginary axis. The unknown coefficient $C_1(k)$ will be determined from matching to the main deck. Applying the no-slip wall boundary condition, the disturbance velocity becomes

$$U_1(Y) = e^{i\pi/6} \frac{C_1(k)}{(k\lambda_0)^{1/3}} \int_{\zeta_0}^{z} Ai(\bar{z})\, d\bar{z} \qquad (2.8a)$$

where

$$\zeta_0 = \frac{S_0 e^{-i\pi/6}}{(k\lambda_0)^{2/3}} \quad . \qquad (2.8b)$$

The pressure in the lower deck can be related to the velocity field by evaluating the momentum equation as $Y \to 0$. This leads to the lower deck pressure in Fourier space

$$P_0(k) = \frac{i}{k} \frac{\partial^2 U_1}{\partial Y^2}\bigg|_{Y=0} = \frac{e^{i\pi/3}(k\lambda_0)^{1/3}}{k} Ai'(\zeta_0) \quad . \qquad (2.9)$$

Turning to the upper deck, the flow field is inviscid and irrotational and the governing equations reduce to Poisson's equation

$$\nabla^2 \phi = q \qquad (2.10a)$$

with

$$\hat{u}_1 = \frac{\partial \phi}{\partial X} \; , \; \hat{v}_2 = \frac{\partial \phi}{\partial \hat{y}} \quad . \qquad (2.10b,c)$$

The source term q depends on the specific physical problem to be investigated. The solution in the upper deck is obtained via Fourier transforms. The matching between the various decks will determine the unknown functions $C_1(k)$ and the (transformed) change of displacement thickness $A_1(k)$. The full solution in physical space can in principle be obtained by inverting the Fourier transform at the end of the analysis. However, here we are only interested in the energy transfer from free-stream disturbance to Tollmien-Schlichting wave. It will be seen that the Tollmien-Schlichting wave appears quite naturally as a pole in Fourier space. Hence, the strength of the generated instability wave is easily obtained by residue theory.

In Section 3, the above analysis is used to model two specific experimental configurations. In Subsection 3.1 a two-dimensional model of Nishioka and Morkovin's (1986) experimental investigation of receptivity to a weak pulsating pressure source in the free-stream is developed, while in Subsection 3.2 Kendall's (1987) rotating rod experiment is considered.

3. Free-stream Unsteady Pressure Sources

Nishioka and Morkovin's (1986) and Kendall's (1987) experiments will be modelled in turn by introducing appropriate pressure sources in the upper deck of the above outlined triple-deck structure.

3.1 Pressure Source in Free-stream

First, let us examine the case of a simple pulsating pressure source in the free-stream. Although a localized three-dimensional source was used in the Nishioka and Morkovin (1986) experiment to generate Tollmien-Schlichting waves, our two-dimensional model is sufficient to elucidate certain general features of the receptivity process.

A time-harmonic pressure source is introduced in the upper deck at $\hat{y} = \hat{y}_0$, $X = 0$, which leads to

$$\nabla^2 \phi = \delta(X)\, \delta(\hat{y} - \hat{y}_0)\, e^{-i\omega t} \quad . \tag{3.1}$$

Transforming (3.1) into Fourier space and solving the resulting differential equation yields

$$\Phi(k) = B\, e^{|k|(\hat{y}_0 - \hat{y})} + \frac{\sinh[|k|(\hat{y}_0 - \hat{y})]}{|k|} \quad \text{for } \hat{y} < \hat{y}_0 \,. \tag{3.2}$$

The integration constant B can be expressed in terms of the transformed change of displacement thickness $A_1(k)$ by matching the normal velocity component of upper (2.10c) and main deck (2.3b) solution. This in turn allows the upper deck pressure

$$\hat{P}(k) = ik\Phi \tag{3.3}$$

for $\hat{y} \to 0$ to be written as

$$\hat{P}(k) = -i\,\mathrm{sgn}(k)\, e^{-|k|\hat{y}_0} + |k|\, A_1(k) \,. \tag{3.4}$$

Matching the main deck velocity $\tilde{u}_1$ as $y \to 0$ with the lower deck velocity U_1 as $Y \to \infty$, the unknown function $C_1(k)$ in (2.8) can be expressed as

$$C_1(k) = \frac{\lambda_0 (k\lambda_0)^{1/3} A_1(k)}{\mathscr{I}}\, e^{i\pi/6} \,. \tag{3.5a}$$

where

$$\mathscr{I} = \int_{\zeta_0}^{\infty} Ai(z)\, dz \,. \tag{3.5b}$$

Equating (2.9) with (3.4) and using (3.5), $A_1(k)$ becomes

$$A_1(k) = \frac{i|k|\, e^{-|k|\hat{y}_0}\, \mathscr{I}}{\gamma\, \Delta} \tag{3.6a}$$

with

$$\gamma = k^2\, \mathrm{sgn}(k) \tag{3.6b}$$

$$\Delta = \mathscr{I} - \frac{S_0 \lambda_0}{\gamma} \frac{Ai'(\zeta_0)}{\zeta_0} \,. \tag{3.6c}$$

This completes the analysis for this case. Inversion of the Fourier transform yields

$$a_1(X) = \frac{1}{2\pi} \int_{-\infty}^{\infty} \frac{i|k|\, e^{-|k|\hat{y}_0}\, \mathscr{I}}{\gamma\, \Delta}\, e^{-ikX}\, dk \,. \tag{3.7}$$

A discussion of the physical implications of this result will be postponed until Section 4.

3.2 Traveling Pressure Field in Free-stream

The traveling pressure field under consideration is the rotating cylinder experiment of Kendall (1987). A sketch of the flow configuration to be analyzed is shown in Fig. 1. The path of the cylinders is such that the trailing wake from each cylinder does not interact with the boundary layer directly (Kendall, 1987).

To leading order, the farfield of a cylinder in a viscous flow can be modelled by a two-dimensional volume source. The strength of the source is related to the wake velocity deficit. Hence, it is appropriate to model Kendall's (1987) experiment by introducing a two-dimensional source in the upper deck of the triple-deck structure. In contrast to the case considered in Subsection 3.1, here the unsteadiness in the boundary layer arises from the cylinders in circular motion with angular velocity Ω above the plate. The dimensions of the experimental setup (Fig. 1) suggest to consider only the influence of the one cylinder passing closest to the boundary layer at each instant. Hence, (2.10a) leads to

$$\nabla^2\phi = \delta(X - X_0)\,\delta(Y - Y_0) \tag{3.8a}$$

$$X_0 = a \sin \Omega t \cong a\Omega t + \ldots \tag{3.8b}$$

$$Y_0 = H + a(1 - \cos \Omega t) \cong H + (\Omega t)^2/2 + \ldots \tag{3.8c}$$

with "a" being the radius of the rotor-like assembly and "H" being the cylinder height at the lowest point of its path.

Transforming (3.8) into Fourier space and solving the resulting differential equation yields

$$\Phi(k,t) = B\, e^{|k|(\hat{y}_0 - \hat{y})} + \frac{e^{ikct} \sinh[|k|(\hat{y}_0 - \hat{y})]}{|k|} \quad \text{for } \hat{y} < \hat{y}_0 \tag{3.9}$$

with $c = \Omega a$ being the circumferential speed of the cylinder.

As in Subsection 3.1, the integration constant B is expressed in terms of $A_1(k,t)$ after matching the normal velocity in the upper and main deck. Using (3.3), the upper deck pressure for $\hat{y} \to 0$ yields

$$\hat{P}(k,t) = -i\, \mathrm{sgn}(k)\, e^{ikct - |k|\hat{y}_0} + |k|\, A_1(k,t) \quad . \tag{3.10}$$

In contrast to the case of the harmonic pulsating pressure source considered in Subsection 3.1, here we explicitly use a Fourier transform in time which we denote with $\mathscr{F}(s)$. The transform of (3.10) yields

$$\hat{\mathscr{P}}(k,s) = -\,i\, \mathrm{sgn}(k) \frac{\sqrt{2a\pi}}{\sqrt{|k|}\; c}\, e^{-|k|H - \frac{(kc\,+\,s)^2 a}{2\,|k| c^2}} \quad . \tag{3.11}$$

Since the lower deck unsteady interaction occurs only for frequencies as given by (2.1), we restrict ourselves to this frequency band. The lower deck solution is thus given by (2.8) as before.

The function $\mathcal{C}_1(k,s)$ follows from matching upper and lower deck and is given by (3.5) upon replacing $C_1(k)$, $A_1(k)$ with $\mathcal{C}_1(k,s)$ and $\mathcal{A}_1(k,s)$, respectively. The function $\mathcal{A}_1(k,s)$ in double Fourier space follows from equating the expressions for lower deck (2.9) and main deck (3.11) pressure to

$$\mathcal{A}_1(k,s) = \frac{-\frac{\sqrt{(2a\pi\,|k|)}}{c}\; e^{-|k|H - \frac{(kc+s)^2 a}{2\,|k| c^2}}\; \mathcal{G}}{\gamma\,\Delta} \tag{3.12}$$

with γ, Δ and $\mathcal{G}$ given by (3.6b), (3.6c) and (3.5b), respectively. Hence,

$$a_1(X,s) = \frac{1}{2\pi}\int_{-\infty}^{\infty} \mathcal{A}_1(k,s)\; e^{-ikX}\, dk \quad . \tag{3.13}$$

4. Results and Discussion

In Section 3, the modification in unsteady displacement thickness $a_1(X)$ was obtained for the two cases investigated. In this Section the results will be related to the Tollmien-Schlichting instability wave. Goldstein (1983) showed that $\Delta = 0$ in (3.7) and (3.13) is the lowest-order approximation to the characteristic equation of classical stability theory. Denoting the unstable Tollmien-Schlichting eigenvalue with $k = -\kappa$ and replacing ζ_0 by η such that

$$\kappa = \frac{S_0^{3/2}\, e^{i3\pi/4}}{\lambda_0\, \eta^{3/2}} \tag{4.1}$$

and

$$\Delta(\eta) = \int_{\eta}^{\infty} Ai(\eta)\, d\eta + i\left[\frac{\eta\, \lambda_0^{3/2}}{S_0}\right]^2 Ai'(\eta) \equiv 0 \tag{4.2}$$

the contribution $a_{TS}(X)$ for the two cases considered is found by residue theory. Following Smith (1979), $a_{TS}(X)$ is related to the streamwise velocity fluctuations and we obtain $u_{TS\text{-}NM}$ for

Nishioka and Morkovin's (1986) and $u_{TS\text{-}K}$ for Kendall's (1987) experiments, respectively:

$$u_{TS\text{-}NM} = \frac{-\beta\epsilon}{\kappa\,\Delta'}\, e^{-|\kappa|\hat{y}_0}\, \mathscr{I}\, e^{i(\kappa X - St)} \tag{4.3}$$

$$u_{TS\text{-}K} = \frac{-i\beta\epsilon\,\sqrt{2a\pi}}{\kappa^{3/2}\,\Delta'}\, e^{-|\kappa|H - \frac{(-\kappa c + S)^2 a}{2|\kappa|c^2}}\, \mathscr{I}\, e^{i(\kappa X - St)} \quad . \tag{4.4}$$

Since a wave packet is generated in Kendall's experiment, (4.4) is the component associated with the most unstable frequency S.

We note that this analysis confirms the generation of unstable Tollmien-Schlichting waves for both experiments modelled. The receptivity strength is linearly related to the strength of the free-stream disturbance field through the parameter β. However, and most importantly, (4.3) and (4.4) show that the absolute strength of the instability wave generated decreases exponentially with the height ($\hat{y}$ or H) of the sources above the boundary layer. It should be pointed out that this exponential dependence on height justifies a posteriori the independent consideration of only the one cylinder passing closest to the boundary layer in Kendall's experiment.

Another result of Kendall's experiment reproduced by this analysis is the dependence of receptivity strength on circumferential speed of the cylinder arrangement (Fig. 2). In Fig. 3 is shown the normalized amplitude of $u_{TS\text{-}K}$ versus speed ratio c/U_∞ (Eq. 4.4). The agreement of the two graphs is evidence of the ability of this model to reproduce the most salient features of the experimental results.

In summary, linearized triple-deck theory can be utilized to model experiments which are often viewed as supporting the concept of distributed receptivity. However, the *local* analysis is able to capture the essential features of the experimental results, which suggests that the experiments under consideration do not support the distributed receptivity concept. The results obtained from this analysis show an exponential decrease of receptivity strength with increasing source height above the boundary layer. This suggests, that receptivity to unsteady free-stream disturbances in an otherwise (locally) parallel boundary layer is exponentially small when compared to mechanisms which involve small surface irregularities (Kerschen, 1989).

Acknowledgements

The authors acknowledge the insightful comments of Mr. M. Choudhari during the course of this study. We also acknowledge the support of the ICASE/LaRC Instability and Transition Workshop. Two of us (RAEH, EJK) would like to acknowledge additional support under NSF grant MEA-8351929.

References

Goldstein, M. E., 1983, "The evolution of Tollmien-Schlichting waves near a leading edge," *J. Fluid Mech.*, Vol. 127, pp. 59-81.

Goldstein, M. E., 1985, "Scattering of acoustic waves into Tollmien-Schlichting waves by small streamwise variations in surface geometry," *J. Fluid Mech.*, Vol. 154, pp. 509-529.

Goldstein, M. E., 1987, "Generation of Tollmien-Schlichting waves on interactive marginally separated flows," *J. Fluid Mech.*, Vol. 181, pp. 485-518.

Goldstein, M. E. and Hultgren, L. S., 1989, "Boundary-layer receptivity to long-wave free-stream disturbances," *Ann. Rev. Fluid Mech.*", Vol. 21, pp. 137-166.

Heinrich, R. A. E., Choudhari, M. and Kerschen, E. J., 1988, "A comparison of boundary layer receptivity mechanisms," AIAA-88-3758, presented at the 1st National Fluid Dynamics Congress, Cincinnati, Ohio.

Kendall, J. M., 1987, "Experimental study of laminar boundary layer receptivity to a travelling pressure field," AIAA-87-1257, presented at the 19th Fluid Dynamics, Plasmadynamics and Lasers Conference, Honolulu, Hawaii, June 8-10.

Kerschen, E. J., 1989, "Boundary-layer receptivity," AIAA-89-1109, presented at the 12th Aeroacoustics Conference, San Antonio, Texas, April 10-12.

Morkovin, M. V., 1969, "Critical evaluation of transition from laminar to turbulent shear layers with emphasis on hypersonically traveling bodies," AFFDL-TR-68-149.

Nishioka, M. and Morkovin, M.V., 1986, "Boundary-layer receptivity to unsteady pressure gradients: experiments and overview," *J. Fluid Mech.*, Vol. 171, pp. 219-261.

Reshotko, E., 1976, "Boundary layer stability and transition," *Ann. Rev. Fluid Mech.*, Vol. 8, pp. 311-349.

Reshotko, E., 1984, "Environment and receptivity," in AGARD Report 709, Special Course on Stability and Transition of Laminar Flow.

Smith, F. T., 1979, "On the non-parallel flow stability of the Blasius boundary layer," *Proc. R. Soc. London A*, Vol 366, pp.91-109.

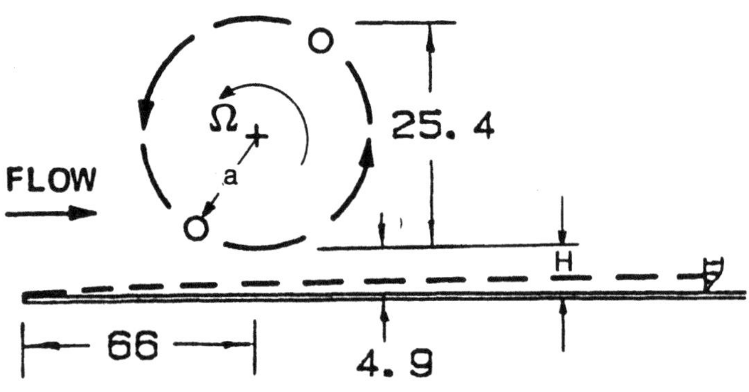

Fig. 1 Kendall's (1987) test configuration showing relation of rotor to boundary layer on plate. Dimensions in cm.

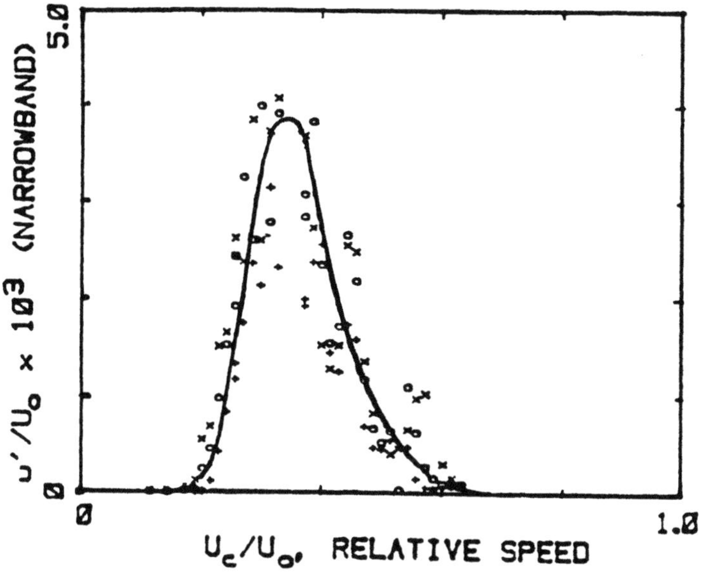

Fig. 2 Induced Tollmien-Schlichting wave amplitude under variation of rotor speed U_c. Measured at vertical location of peak response for various frequencies. From Kendall (1987)

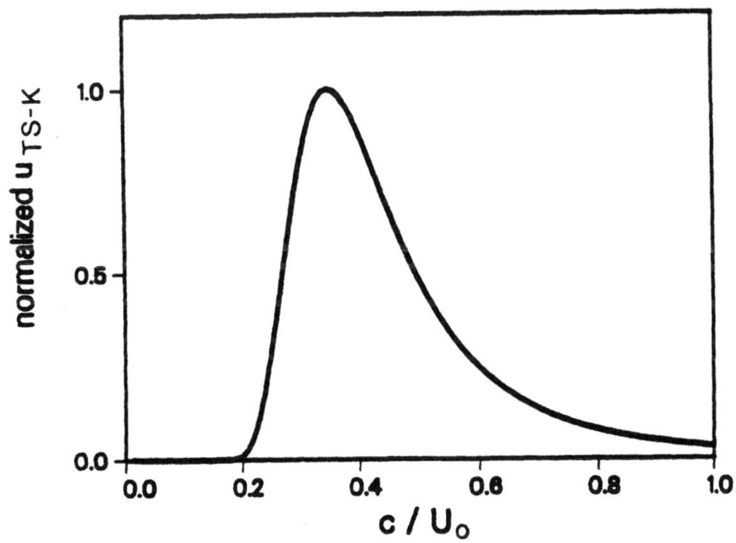

Fig. 3 Normalized amplitude of $u_{TS\text{-}K}$ under variation of rotor speed c (Eq. 4.4) according to the model for Kendall's (1987) experiment developed in this study.

MIX
Papier aus verantwortungsvollen Quellen
Paper from responsible sources
FSC® C105338

If you have any concerns about our products,
you can contact us on
ProductSafety@springernature.com

In case Publisher is established outside the EU,
the EU authorized representative is:
Springer Nature Customer Service Center GmbH
Europaplatz 3, 69115 Heidelberg, Germany

Printed by Libri Plureos GmbH
in Hamburg, Germany